Editorial Advisory Board

Cell Culture and Somatic Cell Genetics of Plants

VOLUME 2

Cell Growth, Nutrition,
Cytodifferentiation, and Cryopreservation

Cell Culture and Somatic Cell Genetics of Plants

VOLUME 2

Cell Growth, Nutrition,
Cytodifferentiation, and Cryopreservation

Edited by

INDRA K. VASIL
Department of Botany
University of Florida
Gainesville, Florida

1985

ACADEMIC PRESS, INC.

Harcourt Brace Jovanovich, Publishers

Orlando   San Diego   New York   Austin
London   Montreal   Sydney   Tokyo   Toronto

ACADEMIC PRESS, INC.
Orlando, Florida 32887

United Kingdom Edition published by
ACADEMIC PRESS INC. (LONDON) LTD.
24–28 Oval Road, London NW1 7DX

Library of Congress Cataloging in Publication Data
(Revised for vol. 2)
Main entry under title:

Cell culture and somatic cell genetics of plants.

 Includes bibliographies and index.
 Contents: v. 1. Laboratory procedures and their
applications.
 1. Plant cell culture—Collected works. 2. Plant
cytogenetics—Collected works. I. Vasil, I. K.
QK725.C37 1984   581'.07'24   83-21538
ISBN 0–12–715002–1 (v. 2 : alk. paper)

PRINTED IN THE UNITED STATES OF AMERICA

85 86 87 88   9 8 7 6 5 4 3 2 1

To my parents,
Pushpalata and the lute
Lal Chand Vasil

Contents

7. Cryopreservation of Cultured Cells and Meristems
LYNDSEY A. WITHERS

Contributors

Numbers in parentheses indicate the pages on which the authors' contributions begin.

M. W. Fowler (103), Wolfson Institute of Biotechnology, University of Sheffield, Sheffield S10 2TN, England

Hiroo Fukuda (149), Department of Biology, Faculty of Science, Osaka University, Toyonaka, Osaka 560, Japan

R. J. Gautheret (1), Paris, France

Wolfgang Hüsemann (213), Department of Plant Biochemistry, University of Münster, D-4400 Münster, Federal Republic of Germany

Atsushi Komamine (149), Biological Institute, Faculty of Science, Tohoku University, Sendai 980, Japan

K. Lindsey (61), Department of Botany, University of Edinburgh, Edinburgh EH9 3JH, Scotland

Peggy Ozias-Akins (129), Department of Botany, University of Florida, Gainesville, Florida 32611

A. H. Scragg (103), Wolfson Institute of Biotechnology, University of Sheffield, Sheffield S10 2TN, England

Indra K. Vasil (129), Department of Botany, University of Florida, Gainesville, Florida 32611

Lyndsey A. Withers (253), Department of Agriculture and Horticulture, School of Agriculture, University of Nottingham, Loughborough, Leicestershire LE12 5RD, England

M. M. Yeoman (61), Department of Botany, University of Edinburgh, Edinburgh EH9 3JH, Scotland

General Preface

Recent advances in the techniques and applications of plant cell culture and plant molecular biology have created unprecedented opportunities for the genetic manipulation of plants. The potential impact of these novel and powerful biotechnologies on the genetic improvement of crop plants has generated considerable interest, enthusiasm, and optimism in the scientific community and is in part responsible for the rapidly expanding biotechnology industry.

The anticipated role of biotechnology in agriculture is based not on the actual production of any genetically superior plants, but on elegant demonstrations in model experimental systems that new hybrids, mutants, and genetically engineered plants can be obtained by these methods, and the presumption that the same procedures can be adapted successfully for important crop plants. However, serious problems exist in the transfer of this technology to crop species.

Most of the current strategies for the application of biotechnology to crop improvement envisage the regeneration of whole plants from single, genetically altered cells. In many instances this requires that specific agriculturally important genes be identified and characterized, that they be cloned, that their regulatory and functional controls be understood, and that plants be regenerated from single cells in which such gene material has been introduced and integrated in a stable manner.

Knowledge of the structure, function, and regulation of plant genes is scarce, and basic research in this area is still limited. On the other hand, a considerable body of knowledge has accumulated in the last fifty years on the isolation and culture of plant cells and tissues. For example, it is possible to regenerate plants from tissue cultures of many plant species, including several important agricultural crops. These procedures are now widely used in large-scale rapid clonal propagation of plants. Plant cell culture techniques also allow the isolation of mutant cell lines and plants, the generation of somatic hybrids by protoplast fusion, and the regeneration of genetically engineered plants from single transformed cells.

Many national and international meetings have been the forums for discussion of the application of plant biotechnology to agriculture. Neither the basic techniques nor the biological principles of plant cell culture are generally included in these discussions or their published proceedings. Following the very enthusiastic reception accorded the two volumes entitled "Perspectives in Plant Cell and Tissue Culture" that were published as supplements to the *International Review of Cytology* in 1980, I was approached by Academic Press to consider the feasibility of publishing a treatise on plant cell culture. Because of the rapidly expanding interest in the subject both in academia and in industry, I was convinced that such a treatise was needed and would be useful. No comprehensive work of this nature is available or has been attempted previously.

The organization of the treatise is based on extensive discussions with colleagues, the advice of a distinguished editorial advisory board, and suggestions provided by anonymous reviewers to Academic Press. However, the responsibility for the final choice of subject matter included in the different volumes, and of inviting authors for various chapters, is mine. The basic premise on which this treatise is based is that knowledge of the principles of plant cell culture is critical to their potential use in biotechnology. Accordingly, descriptions and discussion of all aspects of modern plant cell culture techniques and research are included in the treatise. The first volume describes every major laboratory procedure used in plant cell culture and somatic cell genetics research, including many variations of a single procedure adapted for important crop plants. Two subsequent volumes in preparation are devoted to the nutrition and growth of plant cell cultures and to the important subject of generating and recovering variability from cell cultures. An entirely new approach is used in the treatment of this subject by including not only spontaneous variability arising during culture, but also variability created by protoplast fusion, genetic transformation, etc. Future volumes are envisioned to cover most other relevant and current areas of research in plant cell culture and its uses in biotechnology.

In addition to the very comprehensive treatment of the subject, the uniqueness of these volumes lies in the fact that all the chapters are prepared by distinguished scientists who have played a major role in the development and/or uses of specific laboratory procedures and in key fundamental as well as applied studies of plant cell and tissue culture. This allows a deep insight, as well as a broad perspective, based on personal experience. The volumes are designed as key reference works to provide extensive as well as intensive information on all aspects of plant cell and tissue culture not only to those newly entering the field but also to experienced researchers.

Indra K. Vasil

Preface to Volume 2

The primary objective of these volumes is to provide authoritative, up-to-date, extensive, and in-depth information on all aspects of cell culture and somatic cell genetics of plants. There is a deliberate bias, however, toward science and technology, although applied aspects are discussed and emphasized also where relevant. A key feature is the selection of individuals who have played a leading role in the development of our knowledge in this important area of biotechnology to author chapters in their fields of specialization, so that others may benefit from insight gained during long years of personal experience.

This volume forms a natural bridge between Volume 1, devoted to laboratory procedures and their applications, and Volume 3, devoted to regeneration and genetic variability. Volume 2 thus begins with a detailed historical account, going back more than two centuries, by Professor R. J. Gautheret, and includes many previously unpublished photographs of scientists who made important contributions to the field. Professor Gautheret, along with the late Philip R. White and P. Nobécourt, played a key pioneering and decisive role in the development of the modern science of plant cell and tissue culture. This chapter is followed by others on callus and cell suspension cultures, nutrition and photoautotrophic growth, cytodifferentiation, and cryopreservation. These accounts of the basic and fundamental aspects of plant cell cultures prepare the ground for the next volume.

I thank all the authors for their fine contributions and members of the Editorial Advisory Board for their suggestions and support. It is also a pleasure to thank the editorial staff of Academic Press for their assistance in the preparation of this volume.

Indra K. Vasil

CHAPTER 1

History of Plant Tissue
and Cell Culture:
A Personal Account

R. J. Gautheret

Paris, France

1

I. INTRODUCTION

The success of experimental research is subject to two important factors. One is the outcome of unexpected discoveries and the other results from technical advances. For example, the wide implications of radioactivity and immunology were discovered by casual observations that significantly and fortunately were perceived by scientists of genius. On the other hand, the invention of the electron microscope revolutionized the study of cytology, whereas the availability of radioisotopes allowed the fantastic boom of biochemistry.

The development of the science of tissue culture was more complicated. Interest in it had been foreseen before the middle of the nineteenth century by the promoters of the cell theory. One experimental approach was unsuccessfully attempted during the first years of the twentieth century, but success was reached in only 1912 with animal cells, and another 22 years later with plant tissues.

The technique of cell culture was exploited quickly for animal cells, while in the case of plant tissue culture a long period of stagnation followed the initial establishment of the basic methods. After 30 years of relative indifference, thousands of scientists rushed suddenly toward this "new field" of plant biology, which currently is undergoing considerable expansion with the new name of biotechnology.

II. PREHISTORY OF TISSUE CULTURE

A. Discovery of Callus Formation

The prehistory of plant tissue culture began more than 225 years ago. The discovery of callus formation was made by Duhamel, "Seigneur du Monceau et de Vrigny" (Fig. 1). Like many of the bright people of the eighteenth century, he was an encyclopedist. He published 11 volumes on naval architecture and 18 of a dictionary of sciences and arts. In the manor of his brother, which was located close to Pithiviers (about 50 miles from Paris), he made many observations and experiments on plants, especially on trees. His results were published in two treatises: the dictionary of trees and bushes, and the well-known "La Physique des Arbres" (Duhamel du Monceau, 1756). In this latter book, he described many experiments on sap circulation, grafting, and wound healing. Removing a small ring of cortex from an elm, he observed the development of a swelling above the area of

Fig. 1. Henri-Louis Duhamel du Monceau (1700–1782). General Inspector of the Navy and an agronomist. His pioneering experiments on wound healing identified in 1756 the plant material that was used 178 years later for the first successful tissue cultures.

decortication, while buds developed on the lower part (Fig. 2). Duhamel du Monceau noticed that the buds seemed to come out from the interface between wood and cortex. He had the prescience of cambium when he described "un tissu cellulaire très abreuvé et très délié qui, quand il sera converti en bois, unira l'une a l'autre deux couches très minces de fibres longitudinales" (Duhamel du Monceau, 1756). This very old work was a foreword for the discovery of plant tissue culture. But in 1756 the bacterio logical technique was not invented, asepsis was unknown, the concept of tissue culture had not yet been expressed, and finally nobody was able to appreciate Duhamel's discovery.

B. The Cell Theory

About 160 years after the extensive work of Malpighi (1675) on microscopic anatomy, the celebrated cell theory was expressed independently and almost simultaneously by Schleiden (1838) with respect to plants and

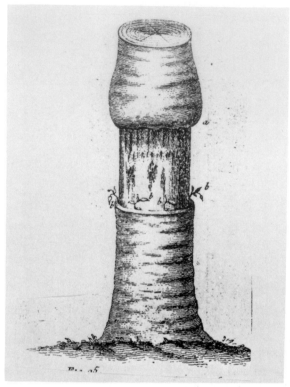

Fig. 2. Wound healing following ring-shaped "debarking" of an elm tree. A swelling developed above the debarked area, while buds appeared on the lower part (Duhamel du Monceau).

by Schwann (1839) considering both animals and plants. In their publications, these two biologists admitted implicitly that the cell is capable of autonomy and even that it is totipotent. This was expressed very clearly by Schwann (1839) in these words:

Among the lower plants, any cell can be separated from the plant and continue to grow. Thus, entire plants may consist of cells whose capacity for independent life can be clearly demonstrated. . . . That not every cell, when separated from the organism, does in fact grow is no more an argument against this theory than is the fact that a bee soon dies when separated from the swarm a valid argument against the individual life of bee [translated from German].

In the case of egg cells of all organisms and spores of haplobiontic plants which are able to divide and form complete individuals, totipotency is obvious. This totipotent behavior occurs similarly in the animal kingdom, for example when a somatic cell of a hydra gives rise to a new individual, and in the case of some plants such as *Begonia*, whose epidermal cells may transform into new begonias (Hartsema, 1926). But in many cases, somatic cells do not produce complete organisms. Their totipotency was postulated by the cell theory, but neither Schleiden nor Schwann proposed any experimental methods capable of proving this consequence of their theory. This limitation of the cell theory was of course distinctly perceived by Schwann (see the previous quotation). The demonstration required two steps: first, the multiplication of a single cell and, second, its trasformation into a complete organism. The first step was reached in 1954 and the second 11 years later. However, this ultimate target was attained by a very long and often illogical, but very interesting, journey.

While Duhamel du Monceau had shown the way, Schleiden and Schwann defined the aim. The race began slowly and initially by the avenue opened by Duhamel. Trécul (1853) performed experiments on callus formation by decorticated trees such as *Robinia, Pawlonia, Ulmus,* and others. While Duhamel had almost neglected the microscope, Trécul used it intensively and published a large number of excellent pictures of callus sections. These proved that scar tissue comes not only from cambium but also from the phloem, and even from medullary rays and the youngest parts of the xylem (Fig. 3). He had really shown what material could lead to tissue culture.

Similar experiments were undertaken by Vöchting (1878). He obtained very luxuriant callus, for example, with explants from *Brassica rapa.* His attention was especially attracted by the polarity that characterizes the development of plant fragments. Thus, the upper portion of a piece of stem always produced buds, while the basal portion produced callus or roots. Even very thin slices showed such a polar development. Thus, this property belongs to the cell itself.

Other experiments carried out by Goebel (1902) confirmed Vöchting's results and supported the notion of totipotency. Vöchting undertook grafting experiments between different species of *Opuntia, Salix, Beta,* and others. These experiments demonstrated that the behavior of a tissue is not altered by contact with another tissue. The dependence of morphogenetic capacity on hereditary internal factors is very strict. A synthetic view of the results provided by experimental morphology led Sachs (1880–1882) to conceive a general theory that suggested the existence of organ-forming substances distributed in a polar fashion. A similar view was proposed by Wiesner (1884) from experiments on roots.

Nine years later, a new approach to tissue culture was used by Rechinger (1893) when he determined the minimal size of explants which permits cell division. For this purpose, he isolated small pieces from buds of *Populus nigra* and *Fraxinus ornus,* from roots of *Beta vulgaris* and *Brassica rapa,* and from stems of *Pothos celatocaulis* and *Coleus arifolia.* The explants were placed at the surface of moistened sand. Pieces thicker than 20 mm were sometimes able to produce buds and even to regenerate entire plants.

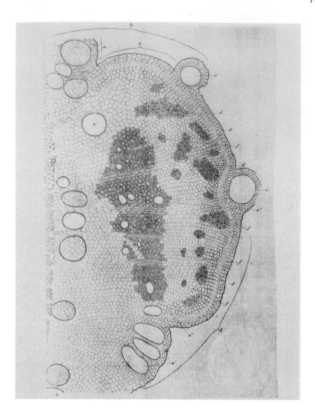

Fig. 3. Section of callus produced by a decorticated stem of *Robinia*. The callus developed mainly from the dedifferentiation and proliferation of the youngest parts of the stem (v', old xylem vessel pushed away by the proliferating callus; v", young xylem differentiated in the callus; Trécul).

Smaller pieces proliferated without organization and this proliferation was facilitated by the presence of vascular tissues in the explants. Finally, when the thickness of the explants were less than 1.5 mm, which represented no more than 21 cell layers, they were not able to grow. These experiments were really outlines of tissue cultures, but Rechinger did not understand the significance of his results, perhaps from the lack of theoretical concepts.

C. The Concept of Cell Culture: Unsuccessful Attempts

A theoretical concept for cell culture was finally imagined in 1902 by the German botanist Haberlandt (Fig. 4). Unfortunately, he was influenced by

Schleiden and Schwann's cell theory rather than by experimental expectation. And he neglected Duhamel's results as well as Vöchting and Rechinger's experiments. His dogmatic attitude and the ignorance of the past explains the failure of his own attempts. But he appreciated very clearly that, when the technical difficulties were removed, the method of cultivating isolated plant cells in nutrient solution should make possible the experimental study of many outstanding problems from new points of view. He, therefore, chose to work with single cells. Appreciating the importance of photosynthesis, he presumed that green cells would be the best material. However, he neglected the fact that green cells of phanerogams are relatively differentiated and cannot recover meristematic competence without stimulating substances which were unknown at the time. He worked with palisade cells of *Lamium purpureum*, pith cells from petioles of *Eichhornia crassipes*, glandular hairs of *Pulmonaria* and *Urtica*, stamen hairs of *Tradescantia* (Fig. 5), stomatal guard cells of *Ornithogalum*, and many other materials. At this time it was recognized that asepsis was absolutely necessary when the culture media are enriched in organic substances metabo-

Fig. 4. Gottlieb Haberlandt (1854–1946). He was the first to formulate clearly the principles of plant cell culture.

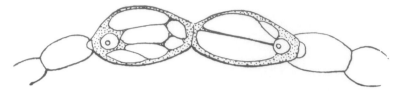

Fig. 5. Haberlandt's (1902) attempt to cultivate stamen hair of *Tradescantia*. The cells increased in volume but did not divide.

lized by microorganisms. Haberlandt's media contained glucose and pep-
ton, he carefully avoided contamination, and his cultures remained free of
microorganisms.

The results, however, were disappointing. The cells survived for several
weeks. They were capable of synthesizing starch and enlarging, but they
never divided. Fifty-six years passed before the realization of Haberlandt's
dogmatic dream. But this physiologist learned the lesson of his failure and
understood that it was necessary to know the requirements for cell division. He began experiments similar to those of Rechinger and after 10 years
of mediations wrote:

> I again attacked the question of the physiological basis of cell division. I no longer experi-
> mented with isolated cells but used rather large cell complexes, i.e., tissue fragments. I sought
> to establish how small the pieces of tissue could be and still be able to detect the cell division
> that led to wound periderm formation or accompany the callus formation of mechanically
> injured organs.

This new orientation was fruitful, and Haberlandt (1913a,b) performed
experiments that prefaced many of the modern investigations. Using
pieces of potato tuber tissues, he discovered that "cell division occurred in
small, thin tissue discs almost without exception when the discs contained
a vascular bundle fragment. It was sufficient when it was comprised of
leptome, i.e., of sieve tubes and their companion cells . . . ." Haberlandt
concluded logically that "the leptome would secrete one or even more
stimuli which diffused through the adjacent storage tissue." Other experiments indicated, in addition, the involvement of a wound stimulus. If
Haberlandt had persevered in these investigations and used in place of
potato tissue some other common material such as willow or carrot, which,
as it is now established, can proliferate without growth substances, he
certainly would have obtained the first true tissue cultures. But he did not
suspect that experiments on callus could be a step toward tissue and even
cell culture. Winkler (1908), who was inspired by Haberlandt's work, tried
to cultivate segments of string bean and observed some cell divisions but
no true proliferation.

More promising experiments were started by Simon (1908; Fig. 6), following the initial observations of Duhamel du Monceau, Trécul, and Vöch-

Fig. 6. Section of callus formed on the upper part of a poplar stem segment. Note formation of shoot bud on upper left (Simon, 1908).

ting. Cultivating poplar stem segments he observed the development of bulky callus, buds, and roots. And like Vöchting he turned his attention to the phenomenon of polarity. If the segments were grown in normal position, that is to say, if their morphologically lower part was placed in the medium and the upper part in the air, the callus that developed on the basal end in the medium produced roots, whereas the callus coming from the upper extremity gave rise to buds. Inversion of the segment's position in respect to medium contact provoked an alteration of polarity expression, but this was not completely reversed. Thus, in effect Simon had established the basis of callus culture and even of micropropagation. However, he did not transfer his cultures and these were not aseptic. Therefore, his results were unfairly neglected.

D. Animal Cell Culture

While Haberlandt's views were clumsily applied by botanists, they suggested wonderful experiments to animal physiologists, experiments that had no direct connections with tissue culture. When Ramon y Cajal conceived the neuron theory, many histologists wondered if this theory had any real basis. Experimental argument seemed to be necessary. This was understood by Harrison (1907), who was a professor at Yale University. He

Fig. 7. Alexis Carrel (1873–1945). A French surgeon who came to the United States and worked at the Rockefeller Institute for Medical Research. He was the first individual to obtain indefinite proliferation of animal cells in culture.

transferred a small piece of neural tube of frog onto a drop of clotted lymph. Axons coming from this explant expanded quickly in the medium, and thus was demonstrated the individuality of the neuron. Following this experiment, he brilliantly demonstrated the possibilities of this new method. Two years later Burrows (1910), a student of Harrison, began to exploit this technique and he joined Carrel (Fig. 7), a French surgeon who came to the United States, where he found good support for his plans to spread out attempts on arterial suture. Together (Burrows and Carrel, 1912) they established the present method of cultivating animal tissues in a nutrient made of blood plasma and embryo juice or their equivalents. The method outlined by Carrel has changed only in minor details.

Many plant physiologists were humiliated by Carrel's success obtained so quickly and without visible efforts. But this mortification did not change their orientation. While Ebeling, following Carrel (1912), developed animal tissue culture, the investigations on plant tissue culture were profoundly influenced for a quarter century by Haberlandt, and between 1917 and 1935 many plant physiologists unsuccessfully tried to obtain multiplication of isolated cells. Bobilioff-Preisser (1917) used, like Haberlandt, palisade cells. He obtained a very long life of such cells from *Viola, Thunbergia*, and other plants. Transferred into various nutritive solutions, their plastids and nuclei changed shape and structure, but they never divided. Knudson (1919), working with root cap cells, remarked that in opposition to classical opinion these cells are still living when they break off from the root. They can live for several weeks in culture but, like palisade cells, they do not divide.

Thielman (1924) undertook attempts with stomatal cells and remarked that they can remain alive for a longer time than other epidermal cells. Börger (1926) tried to cultivate various kinds of tissues from dicotyledons and pteridophytes. The same year, Czech (1926) tried to cultivate cells isolated by a new technique. Believing that the mechanical isolation could injure the cells, he used chemical dissociation with mineral salts such as magnesium chloride. The isolated cells were transferred to media containing tissue extracts. No divisions occurred.

Kunkel (1926) used perianth cells, and Kemmer (1928) worked with leaf epidermis of *Rheo discolor*. Pfeiffer (1931) used cells coming from fleshy fruits. All these attempts were unsuccessful. In the best cases, the cells remained alive for several weeks and were capable of enlarging, but they never divided.

More cheering results were obtained with explants consisting of numerous cells. Thus, Ulehla (1928) observed some cell divisions in explants cut off from cactus. He noticed for the first time that the contact of an aqueous solution injured the cells and that it is better to expose them to atmospheric humidity, which is secured by using a jelly. Mention must be made of original attempts of Lamprecht (1918) with pieces coming from leaves. These pieces were not really isolated but transferred on wounds cut on organs of the same plant. In fact this practice represented grafting, and it is not surprising that the tissues could proliferate. These experiments gave Lamprecht the opportunity to verify Haberlandt's views on the stimulating properties of phloem. Although encouraging, all these attempts did not lead to success, and many other people conducted similar experiments with similar failures.

However, Schmucker (1929) reported that he had successfully grown individual leaf mesophyll cells of *Bocconia* on a medium containing an extract from leaves. He reported his experiments in the following words:

In sehr konzentrierten Reibsaft der *Bocconia* Blätter der durch bakteriendichtes Filter gegangen war, begannen die isolierte Mesophyllzellen sich zu teilen und auszuwachsen und es wurden bereits ansehnliche, mehr oder minder dichte Zellhaufen, zum Teil aus Dutzenden von Zellen aus einer Zelle erhalten. Dieser theoretisch und praktisch wichtige, neuartige Befund lasst vielleicht sogar auf Totalregeneration der ganzen Blutenpflanze aus einer Zelle hoffen.

His statement was very clear and escaped any other interpretation. However, this result was not repeated by Schmucker or others for 30 years. Melchers (1975) reported that he saw the cell divisions in Schmucker's laboratory when he was a young student in 1929. This was in opposition to observations of other workers who reported at this time the toxicity of tissue juices, xylem sap, phloem exudate, etc. But as a matter of fact, it was reported many years later that some plant juices and particularly liquid endosperms stimulate cell division. On the other hand, the result of cell dissociation is usually a mixture of isolated cells with groups of two or more cells so that it is difficult to attest that such groups come from multiplication of a single cell and not from an incomplete dissociation. And it is surprising that a result of such outstanding importance had been announced in a brief notice in *Planta,* without pictures and without repetition. About Schmucker's experiment White (1943) wrote that "details were apparently never published, the work has not been verified by any later worker and the result is so at variance with all other recorded experiments that its correctness is to be doubtful." This criticism expressed in 1943 seems to be harsh but one does not forget that, if White was an excellent scientist, he had not a tender heart. Now, Bergmann (1959) reported that in a puree coming from a suspension of tobacco and kidney bean cells 20% of them were able to divide. But his experiments were somewhat different than those realized by Schmucker. Indeed, Bergmann started from colonies of cells that divided actively, while Schmucker used differentiated cells issuing from leaves. However, Rossini (1969), starting from leaves of *Calystegia sepium,* performed experiments similar to those of Schmucker and obtained—thanks to the mixture of two very powerful stimulating substances like 2,4-dichlorophenoxyacetic acid (2,4-D) and benzyladenine—multiplication in cell suspension cultures.

Now, of course, no discussion of these issues can be expected. Schmucker, White, and Rossini have died, and we will probably never know if Schmucker was 20 years ahead of his contemporaries or if he was wrong in his interpretations.

Such a succession of fruitless work disheartened many, and indeed the famous cytologist Küster (1928) considered the problem hopeless and asked if perhaps the assumption that cells are totipotent might not itself be false.

Concurrently to the disappointing attempts concerning true tissue cultures, successful endeavors were performed on embryo culture in order to obtain the full development of very young embryos outside of the ovule. Between 1904 and 1907, embryos were successfully cultivated by many people, such as Hanning (1904), Stingl (1907), Buckner and Kastle (1917), Andronescu (1919), and Dietrich (1924). These investigations were far from true tissue cultures, but they promoted technical progress, which was helpful.

E. Root Tip Culture

A new approach to tissue culture was conceived simultaneously in Germany by Kotte (1922) and in the United States by Robbins (1922a,b; Fig. 8). They postulated that true *in vitro* culture could be made easier by using meristematic cells such as those that operate in root tips or in buds. Indeed, this kind of material was considered as early as 1908 by Winkler. But his cultures of small pieces of roots exhibited poor growth. The field of organ culture was unknown at the time.

Kotte chose to work with excised root tips. He succeeded in cultivating very short root tips of pea and maize in a variety of nutrients. These contained salts of the Knop's solution, glucose, and several nitrogen com-

Fig. 8. W. J. Robbins (1890–1978). First individual to obtain subcultures of isolated roots (1922).

Fig. 9. Philip R. White (1901–1968). First individual to establish indefinite growth of isolated roots (1934a) and of tumor tissues of plants (1939a).

pounds such as asparagin, alanin, and meat extract. Liebig's meat extract was especially effective, and in the best cases the excised roots would grow for 2 weeks while they retained a normal morphology. No subcultures were attempted. At the same time, Robbins started an important series of experiments with maize roots and successfully maintained them by subcultures. He demonstrated the efficiency of yeast extract, from which a powerful growth substance was extracted 32 years later by Miller *et al.* (1955b). However, Robbins's cultures did not survive indefinitely. In every experiment, their growth gradually diminished and the cultures were ultimately lost.

Robbins's failure resulted from the choice of the living material. If he had worked with dicotyledonous roots, his attempts would probably have been successful, because his compatriot White (Fig. 9) later obtained indefinite cultures using a similar medium with tomato roots (White, 1934a). As for continuous maize root cultures, they were achieved 6 years later by Mc-Clary (1940), thanks to a really aberrant medium containing a high concentration of sodium chloride.

Chambers (1923) tried to cultivate very small root tips but observed only a very slight development. Similar attempts by Mayer (1929), Dauphiné (1929), and Scheitterer (1931) were not any better.

It was at this time that White began his famous experiments on root culture. The circumstances of his success are historically interesting. In 1932, White had a position at the Rockefeller Institute at Princeton. There he met the famous biochemist Wendel Stanley, who had discovered the nature of a plant virus. The multiplication of this virus was secured by experimental infections of whole plants. Stanley gave advice to White to cultivate the virus on isolated roots growing *in vitro*. The first attempt using tobacco roots was unsuccessful. But White (1934a) obtained good results with tomato roots and announced that by means of subculture he was able to obtain indefinite growth of roots of this plant (Fig. 10). He announced the multiplication of the virus of tobacco and aucuba mosaic in his growing excised tomato roots (White, 1934b). This gave Stanley a new method for his investigations on viruses. White observed that in some cases sub-

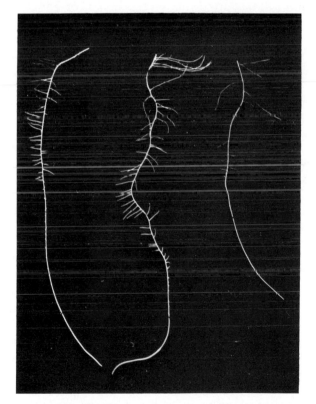

Fig. 10. Isolated roots cultured by White. Left to right: tomato, *Trifolium repens*, and *Trifolium pratense*.

cultures of infected roots gave rise to healthy tissues. In order to maintain the virus in the roots it was necessary to transfer not only the root tip itself but a considerable amount of the older tissue also. The meristems themselves were often virus free. This remarkable fact was explained 18 years later and forms the basis of the best cure for virus diseases, i.e., virus elimination by meristem culture. The successful experiments of White also opened the field of root cultures, which have been explored by many people.

The experiments on root culture raised many interesting questions. First of all, one wondered what could be the origin of the yeast extract activity used by Robbins (1922b) as well as White (1934a). Bonner (1937) demonstrated the importance of thiamine in yeast extract. This was immediately confirmed by Robbins and Bartley (1937) and White himself (1937). And finally it was recognized that the thiamin could be replaced by its two components, thiazole and pyrimidine. Thus, it became possible to cultivate isolated roots on completely synthetic media. Later, Bürstrom, and especially Street and associates, studied the action of mineral salts on root metabolism. Bürstrom (1953), Hannay and Street (1954), and Boll and Street (1951) revealed the importance of trace elements such as copper, manganese, and iodine. Street et al. (1952) established the role of chelating agents, while Sheat et al. (1959) obtained information on the effect of NH_4^+ and amino acids. Delarge (1941) showed that there was considerable variability of roots in culture.

The root culture technique developed vigorously and helped to solve many morphological, physiological, and pathological problems. However, it was realized very early that root tips will always lead to organ cultures and that the accomplishment of Haberlandt's dream would have to come from elsewhere.

F. Bud Culture

Bud cultures initially produced similar results. Robbins (1922a) had obtained limited growth of cotton and pea stem tips. Eleven years later, White (1933) announced equally poor results with bud meristems of *Stellaria media*. Success was obtained only when Loo (1945) achieved excellent cultures of *Asparagus* and Dodder stem tips. Finally, Ball (1946; Fig. 11) published in the *American Journal of Botany* a rather extensive paper in which he identified the exact part of a shoot meristem that was able to give rise to a whole plant and also indicated the kinds of explants that produced only callus. This almost unknown publication prefaced the modern method of vegetative multiplication. Ball is really the father of the so-called

Fig. 11. Ernest A. Ball (1909–   ). He discovered the principle of micropropagation (1946).

micropropagation method, which is now used by thousands of people and applied extensively in important plant propagation industries established in the United States, Europe, and Asia.

In summary it can be said that by 1934 it was well established that attempts involving isolated cells failed constantly while those concerning root tips or buds produced organ cultures. The difficulties faced were compounded by the failure of most plant physiologists to consider the problems of tissue culture. However, the old experiments on callus formation, especially those of Vöchting and Simon, did not justify such pessimism. Furthermore, the illustrations published in 1756 by Duhamel du Monceau (Fig. 2) and in 1853 by Trécul (Fig. 3) had clearly suggested the path to success.

G. The Way to Success

The way to success was discovered again by two physicians, Blumenthal and Meyer (1924), who obtained voluminous callus on carrot root explants treated with lactic acid. Their medical training influenced them in favor of pathological interpretation and they suggested that lactic acid had induced a tumoral transformation. Tissue culture was within their reach but they were not in a position to seize this opportunity to solve the problem. Auler (1925), Bittman (1925), and Nobécourt (1927) performed similar experiments and they too emphasized the pathological aspect of their results. But in 1927 Rehwald, a student of Küster, demonstrated that slices of carrot, *Cochlearia, Scorzonera*, etc., could produce calluses spontaneously, without pathogenic intervention. Unfortunately, he did not attempt to isolate the callus tissues and to transfer them to nutrient media. Thus ended the prehistory of plant tissue culture with this incomplete success and the publication of a photograph representing callus produced by a carrot slice. But the history began without delay.

III. THE FIRST TISSUE CULTURE: AUXIN UTILIZATION

In 1931 I was accepted in Guilliermond's laboratory. Guilliermond was a famous cytologist who had observed for the first time mitochondria in the plant cell and discovered the sexuality of yeasts. He did not share the pessimistic views expressed by his friend Küster about tissue culture and oriented me toward this subject. First, I tried to cultivate roots and obtained the same results as Robbins (Gautheret, 1932). Attempts with root cap cells verified Knudson's failures. After three years of unsuccessful experiments, I too began to doubt. But Guilliermond, who knew what was possible and what was hopeless, urged me to work with cambium. Then I began experiments with the trees of my own garden, *Acer pseudoplatanus, Ulmus campestre, Robinia pseudoacacia,* and *Salix caprea* (Gautheret, 1934, 1935). I settled on a technique of securing aseptic small explants consisting of cambium and phloem. Preliminary attempts with liquid media failed completely. Later, the explants were placed on the surface of a medium solidified with agar. In fact, I did not expect good results. I placed these cultures in a cupboard and inspected them only 2 months later. Then I was very surprised. The surface of the explants was covered with white calluses which seemed very healthy (Fig. 12). Examining these calluses with a microscope, I could recognize perfect living turgescent cells of which many were dividing (Fig. 13). This success had been made possible by Guilliermond's perspicacity. My results were immediately verified by Gioelli (1938) in Italy. However, the activity of my cultures ceased after about 6 months. They were clearly affected by a deficiency.

At this time, Kögl and collaborators (Kögl *et al.*, 1934) had just estab-

Fig. 12. First plant tissue culture (Gautherel, 1934, 1935). A fragment of cambial tissue from the stem of *Acer pseudoplatanus* associated with phloem development.

Fig. 13. Surface of a tissue fragment of cambium from *Alnus glutinosa* (Gautheret). Dividing cells can be seen.

lished that the growth substance discovered by Went (1926) was indoleacetic acid. Snow (1935) demonstrated that it stimulated cambial activity. Thanks to indoleacetic acid, the proliferation of my cambial tissue cultures was enhanced and subcultures became possible. Nevertheless, their activity again ceased after 18 months.

This incomplete success encouraged White to make new attempts. Nobécourt (1937), a French plant pathologist, remembered Rehwald's results on carrot slices. Finally, the possibility to cultivate plant tissues for unlimited periods of time was announced independently and simultaneously by White (1939a), Nobécourt (1939a; Fig. 14), and myself (Gautheret, 1939; Fig. 15).

White had used tissues coming from tumors produced spontaneously by hydrid *Nicotiana glauca* × *N. langsdorffii*, while Nobécourt and I had worked with carrot tissues. Thus ended a long line of failures, but one must not forget that they also set the stage for the final success.

Since the middle of 1938, I was convinced of the perenniality of my

Fig. 14. P. Nobécourt (1895–1961). One of the pioneers, along with Gautheret and White, of plant tissue culture. He obtained sustained growth of carrot tissues in 1939.

Fig. 15 The first unlimited plant tissue culture, a strain of carrot tissue isolated by Gautheret in 1937.

cultures but the French Academy of Science did not accept my results for publication. Indeed Guilliermond had associated me with his extensive work on vital staining and we had published together in the *Comptes Rendus Hebdomadaires des Seances de l'Academie des Sciences* the maximal number of reports that were accepted for a year. For this reason my results were announced at the first meeting in 1939, on January 9. Ten days previously, White (1939a) had presented his work on tobacco tissues at a meeting of the Botanical Society of America, and on February 20, Nobécourt (1939a) announced at the French Society of Biology the perpetuity of his own cultures.

The dates of these announcements are not of interest. The important fact is only that Carrel's results had been transposed to plants. Likewise, the controversies that existed on the one hand between Robbins and White and on the other between Nobécourt and myself are buried in the far past and nowadays nobody rightly remembers them.

The announcements of the successful realization of continuous plant tissue cultures were made a few months before the beginning of World War II. For 6 years American and French scientists worked without knowledge of their mutual results. When peace came again, they could compare their results. Nobécourt was in poor health. For this reason, he did not vigorously follow his initial results and died prematurely. His death went almost unnoticed. This was unfortunate. He was one of the true pioneers of plant tissue culture.

Fig. 16. Georges Morel (1916–1973). He was among the first to culture monocotyledonous tissues, develop the method of meristem culture for the elimination of viruses and the micro-propagation of orchids, and discover the two unique opines in crown gall tissues.

During the war, in spite of material difficulties, I was able to accept some collaborators in my laboratory. One of them, Morel (Fig. 16), was particu-larly able and enthusiastic. While I was working on fleshy organs, es-pecially *Scorzonera,* chicory root, Jerusalem artichoke, and *Brassica napus* tuber (Gautheret, 1942a), Morel established strains of lianas cells such as grape and Virginia creeper and of herbaceous plants such as tobacco and trees, especially hawthorn. I came back to my first material and succeeded in cultivating *Salix* cambium (Gautheret, 1948), while Jacquiot (1964) iso-lated tissue strains from chestnut, *Robinia,* elm, etc. During the war, inves-tigators in the United States and France worked independently and in different directions. White continued his attempts on tumor tissues and began with Braun (White and Braun, 1942) very important experiments on crown gall cells. The culture media used on the two sides of the Atlantic Ocean were very different. White (1943) had proposed a medium containing glycine, nicotinic acid, thiamine and pyridoxine which had been used for root cultures. In fact, the need for these organic compounds by tissue cultures had not been demonstrated and it was certain that tumor tissues did not require auxins. On the contrary, I had noticed very early that auxins stimulated the growth of normal tissues such as those of carrot or chicory and were an absolute requirement by others such as Jerusalem artichoke (Gautheret, 1942b). When I had the opportunity to read the American publications, I was impressed by the contrast between American and French nutritive media.

Interestingly, American and French results were identical on a point that remains enigmatic. It consisted of the fact that in culture normal cells can attain tumoral properties. This transformation was achieved for the first time under the following circumstances.

IV. THE VARIABILITY OF TISSUE CULTURES: ANERGY

Observing carefully many tissue cultures of carrot and *Scorzonera*, I had an opportunity to demonstrate variations of their sensitivity with respect to auxins (Gautheret, 1942b, 1955). The initial explants displayed weak responses to the stimulating action of this growth substance. After a few subcultures, their sensitivity increased considerably and remained at a high level. This was explained by the fact that the multiplication of the parenchymatous cells provided small cells that responded more intensively with regard to auxin. But after many transfers, in some cases, I noticed a stressing of the growth and, concurrently, the disappearance of the cell sensitivity in respect to auxins (Fig. 17) and of their ability to form organs, differentiated cells, and even secondary products. For this common artifact, I proposed the term of "accoutumance," which was translated in English to "habituation." Later, I observed that the loss of reaction in respect to auxin was the main aspect of "accoutumance." I proposed to replace this word with "anergy," which comes from the Greek word denoting lack of action. Kulescha (1952) demonstrated that anergied cells synthesize much more auxin than normal cells. That is a tumoral property. Finally, tumors were obtained by grafting anergied tissues of *Scorzonera* (Camus and Gautheret, 1948b). On the whole, plant cells cultivated out of the organism move toward the tumoral state like animal cells. This evolution raised many disturbing questions at the time concerning the applications of plant tissue culture.

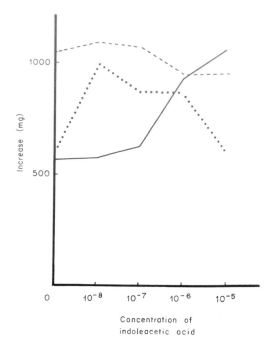

Fig. 17. Demonstration of anergy (Gautheret, 1942). An initial (——) explant displays weak response to the stimulating action of auxin. After a few subcultures, this sensitivity increases (....) because the multiplication of parenchymatous cells provides small cells more sensitive to auxin. After many transfers one observes a stressing of growth and the disappearance of cell sensitivity (----).

V. SUCCESS OF CALLUS CULTURES

Just after the war, several American pathologists were interested in plant tissue culture. Riker and Hildebrandt (1951; Fig. 18) isolated strains of crown gall tissues from *Tagetes erecta, Chrysanthemum frutescens,* and sunflower, while Black (1946) and Nickell (1949, 1954) cultivated tissues from wound virus tumor of sorrel. Riker's group analyzed in detail the nutrition of crown gall tissues, considering successively the action of carbon derivatives, nitrogen compounds, and growth substances. When the conditions of use of auxins were correctly known, the American scientists successfully obtained cultures of normal tissues. The first success was obtained by Ball (1950) with *Sequoia* and shortly afterward by Henderson *et al.* (1952) with sunflower. Nickell isolated tissue strains from several dicotyledons: *Persea americana, Vigna sinensis, Rumex acetosa,* and *Melilotus officinalis.* Following

Fig. 18. A. J. Riker (1894–1981). He and his colleagues, especially A. C. Hildebrandt, exerted a decisive influence on the early development of plant tissue culture media and techniques.

the success of Ball with *Sequoia*, tissues of other gymnosperms were cultivated: cypress by La Rue (1953), *Picea glauca* by Reinert and White (1956), and juniper by Constabel (1958). Many other tissues of dicotyledons were cultivated (safflower, *Crepis, Digitalis, Euphorbia*, tomato, *Phaseolus*, pea, flax, apple, snapdragon, ivy, mint, cotton, ash, string bean, etc.). The anergied alteration was observed by Henderson (1954) and Nickell (1961) in the United States and by Kandler (1952) in Germany.

One must not forget that at this time Thimann (Fig. 19), who carried out outstanding work on plant hormones, exerted an indirect but overwhelmingly positive influence on the progress of plant tissue culture.

The first success had been secured with dicotyledons and this resulted from the sensitivity of cambium to auxins, the sole plant-growth-stimulating substance available at the time. It seemed desirable to generalize the tissue culture method and to consider, for example, monocotyledonous and pteridophyte tissues. Preliminary attempts with the classic media con-

Fig. 19. Kenneth V. Thimann (1904–   ). His outstanding and pioneering work on plant hormones greatly influenced the success of plant tissue cultures.

taining auxin were disappointing, and it became clear that this growth substance would not support the growth of tissue cultures from these plants.

However, some promising results had been obtained on monocotyledons. Curtis and Nichol (1948) had the idea to cultivate orchid seedlings in a medium containing barbiturates. The plantlets developed into shapeless masses similar to tissue colonies. But, in fact, the structure of those masses looked like that of an embryo and they were especially limited by a true epidermis. This highly organized character was also reported by Rao (1963). La Rue (1947, 1949) initiated endosperm cultures of *Zea mays* and obtained subcultures. This was the first successful attempt to cultivate monocotyledonous tissues. The results of La Rue, though very interesting, did not consider the whole problem of tissue culture. The solution came after a tortuous journey which deserves to be reported.

VI. ACTIVITY OF COCONUT MILK:
DISCOVERY OF CYTOKININS

Toward 1940, the famous geneticist Blakeslee (Fig. 20) tried to obtain hybrids between several species of *Datura*. The cross-fertilization was successful, but the hybrid embryos died after a very slight development. Blakeslee supposed that the surrounding ovule was toxic and he thought that the embryos must be cultivated out of their natural environment.

His collaborators Conklin and Van Overbeek suggested that a liquid endosperm such as coconut milk would be a good medium for embryo culture. Experimental testing showed the validity of their hypothesis, and they were delighted to obtain the full development of several *Datura* hybrids (Van Overbeek *et al.*, 1941). This imaginative scientist realized that coconut milk would play a part in tissue culture history, and during my

Fig. 20. Albert Blakeslee (1874–1954). This famous geneticist, with his co-workers, discovered the stimulating properties of coconut milk on embryo development.

first visit to the United States he offered me a position in his laboratory. This was in 1946. Four years earlier I had been elected Professor at the Sorbonne. Several young people were already working in my laboratory, among them Morel, Camus, and Kulescha. I had to stay in France.

The application of coconut milk for the growth of tissue cultures was achieved 2 years later by Caplin and Steward (1948). Working with my principal material, carrot explants, they obtained, thanks to coconut milk, a proliferation much more active than that observed with auxin. Duhamet (1950) reported that coconut milk enhances the development of crown gall and anergied tissues that are not influenced by auxin. Using a medium enriched with this natural extract, Morel (1950a) obtained the indefinite growth of *Amorphophallus rivieri, Sauromatum guttatum,* and other mono-cotyledonous tissues such as those of *Gladiolus, Iris,* and lily. And later the same medium allowed him to cultivate the royal fern tissue (Morel, 1950b).

The tissue culture method was now extended to all groups of vascular plants. The fact that coconut milk enhances the proliferation of tumoral tissues indicated clearly that it contained a stimulating substance that was not an auxin. It seemed naturally credible that other plant extracts would have the same properties as coconut milk. Tests were carried out with different fruits or seed extracts. Investigators did not fail to compare the activity of these extracts with respect to normal and tumoral tissues and to confront them with auxins. In their first paper concerning coconut milk, Caplin and Steward (1948) reported similar action with extracts from young maize albumens. A more detailed attempt was made by Nétien *et al.* (1951). Similar experiments were performed with various other materials such as wheat, oat, and rye caryopses (Steward and Caplin, 1952).

Following La Rue's (1949) pioneering work with tomato juice, extensive tests were made with this extract (Nitsch and Nitsch, 1955; Kovoor, 1954). A comprehensive investigation was carried out by Kovoor (1966) with various materials such as orange juice, grape juice, and extracts from *Elaeis, Borassus, Annona,* and *Allanblackia.* Many other plant extracts were tested, and it appeared that plants contained nonauxinic stimulating substances. At this time, it was necessary to understand the chemical basis for the observed stimulation, but this was out of range for most people.

Inconclusive fractionations were made of coconut milk by De Ropp *et al.* (1952) and by Duhamet and Mentzer (1955). Working on several hundred liters of coconut milk, Shantz and Steward (1955) isolated three active substances and claimed that one of them was 1, 3-diphenylurea. But Nitsch and Nitsch (1957) established that this compound had no activity. Kovoor (1962) reported that the stimulating factor of coconut milk has characteristics of indole compounds as well as nucleic acid bases. Maize endo-sperm yielded similar incomplete chemical information. Nétien and Beau-chesne (1954) and Györffy *et al.* (1955) gave approximate indication about

Fig. 21. Folke Skoog (1908–   ). Leading member of the group that discovered kinetin. He, together with Carlos Miller, discovered the interaction between auxin and cytokinin for the control of organogenesis

the stimulating substances of maize endosperm. A more precise approach was followed by Beauchesne and Jouanneau (1960), and, finally, Beauchesne (1961) suggested that the maize endosperm factor would be a purine, but he was not able to establish its molecular configuration.

The chemical structure of this nonauxinic active substance contained in plant tissues was disclosed indirectly by Skoog (Fig. 21) and collaborators. In 1944, Skoog, visiting White's laboratory, had become interested in an experiment on tobacco tissues (White, 1939b). Maintained on the surface of the medium, these tissues remained unorganized, but immersed in a liquid medium, they produced buds. Skoog considered that the effect of immersion must have a chemical nature and conducted tests with many substances. With the collaboration of Tsui, he demonstrated that adenine stimulates cell division and induces bud formation in tobacco tissues (Skoog and Tsui, 1948).

At this time, the intervention of nucleic acids in cell division began to be understood and Skoog was convinced that they contained substances that were able, like adenine, to induce tissue proliferation. With the collabora-

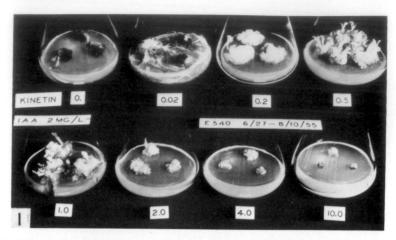

Fig. 22. Hormonal control of organ initiation in tobacco callus tissues (Skoog and Miller, 1957). The relative amounts of auxin and kinetin in the medium induce the formation of callus, root, or shoot bud.

tion of Miller, Okumura, Von Saltza, and Strong, he undertook experiments with nucleic acids extracted from herring sperm, calf thymus, and yeast. Extracts of fresh materials did not induce cell division. But extracts prepared with old material were active, and finally Skoog and collaborators (Miller *et al.*, 1955a,b) isolated from autoclaved yeast extract a derivative of adenine, the 6-furfuryl aminopurine, named kinetin, which, in the presence of indoleacetic acid, induced proliferation of tobacco cells and bud formation (Skoog and Miller, 1957). The bud-promoting properties of kinetin suggested a broad interaction of this substance with auxins (Skoog and Miller, 1957). This realization led to the classic experiments which showed that root and bud initiation were conditioned by a balance between auxin and kinetin, the former substance inducing root formation and the latter bud initiation (Fig. 22).

This remarkable concept could be applied to tobacco and other plants, but its dogmatic generalization was wrong. The discoveries of gibberellin and of abscissic acid compelled the alteration of Skoog's theory, and, of course, innumerable experiments carried out since 1951 have revealed the multiplicity of factors controlling organ formation. Organ formation depends on species, tissues, environment such as physical form of the medium, light, temperature, mineral composition of the nutrient solution, sugar, growth substances, polarity, and other factors. A review of the subject by Thorpe (1980) suggests strongly that the determination of organ formation is very complex and cannot be restricted to hormonal balances only.

The enthusiasm caused by the discovery of kinetin was moderated by the fact that it was an artifact. It possesses the same activity as coconut milk, but its discovery did not solve the problem of natural kinetins called cytokinins.

The first step in this direction was realized when Miller (1961) detected in young maize endosperms a substance with kinetin-like properties. This substance, called zeatin, was isolated by Letham (1963), and with Shannon and McDonald he established its molecular structure (Letham *et al.*, 1964). Finally, he demonstrated that the cytokinin of coconut milk was ribosylzeatin (Letham, 1974), which was verified by Van Staden and Drewes (1975). That brought to fruition the work outlined 34 years earlier by Blakeslee. Auxin and coconut milk had provided a powerful stimulus to future work on plant tissue culture.

VII. IMPROVEMENT OF NUTRIENT MEDIA

Research on tissue cultures was also considerably enhanced by the mineral element content of culture media. The first mineral solutions proposed were related to Knop's solution (Gautheret, 1942d; Nobécourt, 1937) or Uspenski and Uspenskaja (1925), used by White (1943), and supplemented with trace elements. These solutions had been chosen empirically and in practice they gave satisfactory results. But they were not very likely optimal, and soon Hildebrandt *et al.* (1946) proposed new media containing a rather high concentration of Na_2SO_4. Burkholder and Nickell (1949), working on tumor tissues, appreciably modified Duggar's solution. But none of these alterations made any significant difference. Heller (1953) carried out extensive investigations on mineral nutrition, especially of carrot and Virginia creeper tissues, and he prescribed appreciable increase in the concentration of salts.

While the total amount of salts in Gautheret's and White's solutions was less than 200 mg per liter, Heller's solution contained twice the amount. Another mineral solution proposed by Nitsch and Nitsch (1956) contained almost 4 g of salts per liter. Murashige and Skoog (1962) studied the mineral requirements of tobacco tissues and proposed a solution far more concentrated than the classical formula. The concentration of some salts was 25 times that contained in Knop's solution. This medium allowed five to seven times more active growth than the other media. Murashige and Skoog's solution is characterized by the presence of large amounts of NO_3 and NH_4, and it seemed that the activity of NH_4^+ ions was very important. Following this conclusive improvement, other researchers suggested new media, most of them containing rather high concentrations of salts.

Some solutions were proposed for very special purposes. For example, Margara *et al.* (1967) proposed a medium containing a concentration of NO_3^- that is almost 50% more than that proposed by Murashige and Skoog. This medium facilitated flower initiation.

Inorganic micronutrients were similarly considered. The pioneers had used empirically the classical formulations proposed for whole plants. These formulations had very low concentrations of microelements. Heller (1953) discovered the importance of Zn and B and suggested an increase in their concentration. Later, this tendency was strongly accentuated. For example, the Murashige and Skoog medium contained 500 times more Fe^{2+} than Gautheret's solution, 7 times more Mn^{2+}, and 50 times more Zn^{2+}.

In general, Murashige and Skoog's solution, as well as the other very concentrated media, ensured good results for beginners, allowed success of experiments previously out of reach, and consequently favored the extension of tissue culture research to many plants. These new media, however, also had some disadvantages. Large amounts of mineral salts frequently induce anergy, hyperhydric transformation, and other kinds of alterations that are also enhanced by the presence in the medium of some very active auxins, such as 2,4-D. When these artifacts affect tissues, their behavior can no longer be considered normal. The ignorance of this very important factor was often a source of failure or of incorrect interpretations. And very bad surprises sometimes occurred in applied research.

VIII. DEMONSTRATION OF TOTIPOTENCY IN CELL CULTURES

During the first 20 years of plant tissue culture research, starting in 1934, the main focus was on what is usually called callus culture. During this period, the principles expressed by Haberlandt seemed to have been forgotten. Nevertheless, these principles, which were suggested by the cell theory and related to the concept of totipotency, were very important. Their verification required two experimental achievements: first, the multiplication of a single cell and, second, the development of a whole plant from this cell.

In principle, somatic cells, even when they are differentiated, possess the same set of genes as the egg cell (or the spore in the case of Bryophyta or Pteridophyta) and must be potentially totipotent. This totipotency is frequently expressed during the course of vegetative propagation. The best example was recorded a long time ago when *Begonia* leaves were seen to produce numerous buds that were capable of forming roots (Hansen, 1881). Each new bud developed from a single epidermal cell (Hartsema, 1926). But in most cases this totipotency is not expressed spontaneously on account of some inhibitions that must be suppressed by growth substances involved in tissue culture. In 1950, the knowledge acquired about the powerful properties of the new media permitted the conclusion that Haberlandt's failures were caused by the lack of appropriate growth substances. And therefore new hopes were born.

I had perhaps approached somatic totipotency in my experiments in 1942 when I observed the development of buds in elm and of plantlets in chicory tissue cultures (Gautheret, 1942a,c). But histological investigations did not prove that a whole plant was derived from a single cell. Two years later, Buvat (1944, 1945) faced similar doubts. Another approach was made by Levine (1947) with a carrot tissue culture. Cultivated in media containing a high concentration of indoleacetic acid, the tissues proliferated without organization, but the removal of the auxin induced bud formation. Levine consigned his cultures to me because I was interested in carrot tissue differentiation and I undertook histological observations. I was able to see many stages of bud and plantlet formation, some of them composed of a very small number of cells. But there was no evidence that one single cell could develop into a plantlet.

More encouraging attempts were made by Tulecke (1953), not on somatic cells but on pollen grains that normally form male gametes. Cultivating pollen grains in a medium enriched with vitamins and amino acids, he obtained cell colonies. La Rue (1953) carried out similar experiments with *Taxus* pollen. But the pollen grain is the prothallus of the male gametophyte and cannot be considered an ordinary cell. The division of an isolated somatic cell was therefore not yet demonstrated. The same year, Duhamet (1953) tried once more to determine the minimal tissue size compatible with proliferation. Earlier, Sanford *et al.* (1948) had obtained the division of isolated animal cells with conditioned media. The isolated cell was placed in a very thin tube of glass and this was plunged into a medium containing numerous cells. Under these conditions, the isolated cell was under the influence of the unknown substances excreted by the surrounding cells, and these substances induced its multiplication. Muir *et al.* (1954) undertook to transpose this "conditioning principle" to plant cells, but the technique used for animal cells had to be modified. Muir placed a single cell of tobacco or *Tagetes erecta* on a small piece of filter paper that in turn was placed on the top of a callus mass. Eight percent of these isolated cells multiplied and formed cell colonies (Fig. 23). But the following year, De Ropp (1955) claimed that "the technique, however, does suffer from one drawback, namely that the development of the single cell cannot be fol-

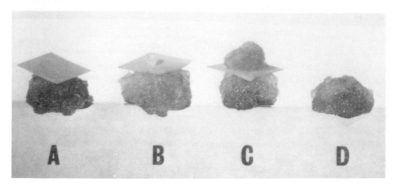

Fig. 23. First example of single-cell culture (Muir *et al.*, 1954). A single cell is placed on a small piece of filter paper, which in turn is placed atop a callus tissue. The isolated cell divides and forms callus tissue.

lowed visually and it seems difficult to rule out the possibility that the cells from the host tissue grow through the filter paper and set up an independent colony." This was a relevant objection, and Muir understood very quickly that he had to repeat his experiment in a different way. De Ropp (1955) had not only expressed criticism but he undertook experiments with small groups and even with isolated cells. He obtained the same results as Marchal and Mazurier (1944). Torrey (1957), who until than had worked not on tissue but on root culture, succeeded in achieving the multiplication of isolated cells placed near active cell colonies. His medium was enriched with powerful stimulating substances such as 2,4-D and yeast extract. Finally, Muir *et al.* (1958), taking into consideration De Ropp's objections, demonstrated in two different ways the success of their technique. First, they placed on the top of cell colonies filter paper pieces without isolated cells. No colonies appeared on such filter papers. Second, they transferred on sunflower colonies fragments of filter paper seeded with isolated cells of *Tagetes*. They observed the development of colonies consisting only of *Tagetes* cells and not of sunflower. These new results refuted De Ropp's doubts, and therefore the prior claim of Muir, Hildebrandt, and Riker was definitely substantiated.

The history of cell culture does not solely contain such irrefutable experiments. Indeed some results could be interpreted in several ways and investigators often chose that which was most favorable to their belief. For example, such uncertain interpretation occurred in observations performed with the shaking apparatus proposed by Steward *et al.* (1952) which allowed dissociation of tissues and led to cell (suspension) cultures. When the culture was established, microscopic observations showed that the medium was crowded with isolated cells or cell clumps. It was tempting to state that the clumps were the result of the multiplication of isolated cells. But the shaking method did not permit the demonstration of this fact, which was possible only with completely isolated single cells as by Muir's method or with observations of the isolated cells made possible by Torrey's technique.

The "plating technique" proposed by Bergmann (1959) permitted following the multiplication of isolated cells in the same way. This method consisted of blending a puree of cells in agar medium not completely cold and plating this mixture in a Petri dish, where it solidifed. The puree consisted of isolated cells and small cell groups, which can be easily located. In Bergmann's experiments, 20% of the isolated cells were able to divide. Lutz (1963) proposed a intermediate method between Muir's and Bergmann's techniques. This method consisted of placing a cell on a solidified medium close to a colony that conditioned the medium. Finally, Vasil and Hildebrandt (1965a) observed the multiplication of completely isolated cells of a tobacco hybrid cultivated in a drop of medium enriched with coconut milk and naphthalene acetic acid. It seemed, and this was not surprising, that the coconut milk could be a substitute for conditioning.

On the whole, it may be pointed out that after Haberlandt's failure, the cell culture problem was neglected for more than 40 years and that suddenly it gave rise to spirited competition, which led to complete success after a few years.

The success achieved had not yet solved the problem of totipotency. The solution of this problem required additional research, allowing the development of a plant from a single cell. With their shaking technique, Steward et al. (1966) observed in suspension cultures of carrot cells the formation of embryos developing into apparently normal plants and they suggested that each embryo came from a single cell. Following this experiment, many researchers reported the phenomenon of somatic embryogenesis and plantlet formation in cell cultures. This occurred particularly in liquid media, and some families of plants such as Umbelliferae were especially favorable. But the cell suspension technique did not allow continuous observations of single cells, and it was not possible to ensure that embryos were derived from isolated cells and not from cell clumps. The uncertainty inherent in the shaking technique was evaded by Vasil and Hildebrandt (1965b). Using their cloning technique, they observed that colonies arising from isolated cells of the hybrid *Nicotiana glutinosa* × *N tabacum* were able to produce plantlets (Fig. 24). The following year, Lutz (1966) verified this result. Thus was demonstrated the totipotency of somatic cells. Backs-Hüsemann and Reinert (1970) later showed the formation of a somatic embryo from a single plated cell of carrot following the formation of a proembryonal mass of cells.

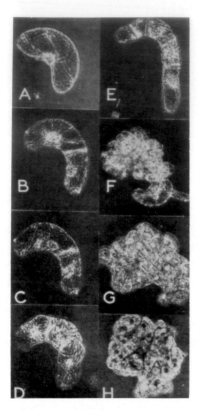

Fig. 24. First demonstration of totipotency. A single isolated cell of the hybrid *Nicotiana glutinosa* × *N. tabacum* cultured in a microchamber gave rise to a cell colony (Figs. A–H, left), a callus, and finally a whole plant (Figs. A–H, right; Vasil and Hildebrandt, 1965b).

IX. DISCOVERY OF ANDROGENESIS

The normal destiny of a gamete is to fuse with another gamete of the opposite sex. The gametes are derived as a result of meiosis and the gametophytic stage prepares the way for fertilization. It was, therefore, surprising that, when Tulecke (1953) obtained cell colonies from *Ginkgo* pollen grains, nobody understood that this work represented a genuine discovery, for it had broken for the first time the rigorous course of the sexual cycle. The haploid pollen grain is actually the male prothallium consisting of a variable number of cells according to the species. In the case of *Ginkgo* this prothallium is made up of five vegetative cells and one reproductive cell. In angiosperms the pollen grain has one vegetative cell

Fig. 25. Formation of somatic embryos from microspores of *Datura innoxia*. Photograph of the first successful anther culture obtained in 1964 by Guha and Maheshwari.

and one reproductive cell, which later gives rise to the two male gametes. In Tulecke's cultures, pollen grains formed unorganized colonies. Similar behavior of gymnosperm pollen grains was observed by La Rue (1953), by Konar (1963), and by others between 1959 and 1970.

Yamada *et al.* (1963) had the original idea to cultivate not pollen grains but whole anthers. They obtained haploid callus and stated without proof that these must have developed from microspore mother cells following meiosis. During the following year, a completely different kind of development was discovered by Guha and Maheshwari (1964). Working with anthers of *Datura innoxia*, a member of the Solanaceae, they observed that the anthers burst and released apparently normal embryos and plantlets (Fig. 25). Although in their initial publication they hesitantly speculated about the possible haploid origin of the plantlets, in a subsequent study they confirmed that the plantlets were haploid and developed from pollen grains (Guha and Maheshwari, 1966). Soon thereafter, Bourgin and Nitsch (1967), using *Nicotiana* anthers, observed that the pollen embryos developed directly into haploid plants. Cytological observations, particularly those of Sunderland (1977), revealed the successive stages of this embryogenesis and established that in most cases embryos were produced by microspores formed as a result of meiosis. The first step frequently consists of an equal division of these cells in place of the normal unequal partition that gives birth to pollen grains. However that may be, Guha and Maheshwari had discovered that totipotency belongs to haploid pollen cells as well as to somatic cells.

Curiously, the results collected experimentally on phanerogams were not compared with those appearing spontaneously in Pteridophyta and Bryophyta. In the case of these cryptogams, diploid and haploid cells are totipotent, but they give rise to different kinds of organisms. On the contrary, in the case of phanerogams, diploid and haploid cells produce the same kind of embryos except with respect to chromosome numbers. More than 60 years had passed and Haberlandt's dream had finally taken shape. Thus ended the fundamental investigations on plant cell and tissue culture.

In the preceding pages I have provided a detailed account of the earliest events in the history of plant tissue and cell culture. There are three reasons for this:

1. I would like to illustrate that this history, submerged in a very distant past, is sometimes forgotten or stated wrongly.

2. It seemed instructive to show that the principal results, which are nowadays so easily verifiable that they are even considered suitable as laboratory exercises for undergraduate students, were really obtained after long, harsh, and disappointing attempts.

3. Finally, the reader will be able to note that in most cases new results were not obtained suddenly, but after a very long period of experimentation involving successive investigators, the last one harvesting without visible difficulty the fruits of his predecessor's labors.

X. APPLICATIONS OF TISSUE AND CELL CULTURE

While the theoretical aspects of *in vitro* culture were reached slowly and step by step, practical applications emerged and developed at first slowly but then in an almost exploding manner. Now the applied field extends more and more quickly. In some cases, the applied investigations were founded on scientific truth, but unfortunately the fundamental facts and principles were often neglected, if not despised. This gave birth to costly disappointments which could have been avoided if rigorously tested concepts had been respected.

Before the final establishment of the techniques, all the efforts had been dominated by two objectives: realization of indefinite growth of cells in culture and demonstration of cellular totipotency. In some cases, the pioneers' determination to reach these objectives was considered with amusement and even with contempt on account of their philosophical aims. But when the solution was in view, many people said, "What can we do now?" Then it appeared that the new methods could be introduced into the main chapters of plant biology and its consequences spread out more and more quickly in different directions. It is not possible to report details of the principal applications of plant tissue culture. These are covered elegantly in different volumes of the this treatise. I will, nevertheless, describe briefly the origins of these applications and their main directions, but must neglect most of the recent publications.

The principal applications of plant cell culture techniques are in the area of morphology, biochemistry, pathology, and genetics.

A. Morphological Applications

The morphological field was the first to be considered. It seemed very easy to understand that tissue culture would be a good means to appreciate factors of cell differentiation and organ formation. It was pointed out that the establishment of a tissue culture involves the process of cell dedifferentiation. The possibility of dedifferentiation was not a new concept. It was indeed suggested 130 years ago by Trécul's histological observations on callus development. Plant tissue culture forced differentiated cells to divide and represented therefore an excellent method for following the morphological events of dedifferentiation. This subject was investigated by Buvat (1944, 1945). He demonstrated that in tissue culture the dedifferentiation of cells consists of two steps: regression to the cambial stage, followed by return to the cytological structure of a primary meristematic cell. The first stage can affect highly differentiated cells. Usually, the primary meristematic stage was accompanied by the organization of bud or root meristems. But the utilization of very powerful growth substances such as 2, 4-D permitted the separation of the cytological meristematic stage from the process of organ formation. Following Buvat's pioneer work, many investigations were carried out on the cytology of cells in callus cultures and in suspension cultures. It was recognized that in established strains chromosome numbers showed very extreme variations and aneuploidy was almost a rule. This resulted from irregular mitosis or endomitosis and led to physiological instability, which could be reduced, but not suppressed, by cloning. In addition, many histological investigations of initial explants and the evolution of their histological structure allowed the comparison of histogenesis *in vitro* and in the whole plant. Such comparison could explain the mechanisms involved in the establishment of anatomical structures. Gautheret's treatise provides a detailed account of the histological development of explants (Gautheret, 1959). More recent information can be found in the review by Phillips (1980).

Most of the histological investigations concerned the physicochemical conditions of differentiation. A few attempts were devoted to tissue interactions. These interactions play a part in the establishment of plant organization. When an explant produces a bud, it is easy to note that differentiation occurred at its bottom (Gautheret, 1942c). This phenomenon could be easily analyzed by means of grafting. Such work was initiated by Camus (1945). He observed the interesting phenomenon of histogenetic induction caused by buds. His results were verified by Wetmore and Sorokin (1955), while Rodkiewicz (1952) and Kuroda (1961) demonstrated that the inducing power belonged not to the bud itself but to the cambium tissue. Thus was confirmed Haberlandt's hypothesis on leptome hormone. Curiously, the grafting technique which led to the original results has scarcely been applied in tissue culture.

For a long time, histological investigations concerned only dicotyledons and gymnosperm tissues. This was due to the fact that tissue culture of monocotyledons was realized 13 years after that of dicotyledonous tissues, on account of the lack of a cambium.

Organ formation in tissue cultures was observed a long time ago with tobacco tissue (White, 1939b), carrot explants (Nobécourt, 1939b, 1946, 1955), *Ulmus* cambium cultures, and Witloof tissues (Gautheret, 1942a). Flower initiation in tissue cultures was reported for the first time by Aghion (1962). The phenomenon of organogenesis exhibited by plant tissue cultures gave rise to innumerable investigations, which suggested several theories. The most popular is that proposed by Skoog and Miller (1957), which demonstrated hormonal control of organogenesis. One of the more interesting cases of organization is somatic embryogenesis (Reinert, 1959; Steward *et al.*, 1966; Vasil, 1983, 1985).

Another important application of plant tissue culture is micropropagation. The meristem culture work initiated by Ball (1946) was applied by Morel (1963) to orchids, especially *Cymbidium* and *Odontoglossum*. Morel's method has been exploited all over the world and he became celebrated for this discovery. Later, the method was extended to dicotyledons, gymnosperms, and many monocotyledons. Now this new method of vegetative propagation is exploited intensively in horticulture and the nursery industry for the rapid clonal propagation of plants. More than 1000 species can be cloned by this method, and nowadays the "micropropagation" represents the "voie royale" of plant tissue culture (see the review by Vasil and Vasil, 1980).

B. Production of Metabolites by Cell Cultures

Another possible application of plant tissue culture may be the industrial production of secondary metabolites of economic value. Preliminary experiments by the pioneers established many years ago that tanins (Ball,

1950) or alkaloids (Telle and Gautheret, 1947) disappeared quickly, although not completely, during the first stages of a tissue culture. In those instances where secondary metabolites were produced by specific organs, it was recognized (Paupardin, 1971) that their synthesis was not obtained in callus, which did not undergo such organization. However, when the callus produced buds and specific structures associated with secondary metabolites, terpene synthesis started again. Finally, Benveniste *et al.* (1964) observed that in tobacco tissue cultures there is a deviation of steroid synthesis, which is increased and partly oriented toward new steroids unknown in this plant.

The first attempt for the industrial production of secondary metabolites was made between the 1950s and the 1960s by the Pfizer Company. But in spite of the assistance of Nickell (Fig. 26), distinguished expert of plant tissue culture, this attempt failed and individuals interested in applied work involving plant tissue culture were dissuaded for many years. The problems remained unresolved for many years, but these early disappointments laid the foundation and the fundamental basis of secondary metabo-

Fig. 26. Louis G. Nickell (1921–). A pioneer in the early development of plant tissue cultures, especially their ability to produce secondary metabolites.

lite production by plant tissue cultures. This subject was almost completely forgotten for many years, except with respect to technical improvements that were in advance of profitable production. Clearly, the industrial production of secondary metabolites required equipment allowing suspension cultures of very large masses of cells, and many kinds of apparatus were conceived for this purpose. The available technology was ready to industrially produce secondary products, although the economic aspects were not encouraging.

Then, from 1975, the fundamental aspects of the subject were explored carefully and very intensively in order to find the way to success. Rapid progress in this area attracted many new workers (Bohm, 1980.1980; Staba, 1980, 1982).

Important results were obtained with many species. These investigations were very useful for understanding the fundamental aspects of secondary metabolite production. But, concerning the industrial point of view, the main problem still remains the production *in vitro* of substances more cheaply than they can be obtained from whole plants. Until now, the instances of success have been really exceptional and two opposite points of view have been stated: the first by Melchers (1980), who considered that "confidence in this type of work became very poor in the industrial world," and the second by Staba (1982), who asserted at the Fifth Congress of Plant Tissue and Cell Culture that this method "will be successfully applied by the industry to produce compounds within the decade."

C. Virus Eradication

There have been many valuable contributions of tissue culture to problems concerning pathology. One outstanding success concerned virus eradication. The history of virus eradication began when White (1934b) observed that subcultures of virus-infected isolated roots led frequently to virus-free cultures. Studying thoroughly this peculiarity, he demonstrated that the meristem was deprived of virus and that the virus could be maintained only if the transferred part of the root included a sufficient amount of old tissue.

Two pathologists, Limasset and Cornuet (1949), verified that the lack of viruses in the meristematic cells affects not only the root tips but also buds. They suggested to their colleagues Morel and Martin to cultivate shoot meristems of infected plants. The attempts were a great success and this method allowed recovery of healthy plants from *Dahlia* (Morel and Martin, 1952) and potato (Morel and Martin, 1955). But in both cases the shoots produced no roots and their propagation required grafting on healthy seedlings. Rooting of these shoots was obtained after additional investigations by Quack (1961) and Hollings and Stone (1968). This technique is universally used for virus eradication and has had important consequences at the economic level.

D. Tissue Culture and Plant Tumors

The second success of tissue culture in plant pathology is the result of its application to the problem of plant tumors, especially crown gall. Smith and Townsend (1907) discovered that this tumor was induced by a bacterium, *Agrobacterium tumefaciens,* and stated that it was a cancer. Now, a cancer involves a hereditary change at the cell level and this was not demonstrated for crown gall. This demonstration was obtained when Braun (1941; Fig. 27) showed that in sunflower *Agrobacterium* could induce

Fig. 27. Armin C. Braun (1911–   ). His pioneering and outstanding work on crown gall was the beginning of the modern work on genetic transformation by the tumor-inducing plasmid of *Agrobacterium tumefaciens.*

tumors not only at the inoculated point but at a considerable distance, where secondary tumors free of bacteria are formed (Braun and White, 1943). Cells of these secondary tumors could be cultivated under axenic conditions (Braun and White, 1943) and multiplied on a medium deprived of growth substances, whereas normal tissues required auxins (Gautheret, 1942b) and in many cases cytokinins (Skoog and Miller, 1957). Crown gall tissues deprived of bacteria gave rise to tumors by graftings. The same property belongs to anergied tissues, and in both cases Kulescha (1952) established that they synthesized more auxin than normal tissues. Thanks to a series of remarkable experiments, Braun demonstrated that crown gall tissues deprived of bacteria contain a tumor-inducing principle (T.I.P.), which may be a macromolecule (Braun, 1950).

Grafts made by De Ropp (1948) and by Camus and Gautheret (1948a,b) suggested that the T.I.P. could be transmitted from cell to cell, but this statement was questionable, for it was possible that in these grafts there was no transformation of host cells but only multiplication of the trans- formed tumoral cells. The demonstration was completed by Aaron-da Cunha (1969). She grafted on a stem of piece of axenic crown gall tissue previously treated by a high dose of X rays, in order to suppress their capacity to proliferate. Nevertheless, tumors appeared that were produced by the host cells.

The biochemistry of the crown gall problem was studied by Lioret (1957) who discovered an amino acid called lysopin (Biemann et al., 1960), which seemed specific for this tumor. Ménagé and Morel (1965), Goldmann-Mé- nagé (1970), and Morel (1971) isolated two substances of the same chemical family, octopine and nopaline, and remarked that these opines were not characteristic of the plant but of the bacterium that had induced the tumor. Crown galls induced by some strains of A. tumefaciens synthesize octopin, while tumors produced by other strains elaborate nopaline. Following these observations, Morel (1971) suggested that the synthesis of these opines depends on the presence in the tumor cells of genes coming from the bacterium. In other words, the T.I.P. would consist of DNA. Kerr (1969) reported that the T.I.P. could be transferred from a virulent to a nonvirulent bacterium. Hamilton and Fall (1971) established that some strains lost the T.I.P. when grown at 36°C. Finally, Zaenen et al. (1974) discovered the segment of bacterial DNA which is responsible for the tumoral transformation and opine synthesis. This segment belongs to a large plasmid. Only a small part (about 8%) of the plasmid is stably incor- porated and replicated in plant cells.

Schematically, we can say that this transferred DNA (T-DNA) contains the genetic information which promotes the tumoral transformation and gene coding for octopine or nopaline. Working with cells of teratomas, a kind of tumor capable of producing buds, Braun (1959) reported that in some circumstances the tumoral power is partially lost. Later, it was found that this recovery from the tumorous condition is complete when the shoots developed from teratomas blossom and produce seeds. In this case, the transferred DNA disappears.

Finally, geneticists and molecular biologists were able to establish the map of the crown gall plasmid. Positions of the octopine (or nopaline) and other genes were determined. But the basic mechanism of tumoral transformation has not been completely clarified and presently Braun's work remains the secure foundation of the crown gall problem.

E. Genetics and Tissue Culture

The last and most promising application of plant tissue culture concerns genetics. Hereditary variations can be observed in cell colonies or regenerated plants. In the later case, they are expressed at the time of vegetative multiplication or sexual reproduction.

1. Variability of Cells and Tissues in Culture

Variations concerning cell colonies were first detected by Gautheret (1942b) in the case of anergy. Such alteration had no similarity with the classical genetic process, and therefore it was almost neglected. Nevertheless, it is the most common variation occurring in tissue cultures. Recently, Melchers (1980) wrote that "we still know little about what, since Gautheret, has been called 'accoutumance' (habituation)." Similarly, the first mutation observed by Eichenberger (1951) in a carrot strain has not been explained.

Variability of meristem initiation in callus culture was reported many years ago (Gautheret, 1955; Nobécourt, 1955; Jacquiot, 1964). Following these pioneering investigations, many people described variations in organ formation exhibited by callus cultures. In addition, plants coming from these newly formed buds exhibited many more variations than those observed in the case of sexual reproduction. It is clear that tissue colonies are mosaics of cells of different kinds. Cultivated separately, some cells produced unorganized colonies, while others could regenerate plants frequently affected by morphological anomalies.

Cloning reduces variability but does not suppress it. This indicates that *in vitro* culture is a direct source of variations. Many people took advantage of this characteristic feature, but in return they had to face the almost spontaneous instability of tissue cultures. Caryological observations that were carried out first by Gautheret (1959) revealed abnormal mitosis in tissue culture leading to aneuploid cells. Careful counting performed by De Torok and White (1960) as well as Fox (1963) confirmed the instability of chromosome numbers, and thus was explained the variability of tissue cultures. Innumerable investigations were carried out on this variability and its relation with caryological instability (see, for example, reviews of Sunderland, 1973; Bayliss, 1980).

2. Haploids Produced by Anther Culture

The discovery of Guha and Maheshwari opened the androgenesis field. The first success achieved with *Datura innoxia* was transferred to many Solanaceae and by this technique haploid plants coming from 160 species belonging to about 25 families were obtained (see Vasil, 1980). These haploid plants could produce flowers but no seeds. However, in some cases androgenesis gave rise to diploid and even to polyploid and aneuploid plants. By means of colchicine treatments, it was possible to double the chromosome number and secure fertile plants from haploids. This can be obtained less directly by tissue cultures starting from haploid plants; endomitosis occurring in these haploid tissue cultures sometimes led to diploid buds.

Theoretically, starting from a hybrid, androgenesis followed by chromosome doubling would immediately produce many different stabilized plants. But practically, the drift that so frequently affected the *in vitro* cultures led to erratic and sporadic results which are really disturbing.

Some improvements have been obtained by the haploid breeding technique, for example, with *Asparagus, Triticale,* rice, and tobacco. But this fascinating technique is far from supplanting the classical methods of plant breeding.

3. Hybridization by Protoplast Fusion

Other attempts have been made to escape from the limits of classical genetics. The more promising of these is the culture and fusion of protoplasts. As a matter of fact, protoplast isolation was performed very early by Klercker (1892). The first fusion was achieved by Küster (1909; Fig. 28), who established the facilitating action of some salts and was convinced that, in the future, hybridization by fusion of protoplasts would become possible. It was reasonable to predict that this fusion will enable the breaking of barriers of sexual incompatibility. It is alsmost unknown that Haberlandt (1919) too had observed the multiplication of protoplasts not isolated but confined in plasmolyzed cells. Küster and Haberlandt were ahead of their time. Their results were neglected but were indirectly brought to light by Barski *et al.* (1960), who announced for the first time the

Fig, 28. E. Küster (1874–1953). Famous cytologist who was the first individual to fuse plant protoplasts.

fusion of cells in animal tissue cultures. This work impressed the geneticist Ephrussi (1972) to such an extent that he immediately recovered his old inclination for tissue culture and developed the hybridization of animal somatic cells. But such investigations would be necessarily apart from classical genetics for, in the animal kingdom, a somatic cell will not develop into a complete organism. At that time, the totipotency of plant cells was clearly known and there was no theoretical opposition to achieving somatic hybridization. Simply, some technical barriers had to be removed. Thanks to cellulase, which destroyed the cell wall, Cocking (1960) succeeded in preparing protoplasts, but no other important progress was secured for ten years. The formation of a new cell wall was first observed by Pojnar *et al.* (1967). Soon divisions in the regenerated cell were reported (Nagata and Takebe, 1970; Kao *et al.*, 1970). The next year, plants were obtained from protoplasts by Nagata and Takebe (1971) as well as by Takebe *et al.* (1971). This result was verified very quickly by many people who suddenly rushed to this field of research.

The facilitating action of cations in protoplast fusion was verified by

Fig. 29. George Melchers (1906–   ), with his famous somatic hybrids of potato and tomato obtained by protoplast fusion.

Keller and Melchers (1973), while Wallin *et al.* (1974) and Kao *et al.* (1974) discovered that this process is considerably improved by high-molecular-weight polyethylene glycol (PEG). Protoplasts coming from very distant species can be fused easily with this method.

Successful protoplast hybridization must lead to a plant. This result was obtained for the first time by Carlson *et al.* (1972), by fusion of protoplasts *Nicotiana glauca* with *N. langsdorffii*. But it was not demonstrated that the hybrids were really amphidiploids. Moreover, these hybrids could be obtained by means of sexual crosses as well. However, a true escape from sexual incompatibility was realized by Melchers *et al.* (1978). By fusion of protoplasts from potato and tomato, they obtained hybrids (Fig. 29), even though sexual crossbreeding is not possible between these plants. This result was really a jump over limits of sexual compatibility. But this jump was short, for the potato × tomato hybrid could not be backcrossed with tomato or potato or other Solanaceae.

4. DNA Transfer

The last and very aggressive application of plant tissue and cell culture in genetics is represented by DNA uptake. This consists of four steps: insertion, integration, expression, and replication of foreign DNA into a host cell. These criteria exist in the case of viruses, but cell function is so disturbed that death takes place very quickly. Finally, since 1976, several researchers reported that various eukaryotic genes introduced in bacteria express their specific activity. This represented an extension of transformation and suggested new experiments in order to obtain similar transformation of eukaryotic cells.

This was the basis of the so-called genetic manipulations. And a long and deceptive series of investigations was performed for more than 15 years (see review by Kado and Kleinhofs, 1980).

However, in one case, the transformation of plant cells by DNA uptake was undoubtedly demonstrated. This was in the course of studies on crown gall, and therefore it is exceptional and took place in the last stage of the very long history of tumoral transformation. It must be remembered that *Agrobacterium tumefaciens* can transfer genetic information to plant cells in the form of a plasmid that promotes tumoral transformation and carries a gene coding for one opine.

It has been demonstrated that the transferred part of the plasmid could lose its "oncogenic and opine regions," and this suggested its use as a gene vector. It was possible to introduce in the transferred DNA a gene coming from a yeast that is coding for alcohol dehydrogenase. The modified plasmid was accepted by tobacco cells but the foreign gene showed no activity. However, in similar experiments, a bacterial gene conferring resistance to kanamycin could express this property in plant cells.

Thus was discovered the first key for transferring bacterial genes to eukaryotic cells. Until now the *Agrobacterium* plasmid seems to be the sole vector capable of this transfer, and it would be advisable to avoid excessive speculations.

XI. CONCLUDING REMARKS

It is time to summarize this long history and to recall its principle steps. The main fundamental results were:

1. The establishment of true tissue cultures.
2. The demonstration of plant cell totipotency.
3. The discovery that pollen cells could develop into plants.
4. Overcoming sexual barriers by means of protoplast fusion.

Tissue cultures also revealed the astonishing instability of plant cells when separated from the organism and the fact that the fixed chromosome number is maintained only with meiosis and sexual fusion. Some somatic alterations such as the common anergy are enigmatic. On the whole, classic rules of genetics cannot be applied in tissue culture. This was the origin of many disappointments but opened the way toward a new and promising genetics, the laws of which are still unknown.

To conclude, it can be said that the progress of plant tissue culture was made possible by only a few genuine discoveries. These discoveries did not appear suddenly but after a long and slow journey, unpretentiously covered by pioneers. Most of them are dead. They were my friends, and this historical sketch is dedicated to their memory.

REFERENCES

Aaron-da Cunha, M. I. (1969). Sur la libération par les rayons X d'un principe tumorigène contenu dans les tissus de crown-gall de tabac. C. R. Hebd. Seances Acad. Sci., Ser. D **268**, 318–321.

Aghion, D. (1962). Conditions expérimentales conduisant à l'initiation et au développement de fleurs à partir de la culture stérile de fragments de tige de tabac. C. R. Hebd. Seances Acad. Sci. **255**, 993–995.

Andronescu, D. J. (1919). Germination and further development of the embryo of Zea mays separated from the endosperm. Am. J. Bot. **6**, 443–453.

Auler, H. (1925). Über chemische und anaerobe Tumorbildung bei Pflanzen. Z. Krebsforsch. **22**, 393–403.

Backs-Hüsemann, D., and Reinert, J. (1970). Embryobildung durch isolierte Einzelzellen aus Gewebe Kulturen von Daucus carota. Protoplasma **70**, 49–60.

Ball, E. (1946). Development in sterile culture of stems tips and subjacent regions of Tropaeolum majus L. and of Lupinus albus L. Am. J. Bot. **33**, 301–318.

Ball, E. (1950). Differentiation in callus culture of Sequoia sempervirens. Growth **14**, 295–325.

Barski, G., Sorieul, S., and Cornefert, F. (1960). Production dans les cultures in vitro de deux souches cellulaires en association de cellules de caractère "hybride." C. R. Hebd. Seances Acad. Sc. **251**, 1825–1827.

Bayliss, M. W. (1980). Chromosomal variation in plant tissues in culture. Int. Rev. Cytol. Suppl. **11A**, 113–139.

Beauchesne, G. (1961). Recherches sur les substances de croissance du maïs immature. Ph.D. Thesis, Paris.

Beauchesne, G., and Jouanneau, J. P. (1960). Progrès dans la technique de séparation et l'etude de la fraction neutre des substances de croissance des graines de Maïs immatures. C. R. Hebd. Seances Acad. Sc. **251**, 2396–2398.

Benveniste, P., Hirth, L., and Ourisson, G. (1964). La biosynthèse des stérols dans les cultures de tissue de tabac. Identification du cycloarténol et du méthylène 2-4 cycloarténol. C. R. Hebd. Seances Acad. Sc. **259**, 2284–2287.

Bergmann, L. (1959). A new technique for isolating and cloning cells of higher plants. Nature (London) **184**, 648–649.

Biemann, K., Lioret, L., Asselineau, J., Lederer, F., and Polonsky, J. (1960). Sur la structure chimique de la lysopine, nouvel acide aminé isolé de tissus de crown-gall. Bull. Soc. Chim. Biol. **42**, 979–991.

Bittmann, O. (1925). Ein Beitrag zur künstlichen Erzeugung atypischen Zellenproliferation bei den Pflanzen. Z. Krebsforsch. **22,** 291–296.

Black, L. M. (1946). Plant tumors induced by the combined action of wounds and virus. *Nature (London)* **158,** 56–58.

Blumenthal, F., and Meyer, P. (1924). Über durch Acidum lacticum erzeugte Tumoren auf Mohrrübenscheiben. Z. Krebsforsch. **21,** 250 252.

Bobilioff-Preisser, W. (1917). Beobachtungen an isolierten Palisaden und Schwamparenchymzellen. *Beih. Z. Bot. Zbl.* **33,** 248–274.

Bohm, H. (1980). The formation of secondary metabolites in plant tissue and cell cultures. *Int. Rev. Cytol. Supp.* **11B,** 183–208.

Boll, W. G., and Street, H. E. (1951). Studies on the growth of excised roots. I. The stimulatory effect of molybdenum on the growth of excised tomato roots. *New Phytol.* **50,** 50–75.

Bonner, J. (1937). Vitamin B_1, a growth factor for higher plants. *Science* **85,** 183–184.

Börger, H. (1926). Über die Kultur von isolierten Zellen und Gewebefragmenten. *Arch. Exp. Zellforsch. Besonders Gewebezuecht.* **2,** 123–190.

Bourgin, J. P., and Nitsch, J. P. (1967). Obtention de *Nicotiana* haploides à partir de'étamines cultivées *in vitro*. *Ann. Physiol. Vég.* **9,** 377–382.

Braun, A. C. (1941). Development of secondary tumor and tumor strands in the crown-gall of sunflowers. *Phytopathology* **31,** 135–149.

Braun, A. C. (1950). Thermal inactivation studies on the tumor inducing principle in crowngall. *Phytopathology* **40,** 3.

Braun, A. C. (1959). A demonstration of the recovery of the crown-gall tumor cell with the use of complex tumors of single cell origin. *Proc. Natl. Acad. Sci. U.S.A.* **45,** 932–938.

Braun, A. C., and White, P. R. (1943). Bacteriological sterility of tissues derived from secondary crown-gall tumors. *Phytopathology* **33,** 85–100.

Buckner, J. D., and Kastle, J. H. (1917). The growth of isolated embryos. *J. Biol. Chem.* **29,** 209–213.

Burkholder, P. R., and Nickell, L. G. (1949). Atypical growth of plants. I. Cultivation of virus tumors of *Rumex* on nutrient agar. *Bot. Gaz. (Chicago)* **110,** 426–437.

Burrows, M. T. (1910). The cultivation of tissues of the chick embryo outside the body. *JAMA, J. Am. Med. Assoc.* **55,** 2057–2058.

Burrows, M. T., and Carrel, A. (1912). On the permanent life of tissues outside of the organism. *J. Exp. Med.* **15,** 516–528.

Bürstrom, H. (1953). Physiology of root growth. *Annu. Rev. Plant Physiol.* **4,** 237–252.

Buvat, R. (1944). Recherches sur la dédifférenciation des cellules végétales. *Ann. Sci. Nat. Bot. Biol. Veg.* **5,** 1–130.

Buvat, R. (1945). Recherches sur la dédifférenciation des cellules végétales. *Ann. Sci. Nat. Bot. Biol. Veg.* **6,** 1–119.

Camus, G. (1945). Mise en évidence de l'action différenciatrice des bourgeons d'Endive par la méthode des greffes. *C. R. Hebd. Seances Acad. Sc.* **221,** 570–572.

Camus, G., and Gautheret, R. J. (1948a). Sur la transmission par greffage des propriétés tumorales des tissus de crown-gall. *C. R. Seances Soc. Biol. Ses Fil.* **142,** 15–16.

Camus, G., and Gautheret, R. J. (1948b). Sur le repiquage des proliférations induites sur des fragments de racines de Scorsonère par les tissus de crown-gall et les tissus ayant subi le phénomène d'accoutumance aux hétéro-auxines. *C. R. Seances Soc. Biol. Ses Fil.* **142,** 771–773.

Caplin, S. M., and Steward, F. C. (1948). Effect of coconut milk on the growth of explants from carrot root. *Science* **108,** 655–657.

Carlson, P. S., Smith, H. H., and Dearing, R. D. (1972). Parasexual interspecific plant hybridization. *Proc. Natl. Acad. Sci. U.S.A.* **69,** 2292–2294.

Carrel, A. (1912). On the permanent life of tissues outside of the organism. *J. Exp. Med.* **15**, 516–528.

Chambers, W. H. (1923). Culture of plant cells. *Proc. Soc. Exp. Biol. Med.* **21**, 71–72.

Cocking, E. C. (1960). A method for the isolation of plant protoplasts and vacuoles. *Nature (London)* **187**, 927–929.

Constabel, F. (1958). La culture des tissus de *Juniperus communis*. *Rev. Gen. Bot.* **65**, 390–396.

Cronenberger, L., Vallet, C., and Nétien G. (1955). Sur les flavanediols des cultures de tissus végétaux. *C. R. Hebd. Seances Acad. Sc.* **241**, 1161–1163.

Curtis, J. T., and Nichol, M. A. (1948). Culture of proliferating orchid embryos *in vitro*. *Bull. Torrey Bot. Club* **75**, 358–373.

Czech, H. (1926). Kultur von pflanzichen Gewebezellen. *Arch. Exp. Zellforsch. Besonders Gewebezuecht.* **3**, 176–199.

Dauphiné, A. (1929). Sur le développement d'organes embryonnaires isolés. *C. R. Seances Soc. Biol. Ses Fil.* **102**, 652.

Delarge, L. (1941). Etude de la croissance et de la ramification des racines *in vitro*. *Mem. Soc. R. Sci. Liege* **5**, 1–221.

De Ropp, R. S. (1948). The interaction of normal and crown-gall tumor tissue in *in vitro* grafts. *Am. J. Bot.* **35**, 372–377.

De Ropp, R. S. (1955). The growth and behaviour *in vitro* of isolated plant cells. *Proc. R. Soc. London, Ser. B.* **144**, 86–93.

De Ropp, R. S., Vitucci, J. C., Hutchings, B. J., and Williams, J. H. (1952). Effect of coconut milk fractions on growth of carrot tissue. *Proc. Soc. Exp. Biol. Med.* **81**, 704–705.

De Torok, D., and White, P. R. (1960). Cytological instability in tumors of *Picea glauca*. *Science* **131**, 730–732.

Dietrich, K. (1924). Über die Kultur von embryonen ausserhalb der Samen. *Flora (Jena 1818–1965)* **17**, 379–417.

Duhamel du Monceau, H. L. (1756). "La Physique des Arbres, où Il Est Traitê de l'Anatomie des plantes et de l'Économie Végétale pour Servir d'Introduction au Traité Complet des Bois et des Forêts." P. H. L. Guérin Pub.

Duhamet, L. (1950). Action du lait de coco sur la croissance des tissus de crown gall de Scorsonère cultivés *in vitro*. *C. R. Hebd. Seances Acad. Sc.* **230**, 770–771.

Duhamet, L. (1953). Recherches préliminaires sur les variations du pouvoir de prolifération de cultures de tissus végétaux en fonction du poids de l'explantat ensemencé. *C. R. Seances Soc. Biol. Ses Fil.* **147**, 81–83.

Duhamet, L., and Mentzer, C. (1955). Essais d'isolement des substances excito-formatrices du lait de coco. *C. R. Hebd. Seances Acad. Sc.* **241**, 86–88.

Eichenberger, M. E. (1951). Sur une mutation survenue dans une culture de tissus de carotte. *C. R. Seances Soc. Biol. Jes Fil.* **145**, 239–240.

Ephrussi, B. (1972). "Hybridization of Somatic Cells." Princeton Univ. Press, Princeton, New Jersey.

Fox, J. E. (1963). Growth factor requirements and chromosome number in tobacco tissue cultures. *Physiol. Plant.* **16**, 793–803.

Gautheret, R. J. (1932). Sur la culture d'extrémités de racines. *C. R. Seances Soc. Biol. Ses Fil.* **109**, 1236–1238.

Gautheret, R. J. (1934). Culture du tissus cambial. *C. R. Hebd. Seances Acad. Sc.* **198**, 2195–2196.

Gautheret, R. J. (1935). Recherches sur la culture des tissus végétaux. Ph.D. Thesis, Paris.

Gautheret, R. J. (1939). Sur la possibilité de réaliser la culture indéfinie des tissus de tubercules de carotte. *C. R. Hebd. Seances Acad. Sc.* **208**, 118–120.

Gautheret, R. J. (1942a). Recherches sur le développement de fragments de tissus végétaux cultives *in vitro*. *Rev. Cytol. Cytophys. Veg.* **6**, 87–180.

Gautheret, R. J. (1942b) Hétéro-auxines et cultures de tissus végétaux. *Bull. Soc. Chim. Biol.* **24**, 13–41.

Gautheret, R. J. (1942c). Le bourgeonnement des tissus végétaux en culture. *Science* **40**, 95–128.

Gautheret, R. J. (1942d). "Manuel Technique de Culture de Tissus Végétaux." Masson Publ., Paris.

Gautheret, R. J. (1948). Sur la culture indéfinie des tissus de *Salix caprea*. *C. R. Seances Soc. Biol. Ses Fil.* **142**, 807–808.

Gautheret, R. J. (1955). Sur la variabilité des propriétés physiologiques des cultures de tissus végétaux. *Rev. Gén. Bot.* **62**, 5–112.

Gautheret, R. J. (1959). La culture des tissus végétaux. Masson Publ., pp. 483–490.

Gautheret, R. J. (1966). Factors affecting differentiation of plant tissue grown *in vitro*. In "Cells, Differentiation and Morphogenesis," pp. 55–95, Intern. Lectures, Wageningen.

Gioelli, F. (1938). Morfologia, istologia, fisiologia e fisiopatologia di meristemi secondari *in vitro*. *Atti. Accad. Sci. Ferrara* **16**, 1–87.

Goebel, F. (1902). Über-regeneration in Pflanzenreich. *Biol. Zentralbl.* **22**, 385–397, 417–438, 481–505.

Goldmann Ménagé, A. (1970) "Recherches sur le Métabolisme Azoté de Tissue de Crown Gall Cultivés *in vitro*." Masson Publ., Paris.

Guha, S., and Maheshwari, S. C. (1964). *In vitro* production of embryos from anthers of *Datura*. *Nature (London)* **204**, 497.

Guha, S., and Maheshwari, S. C. (1966). Cell division and differentiation of embryos in the pollen grains of *Datura in vitro*. *Nature (London)* **212**, 97–98.

Györffy, R., Rédei, G., and Rédei, G. (1955). La substance de croissance du Maïs laiteux. *Acta Bot. Acad. Sci. Hung.* **11**, 57–76.

Haberlandt, G. (1902). Kulturversuche mit isolierten Pflanzenzellen. *Sitzungsber. Akad. Wiss. Wien.; Math.–Naturwiss. Kl., Abt. 1* **111**, 69–92.

Haberlandt, G. (1913a). Zur Physiologie der Zellteilung. *Sitzungsber. K. Preuss. Akad. Wiss. (Berlin)* **16**, 318–345.

Haberlandt, G. (1913b). Zur Physiologie der Zellteilung. *Sitzungsber. K. Preuss. Akad. Wiss. (Berlin)* **16**, 1095–1111.

Haberlandt, G. (1919). Zur Physiologie der Zellteilung 3 Mitteilung: Über Zellteilung nach Plasmelyse. *Sitzungsber. K. Preuss. Akad. Wiss. (Berlin)* **20**, 322–348.

Haberlandt, G. (1922). Über Zellteilung Hormone und Adventivembryonie: Ihre Beziehung zur Wundheilung, Befruchtung, Parthenogenese und Adventivembryonie. *Biol. Zentralbl.* **42**, 145–172.

Hamilton, R. H., and Fall, M. Z. (1971). The loss of tumor-initiating ability in *Agrobacterium tumefaciens* by incubation at high temperature. *Experientia* **27**, 229–230.

Hannay, J. W., and Street, H. E. (1954). Studies on the growth of excised root. III. Molybdenum and manganese requirements of excised tomato roots. *New. Phytol.* **2**, 297–302.

Hanning, B. (1904). Über die Kultur von Cruciferen-Embryonen ausserhalb des Embryosachs. *Bot. Zeitung* **62**, 45–80.

Hansen, A. (1881). Vergleichende Untersuchungen über Adventivbildungen bei den Pflanzen. *Abh. Herausg. V. d.Sencrenb. Naturf. Ges.* **12**, 147–198.

Harrison, R. G. (1907). Observations on the living developing nerve fiber. *Proc. Soc. Exp. Biol. Med.* **4**, 140–143.

Hartsema, A. M. (1926). Anatomische und experimentelle Untersuchungen über das Auftreten von Neubildungen an Blattern von *Begonia Rex*. *Rec. Trav. Bot. Neerl.* **23**, 205–361.

Heller, R. (1953). Recherches sur la nutrition minérale des tissus végétaux cultivés *in vitro*. *Ann. Sci. Nat. Bot. Biol. Veg.* **14**, 1–223.

Henderson, J. H. M. (1954). The changing nutritional pattern from normal to habituated sunflower callus tissue *in vitro*. *Ann. Biol.* **30**, 329–348.

Henderson, J. H. M., Durrel, M. E., and Bonner, J. (1952). The culture of normal sunflower stem callus. *Am. J. Bot.* **39**, 467–473.

Hildebrandt, A. C. (1951). *In vitro* experiments on tissues of pathological origin. *In* "Plant Growth Substances," (pp. 391–404). Univ. of Wisconsin Press, Madison.

Hildebrandt, A. C., Riker, A. J., and Duggar, B. M. (1946). The influence of the composition of the medium on growth *in vitro* of excised tobacco and sunflower tissue cultures. *Am. J. Bot.* **33**, 591–597.

Hollings, M., and Stone, A. M. (1968). Techniques and problems in the production of virus tested planting material. *Sci. Hortic. (Canterbury, Engl.)* **20**, 57–72.

Jacquiot, C. (1964). Application de la technique de culture des tissus végétaux à l'étude de quelques problèmes de la physiologie de l'arbre. *Ann. Sci. For.* **21**, 317–473.

Kado, C. I., and Kleinhofs, A. (1980). Genetic modifications of plant cells through uptake of foreign DNA. *Int. Rev. Cytol. Suppl.* **11B**, 47–80.

Kandler, O. (1952). Über eine physiologische Umstimmung von Sonnenblummenstengelgewebe durch Dauereinwirkung von beta indolylessig säure. *Planta* **40**, 346–349.

Kao, K. N., Keller, W. A., and Miller, R. A. (1970). Cell division in newly-formed cells from protoplasts of soybean. *Exp. Cell Res.* **62**, 338–340.

Kao, K. N., Constabel, F., Michayluk, M. R., and Gamborg, O. L. (1974). Plant protoplast fusion and growth of intergeneric hybrid cells. *Planta* **120**, 215–227.

Keller, W. A., and Melchers, G. (1973). The effect of high pH and calcium on tobacco leaf protoplast fusion. *Z. Naturforsch. C: Biochem., Biophys., Biol.* **280**, 737–741.

Kemmer, E. (1928). Beobachtungen über Lebensdauer isolierter Epidermen. *Arch. Exp. Zellforsch. Besonders Gewebezuecht.* **7**, 1–68.

Kerr, A. (1969). Transfer of virulence between isolates of *Agrobacterium. Nature (London)* **223**, 1175–1176.

Klercker, I. A. F. (1892). Eine Methode zur Isolierung lebender Protoplasten. *Oefvers. K. Vetensk. Akad. Foerch. Stockholm* **9**, 463–471.

Knudson, L. (1919). Viability of detached root cap cells. *Am. J. Bot.* **6**, 309–310.

Kögl, F., Haagen-Smit, A. J., and Erxleben, M. (1934). Über ein neues Auxin ("Hetero-Auxin") aus Harn. *Hoppe-Seyler's Z. Physiol. Chem.* **228**, 90–103.

Konar, R. N. (1963). A haploid tissue from the pollen of *Ephedra foliata* Boiss. *Phytomorphology* **13**, 170–174.

Kotte, W. (1922). Kulturversuche mit isolierten Wurzelspitzen. *Beitr. Allg. Bot.* **2**, 413–434.

Kovoor, A. (1954). Action de quelques substances stimulantes d'origine naturelle sur le développement des tissus végétaux cultivés *in vitro Ann. Biol.* **30**, 417–429.

Kovoor, A. (1962). Essai d'isolement d'un régulateur de croissance présent dans le lait de coco. *C. R. Hebd. Seances Acad. Sc.* **255**, 1991–1993.

Kovoor, A. (1966). Recherches sur l'action de quelques extraits d'origine végétale sur la croissance et le développement des tissus végétaux cultivés *in vitro*. *Ann. Sci. Nat. Bot. Biol. Veg.*, 12e serie **7**, 219–352.

Kulescha, Z. (1952). Recherches sur l'élaboration de substances de croissance par les cultures de tissus de quelques végétaux. *Rev. Gén. Bot.* **59**, 92–111, 127–157, 195–208, 241–264.

Kunkel, W. (1926). Über die Kultur von Perianthgeweben *Arch. Exp. Zellforsch. Besonders Gewebezuecht.* **3**, 405–428.

Kuroda, K. (1961). Recherches sur les phénomènes d'induction histogénétiques provoqués par des tissus normaux et tumoraux des plantes. *Annu. Rep. Sci. Works Fac. Sci. Osaka Univ.* **9**, 29–50.

Küster, E. (1909). Über die Verschmelzung nachter Protoplasten. *Ber. Dtsch. Bot. Ges.* **27**, 589–598.

Küster E. (1928). Das Verhalten pflanzenlicher Zellen *in vitro* und *in vivo*. *Arch. Exp. Zellforsch. Besonders Gewebezuecht.* **6,** 28–41.

Lamprecht, W. (1918). Über die Kultur und Transplantation kleiner Blattstückchen. *Beitr. Allg. Bot.* **I,** 353–398.

La Rue, C. D. (1947). Growth and regeneration of the endosperm of maize in culture. *Am. J. Bot.* **34,** 585–586.

La Rue, C. D. (1949). Culture of the endosperm of maize. *Am. J. Bot.* **36,** 798.

La Rue, C. D. (1953). Studies on growth and regeneration in gametophytes and sporophytes of gymnosperms. Abnormal and pathological plant growth. *Brookhaven Symp. Biol.* **6,** 187–208. Letham, D. S. (1963). Zeatin, a factor inducing cell division isolated from *Zea mays. Life Sci.* **2,** 569–579.

Letham, D. S. (1974). Regulators of cell division in plant tissues. The cytokinins of coconut milk. *Physiol. Plant.* **32,** 66–70.

Letham, C. D., Shannon, N. N., and McDonald, I. R. (1964). The structure of Zeatin, a factor inducing cell division. *Proc. Chem. Soc. London* **280,** 230–231.

Levine, M. (1947). Differentiation of carrot root tissue grown *in vitro*. *Bull. Torrey Bot. Club* **74,** 321–328.

Limasset, P., and Cornuet, P. (1949). Recherche du virus de la mosaïque du Tabac dans les méristèmes des plantes infectées. *C. R. Hebd. Seances Acad. Sci.* **228,** 1971–1972.

Lioret, L. (1957). Les acides aminés libres des tissus de crown-gall cultivés *in vitro*. Mise en évidence d'un acide particulier à ces tissus. *C. R. Hebd. Seances Acad. Sci.* **244,** 2171–2174.

Loo, S. W. (1945). Cultivation of excised stem tips of *Asparagus in vitro Am. J. Bot.* **32,** 13–17.

Lutz, A. (1963). Description d'une technique d'isolement cellulaire en vue de l'obtention de cultures de tissus végétaux provenant d'une cellule unique. *C. R. Hebd. Seances Acad. Sci.* **256,** 2676–2678.

Lutz, A. (1966). Obtention de plantes de Tabac a partir de cultures unicellulaires provenant d'une souche anergiée. *C. R. Hebd. Seances Acad. Sc.* **262,** 1856–1858.

Lutz, A. (1969). Etude des aptitudes morphogénétiques des cultures de tissus. Analyse par la méthode des clones d'origine unicellulaire. *Rev. Gen. Bot.* **76,** 309–359.

McClary, J. E. (1940). Synthesis of thiamin by excised roots of maize. *Proc. Natl. Acad. Sci. U.S.A.* **26,** 581–587.

Malpighi, M. (1675). Anatome plantarum. Regiae Societati Londoni ad Scientiam Naturalem Promovendem Institutae Dicta. Johannis Martyn Impensis, London.

Marchal, J. G., and Mazurier, A. (1944). Culture *in vitro* de cellules végétales isolées. *Bull. Soc. Bot. Fr.* **91,** 76–77.

Margara, J., Rancillac, M., and Bouniols, A. (1967). La néoformation *in vitro* de bourgeons inflorescentiels chez *Cichorium intybus* L. Etude méthodologique. *Colloq. Int. C. N. R. S.* **167,** 71–82.

Mayer, G. G. (1929). Der Einfluss verschiedenen Nährstoffzuführung auf das Lägenwachstum isolierter Wurzeln. Ph D. Thesis, Giessen.

Melchers, G. (1975). Theodor Schmucker, 1894–1970. *Ber. Dtsch. Bot. Ges.* **88,** 473–484.

Melchers, G. (1980). The future. *Int. Rev. Cytol. Suppl.* **11B,** 241–253.

Melchers, G., Sacristan, M. D., and Holder, A. (1978). Somatic hybrid plants of potato and tomato regenerated from fused protoplasts. *Carlsberg Res. Commun.* **43,** 203–218.

Ménagé, A., and Morel, G. (1965). Sur la présence d'un acide aminé nouveau dans les tissus de crown-gall. *C. R. Hebd. Seances Acad. Sci.* **261,** 2001–2002.

Miller, C. (1961). A kinetin-like compound in maize. *Proc. Natl. Acad. Sci. U.S.A.* **47,** 170–174.

Miller, C., Skoog, F., Okumura, F. S., Von Saltza, M. H., and Strong, F. M. (1955a). Structure and synthesis of kinetin. *J. Am. Chem. Soc.* **77,** 2662–2663.

Miller, C., Skoog, F., Von Saltza, M. H., and Strong, F. M. (1955b). Kinetin, a cell division factor from desoxyribonucleic acid. *J. Am. Chem. Soc.* **77,** 1392.

Morel, G. (1950a). Sur la culture des tissus de deux monocotylédones. *C. R. Hebd. Seances Acad. Sci.* **230**, 1099–1101.

Morel, G. (1950b). Sur la culture des tissus d'*Osmunda cinnamonea*. *C. R. Hebd. Seances Acad. Sc.* **230**, 2318–2320.

Morel, G. (1963). La culture *in vitro* du méristème apical de certaines orchidées. *C. R. Hebd. Seances Acad. Sc.* **256**, 4955–4957.

Morel, G. (1971). Déviations du métabolisme azoté des tissus de crown-gall. *Colloq. Int. C. N. R. S.* **193**, 463–471.

Morel, G., and Martin, C. (1952). Guérison de dahlias atteints d'une maladie á virus. *C. R. Hebd. Seances Acad. Sc.* **235**, 1324–1325.

Morel, G., and Martin, C. (1955). Guérison de pommes de terre atteintes de maladies à virus. *C. R. Seances Acad. Agri. Fr.* **41**, 472–475.

Muir, W. H., Hildebrandt, A. C., and Riker, A. J. (1954). Plant tissue cultures produced from single isolated plant cells. *Science* **119**, 877–878.

Muir, W. H., Hildebrandt, A. C., and Riker, A. J. (1958). The preparation, isolation and growth in culture of single cells from higher plants. *Am. J. Bot.* **45**, 585–597.

Murashige, T., and Skoog, F. (1962). A revised medium for rapid growth and bioassays with tobacco tissue cultures. *Physiol. Plant.* **15**, 473–497.

Nagata, T., and Takebe, I. (1970). Cell wall regeneration and cell division in isolated tobacco mesophyll protoplasts. *Planta* **92**, 301–308.

Nagata, T., and Takebe, I. (1971). Plating of isolated tobacco mesophyll protoplasts on agar medium. *Planta* **99**, 12–20.

Nétien, G., and Beauchesne, G. (1954). Essai d'isolement d'un facteur de croissance présent dans un extrait laiteux de caryopses de maïs immatures. *C. R. Seances Soc. Biol. Ses Fil.* **30**, 437–443.

Nétien, G., Beauchesne, G., and Mentzer, C. (1951). Influence du lait de maïs sur la croissance des tissus de carotte *in vitro*, *C. R. Hebd. Seances Acad. Sci.* **233**, 92–93.

Nickell, L. G. (1949). The effect of certain plant hormones on the growth and respiration of virus tumors of *Rumex. Am. J. Bot.* **36**, 826–827.

Nickell, L. G. (1954). Nutritional aspects of virus tumor growth. *Brookhaven. Symp. Biol.* **6**, 174–186.

Nickell, L. G. (1961). Sur la perte des besoins en vitamine B1 par des tissus végétaux cultivés *in vitro. C. R. Hebd. Seances Acad. Sc.* **253**, 182–184.

Nitsch, J. P., and Nitsch, C. (1955). Action synergique des auxines et du jus de tomate sur la croissance de tissus végétaux cultivés *in vitro. Bull. Soc. Bot. Fr.* **102**, 519–532.

Nitsch, J. P., and Nitsch, C. (1956). Auxin-dependent growth of excised *Helianthus tuberosus* tissues. *Am. J. Bot.* **43**, 839–851.

Nitsch, J. P., and Nitsch, C. (1957). Action de l'urée et de ses dérivés sur la croissance de fragments de plantes. *Bull. Soc. Bot. Fr.* **104**, 24–33.

Nobécourt, P. (1927). Contribution à l'étude de l'immunité chez les végétaux. Ph.D. Thesis, Université de Lyon, Lyon.

Nobécourt, P. (1937). Cultures en série de tissus végétaux sur milieu artificiel. *C. R. Hebd. Seances Acad. Sc.* **200**, 521–523.

Nobécourt, P. (1939a). Sur la pérennité et l'augmentation de volume des cultures de tissus végétaux. *C. R. Seances Soc. Biol. Ses Fil.* **130**, 1270–1271.

Nobécourt, P. (1939b). Sur les radicelles naissant des cultures de tissus végétaux. *C. R. Seances Soc. Biol. Ses Fil.* **130**, 1271–1272.

Nobécourt, P. (1946). Productions de nature caulinaire et foliaire sur des cultures de racines de carotte. *C. R. Seances Soc. Biol. Ses Fil.* **140**, 953–954.

Nobécourt, P. (1955). Variations de la morphologie et de la structure de cultures de tissus végétaux. *Ber. Schweiz. Bot. Ges.* **65**, 475–480.

Paupardin, C. (1971). Sur l'évolution des huiles essentielles dans les tissus de péricarpe de citron (*Citrus liminia* Obseck) cultivés *in vitro*. *C. R. Hebd. Seances Acad. Sci.* **273**, 1690–1693.

Pfeiffer, H. (1931). Beobachtungen an Kulturen nakter Zellen aus pflanzlichen Beerenperikarpien. *Arch. Exp. Zellforsch. Besonders Gewebezuecht.* **11**, 424–432.

Phillips, R. (1980). Cytodifferentiation. *Int. Rev. Cytol. Suppl* **11A**, 55–70.

Pojnar, E. Willison, J. H. M., and Cocking, E. C. (1967). Cell wall regeneration by isolated tomato fruit protoplasts. *Protoplasma* **64**, 460–475.

Quack, F. (1961). Heat treatment and substances inhibiting virus multiplication in meristem culture to obtain virus-free plants. *Adv. Hortic. Sci. Their Appl., Proc. Int. Hortic. Congr., 15th, 1958* **1**, 144–148.

Rao, A. N. (1963). Organogenesis in callus cultures of Orchid seeds. *In* "Plant Tissue and Organ Culture", pp. 332–344. Symposium Intern. Soc. of Plant Morphologists. Dehli.

Rechinger, C. (1893). Untersuchungen über die Grenzen der Teilbarkeit im Pflanzenreich. *Abh. Zool. Ges. Wien* **43**, 310–334.

Rehwald, C. (1927). Über pflanzliche Tumoren als vermeintliche Wirkung chemischen Reizung. *Z. Pflanzenkr. Pflanzenschutz* **37**, 65–86.

Reinert, J. (1959). Über die Kontrolle der Morphogenese und die Induktion von Adventivembryonen an Gewebekulturen aus Karotten. *Planta* **53**, 318–333.

Reinert, J., and White, P. R. (1956). The cultivation *in vitro* of tumor tissues and normal tissues of *Picea glauca. Physiol. Plant.* **9**, 177–189.

Reinert, J., Bachs-Hüsemann, D., and Zerban, H. (1971). Determination of embryo and root formation in tissue cultures from *Daucus carota. Colloq. Int. C.N.R.S.* **193**, 261–268.

Riker, A. J., and Hildebrandt, A. C. (1951). Pathological plant growth. *Annu. Rev. Microbiol.* **5**, 223–240.

Robbins, W. J. (1922a). Cultivation of excised root tips and stem tips under sterile conditions. *Bot. Gaz. (Chicago)* **73**, 376–390.

Robbins, W. J. (1922b). Effect of autolysed yeast and peptone on growth of excised corn root tips in the dark. *Bot. Gaz. (Chicago)* **74**, 59–79.

Robbins, W. J., and Bartley, M. A. (1937). Vitamin B_1 and the growth of excised tomato roots. *Science* **85**, 246–247.

Rodkiewicz, B. (1952). Observations sur la soudure des tissus de la racine de carotte cultivés *in vitro. Acta. Soc. Bot. Pol.* **21**, 789–901.

Rossini, L. (1969). Une nouvelle méthode de culture *in vitro* de cellules parenchymateuses séparées de feuilles de *Calystegia sepium* L. *C. R. Hebd. Seances Acad. Sci.* **268**, 683–685.

Sachs, J. (1880–1882). Stoff und der Pflanzen Organe. *Arch. Bot. Inst. Wurzburg* **2**, 453–689.

Sanford, K. K., Earle, W. R., and Likely, G. D. (1948). The growth *in vitro* of single isolated tissue cells. *JNCI, Natl. Cancer Inst.* **9**, 229–246.

Scheitterer, H. (1931). Versuche zur Kultur von Pflanzengeweben. *Arch. Exp. Zellforsch. Besonders Gewebezuecht.* **12**, 141–176.

Schleiden, M. J. (1838). Beiträge zur Phytogenesis. *Müllers Arch. Anat. Physiol.*, pp. 137–176.

Schmucker, T. (1929). Isolierte Gewebe und Zellen von Blütenpflanzen. *Planta* **9**, 339–340.

Schwann, T. (1839). "Mikroscopische Untersuchungen über die Übereinstimmung in der Struktur und dem Wachstum der Thiere und Pflanzen," No. 176. Oswalds, Berlin.

Shantz, E. M. and Steward, F. C. (1955). The identification of compound A from coconut milk as 1,3-diphenylurea. *J. Am. Chem. Soc.* **77**, 6351–6353.

Sheat, D. E. G., Flechter, B. H., and Street, H. E. (1959). Studies on the growth of excised roots. VIII. The growth of excised tomato roots supplied with various inorganic sources of nitrogen. *New Phytol.* **58**, 128–141.

Simon, S. (1908). Experimentelle Untersuchungen über die Differenzierung Vorgäangen Callusgewebe von Holzgewächsen. *Jahrb. Wiss. Bot.* **45**, 351–478.

58   R. J. Gautheret

Skoog, F., and Miller, C. O. (1957). Chemical regulation of growth and organ formation. *Symp. Soc. Exp. Biol.* **11**, 118–131.
Skoog, F., and Tsui, C. (1948). Chemical control of growth and bud formation in tobacco stem segments and callus cultured *in vitro*. *Am. J. Bot.* **35**, 782–787.
Smith, E. F., and Townsend, C. O. (1907). A plant tumor of bacteriological origin. *Science* **25**, 671–673.
Snow, R. (1935). Activation of cambial growth by pure hormones. *New Phytol.* **34**, 347–360.
Staba, J. (ed.) (1980). "Plant Tissue Culture as a Source of Biochemicals." CRC Press, Boca Raton, Florida.
Staba, J. (1982). Production of useful compounds from plant tissue cultures. *In* "Plant Tissue Culture" (A. Fujiwara, ed) pp. 25–30. Maruzen, Tokyo.
Steward, F. C., and Caplin, S. M. (1952). Investigation on growth and metabolism of plant cells. IV. Evidence on the role of the coconut milk factor development. *Ann. Bot. (London)* **16**, 491–504.
Steward, F. C., Caplin, M., and Millar, F. K. (1952). Investigations on growth and metabolism of plant cells. *Ann. Bot. (London)* **16**, 57–77.
Steward, F. C., Kent, A. E., and Mapes, M. O. (1966). The culture of free plant cells and its signification for embryology and morphogenetics. *In* "Current Topics in Developmental Biology" (A. A. Moscona and A. Monroy, eds.), pp. 113–154. Academic Press, New York.
Stingl, G. (1907). Experimentelle Studie über die Ernährung von pflanzlichen Embryonen. *Flora (Jena, 1818–1965)* **97**, 308–331.
Street, H. E., McGonagle, M. P., and McGregor, S. M. (1952). Observations on the "staling" of White medium by excised tomato roots. II. Iron availability. *Physiol. Plant* **5**, 243–276.
Sunderland, M. (1973). Nuclear cytology. *In* "Plant Tissue and Cell Culture" (H. E. Street, ed)., pp. 161–190. Blackwell Publications.
Sunderland, N. (1977). Observations on anther culture of ornamental plants. *In* "La Culture des Tissus et des Cellules des Végétaux. Travaux Dédiés à la Mémoire de Georges Morel" (R. J. Gautheret, ed.), pp. 34–46. Masson Publ., Paris.
Takebe, I., Labib, C., and Melchers, G. (1971). Regeneration of whole plants from isolated mesophyll protoplasts of tobacco. *Naturwissenschaften* **58**, 318–320.
Telle, J., and Gautheret, R. J. (1947). Sur la culture indéfinie des tissus de la racine de Jusquiame (*Hyoscyamus niger* L.). *C. R. Hebd. Seances Acad. Sci.* **224**, 1653–1654.
Thielman, M. (1924). Über Kulturversuche mit Spaltöffnungszellen. *Ber. Dtsch. Bot. Ges.* **42**, 429–433.
Thorpe, T. A. (1980). Organogenesis *in vitro*: Structural, physiological and biochemical aspects. *Int. Rev. Cytol. Suppl.* **11A**, 71–111.
Torrey, J. G. (1957). Cell division in isolated single plant cells *in vitro*. *Proc. Natl. Acad. Sci. U.S.A.* **43**, 887–891.
Trécul, A. (1853). Accroissement des végétaux dicotylédones ligneux (reproduction du bois et de l'écorce par le bois décortiqué). *Ann. Sci. Nat. Bot. Biol. Veg.* **29**, 157–192.
Tulecke, W. R. (1953). A tissue derived from the pollen of *Ginkgo biloba*. *Science* **117**, 599–620.
Ulehla, V. (1928). Vorversuche zur Kultur des Pflanzengewebes. I. Das Wasser als Faktor des Gewebekultur. *Arch. Exp. Zellforsch. Besonders Gewebezuecht.* **6**, 370–417.
Uspenski, E. E., and Uspenskaja, W. J. (1925). Reinkultur und ungeschlechtliche Fortpflanzung des *Volvox minor* und *Volvox globator* in einer synthetischen Nährlösung. *Z. Bot.* **17**, 273–308.
Van Overbeek, J., Conklin, M. E., and Blakeslee, A. F. (1941). Factors in coconut milk essential for growth and development of very young *Datura* embryos. *Science* **94**, 350–351.
Van Staden, J., and Drewes, S. E. (1975). Identification of zeatin and zeatin riboside in coconut milk. *Physiol. Plant* **34**, 106–109.

Vasil, I. K. (1980). Androgenic Haploids. *Int. Rev. Cytol. Suppl* **11A**, 194–223.

Vasil, I. K. (1983). Regeneration of plants from single cells of cereals and grasses. *In* "Genetic Engineering in Eukaryotes" (P. F. Lurquin and A. Kleinhofs, eds.), pp. 233–252. Plenum Press, New York.

Vasil, I. K. (1985). Somatic enbryogenesis and its consequences in the Gramineae. *In* "Propagation of Higher Plants Through Tissue Culture: Development and Variation" (R. Henke, K. Hughes, and A. Hollaender, eds.). Plenum Press, New York.

Vasil, V., and Hildebrandt, A. C. (1965a). Growth and tissue formation from single isolated tobacco cells in microculture. *Science* **147**, 1454–1455.

Vasil, V., and Hildebrandt, A. C. (1965b). Differentiation of tobacco plants from single isolated cells in microcultures. *Science* **150**, 889–890.

Vasil, I. K., and Vasil, V. (1980). Clonal propagation. *Int. Rev. Cytol. Suppl.* **11A**, 145–179.

Vöchting, II. (1878). "Über Organbildung im Pflanzenreich." Max Cohen Publ., Bonn.

Wallin, A., Glimelius, K., and Eriksson, T. (1974). The induction of aggregation and fusion of *Daucus carrota* protoplasts by polyethylene glycol. *Z. Pflanzenphysiol.* **74**, 64–80.

Went, F. W. (1926). On growth accelerating substances in the coleoptile of *Avena sativa*. *Proc. K. Ned. Akad. Wet., Ser. C* **30**, 10.

Wetmore, R. H., and Rier, J. P. (1963). Experimental induction of vascular tissue in callus of angiosperms. *Am. J. Bot.* **50**, 418–430.

Wetmore, R. H. and Sorokin, S. (1955). On the differentiation of xylem. *J. Arnold Arbor.* **36**, 305–317.

White, P. R. (1933). Results of preliminary experiments on the culturing of isolated stem-tips of *Stellaria media. Protoplasma* **19**, 97–116.

White, P. R. (1934a). Potentially unlimited growth of excised tomato root tips in a liquid medium. *Plant Physiol.* **9**, 585–600.

White, P. R. (1934b). Multiplication of the viruses of tobacco and Aucuba mosaics in growing excised tomato root tips. *Phytopathology* **24**, 1003–1011.

White, P. R. (1937). Vitamin B1 in the nutrition of excised tomato roots. *Plant Physiol.* **12**, 803–811.

White, P. R. (1939a). Potentially unlimited growth of excised plant callus in an artificial nutrient. *Am. J. Bot.* **26**, 59–64.

White, P. R. (1939b). Controlled differentiation in a plant tissue culture. *Bull. Torrey Bot. Club* **66**, 507–513.

White, P. R. (1943). "A Handbook of Plant Tissue Culture." Jacques Cattel Press.

White, P. R., and Braun, A. C. (1942). A cancerous neoplasm of plants. Autonomous bacteria-free crown-gall tissue. *Proc. Am. Philos. Soc.* **86**, 467–469.

Wiesner, J. (1884). Untersuchungen über die Wachstumsbewegungen der Wurzeln. *Sitzungsber. Akad. Wiss. Wien., Math.-Naturwiss Kl., Abt 1* **89**, 223–302.

Winkler,H. (1908). Besprechung der Arbeit G. Haberlandt's Kultur Versuche mit isolierten Pflanzenzellen. *Bot. Z.* **60**, 262–264.

Yamada, T., Shoji, T., and Sinoto, Y. (1963). Formation of calli and free cells in a tissue culture of *Tradescantia reflexa. Bot. Mag.* **76**, 332–339.

Zaenen, I., van Larebeke, N., Touchy, H., Van Montagu, M., and Schell, J. (1974). Supercoiled circular DNA in crown-gall inducing *Agrobacterium* strains. *J. Mol. Biol.* **86**, 109–127.

CHAPTER 2

Dynamics of Plant Cell Cultures

K. Lindsey
M. M. Yeoman

Department of Botany
University of Edinburgh
Edinburgh, Scotland

I. INTRODUCTION

The wounding of plant tissues during explantation initiates a series of metabolic changes in those tissues, which can be sustained and to some extent regulated by the transfer of the cells to appropriate culture conditions (Yeoman and Forche, 1980). As the tissues proliferate in a more or less disorganized manner, the expression of primary and secondary metabolism is disturbed (Aitchison *et al.*, 1977). By manipulation of the chemical and physical environments so that reorganization of cells is encouraged, as for example in embryogenesis, root and shoot development, and even by immobilizing cells, "normal" metabolic activity characteristic of the intact

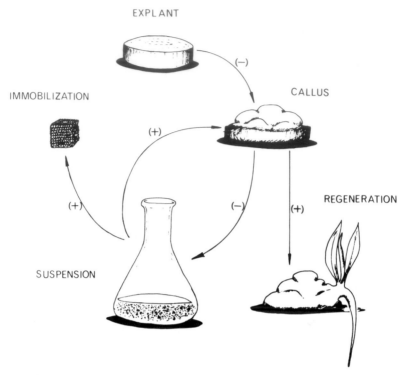

Fig. 1. A schematic diagram representing the dynamics of morphology and secondary metabolic activity in plant cell cultures. A relative increase in the levels of secondary compounds is indicated by a plus and a relative decrease by a minus.

plant may be restored (Fig. 1). Secondary metabolism can be considered to be an aspect of cell differentiation (Luckner, 1980; Yeoman *et al.*, 1982b), and the study of its expression during the transition from explant to cell culture to regenerated plant provides us with an insight into the more complex dynamics of differentiation and development.

Even when maintained under apparently stable culture conditions, cultured cells exhibit variation in morphology and metabolism which is the product of the physiological and genetic constitution of the original (explanted) cells and the "developmental stimuli" supplied in the form of the culture conditions. Street (1977) has described the growth and development of a cell culture as comprising a succession of physiological states, each characterized by distinctive structural and biochemical features. In this article, therefore, we will consider the nature and origins of such heterogeneity, with particular reference to the dynamics of secondary metabolic activity, and attempt to define some of the factors which appear to play roles in the regulation of biochemical aspects of cell differentiation.

II. SPATIAL HETEROGENEITY IN CELL CULTURES

A. The Expression of Variation

Throughout a single population of cultured cells there can invariably be found differences in structure, in size, in karyotype, and in metabolic behavior. Such variability can be termed "spatial" heterogeneity, and the range of variation largely depends on the stage of the growth cycle and on the culture conditions. Structural differences have been reviewed by, for example, Yeoman and Street (1977). Differences between cells in their karyotype (Vanzulli et al., 1980) and particularly in their ability to accumulate secondary metabolites have been the subject of much interest. Davey et al. (1971) have studied various characteristics of clones isolated from cell cultures of *Atropa belladonna* and found stable differences in growth rate, nutrient requirements, callus and cell morphology, cellular fine structure, and pigmentation. Tabata and Hiraoka (1976) and Ogino et al. (1978) have described differences in the growth characteristics and nicotine contents of cultured tobacco cells and have indicated how cell line selection techniques can be used to produce high-yielding cultures. Mok et al. (1976) have shown how, from red roots of carrot (*Daucus carota*), cultures could be obtained which were yellow and orange, and carotenoid synthesis could be manipulated by kinetin and 2,4-dichlorophenoxyacetic acid (2,4-D). The anthocyanin content of cultured cells of carrot and other species is also variable (Fig. 2) and is similarly affected by exogenous cytokinin levels (Kinnersley and Dougall, 1980a). Furthermore, variability arises in single-cell clones, indicating that a certain drift in biosynthetic stability occurs (Dougall et al., 1980). Zenk et al. (1977) have described variation in the capacity to accumulate indole alkaloids in *Catharanthus roseus* cultures, and detailed studies of cell lines of this species are being undertaken in Constabel's laboratory (Kutney et al., 1980; Kurz et al., 1981).

It will be apparent, therefore, that cell cultures provide a valuable system for studying biosynthetic and, in general, developmental variation and stability, and we shall consider some of the factors which influence the establishment and proliferation of heterogeneity in cell cultures.

B. The Tissue Explant as a Source of Heterogeneity

The explant used for callus initiation is usually heterogeneous in cell type. Even in apparently homogeneous tissue, however, there is evidence for genetic and epigenetic differences between cells of a single tissue, and

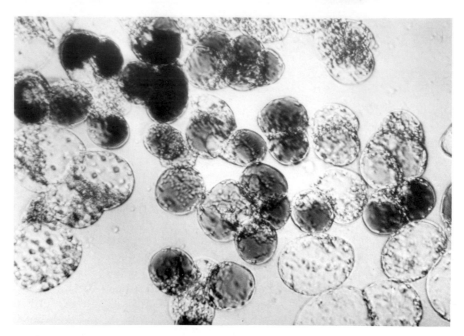

Fig. 2. Heterogeneity in the anthocyanin content of cultured cells of *Catharanthus roseus*. Accumulation is restricted to the stationary phase of the growth cycle, when the largest proportion of accumulating cells is observed.

these will be considered in detail below. With such heterogeneity in source material it would be expected that explant cells would respond differently to the developmental stimuli to which they are subjected *in vitro*. Thus, for example, Yeoman and Davidson (1971) have studied the initiation of division in cultured explants of Jerusalem artichoke (*Helianthus tuberosus*) tuber. They found that division is restricted to the peripheral cells of the explant but the proportion of dividing cells can be altered by changing the light regime. Other factors determining peripheral cell division have been reviewed previously (Aitchison *et al.*, 1977) and include the wound response at the cut surface, differences in availability of atmospheric gases and nutrients between cells of the explant, and differences in the release of inhibitors of division. Torrey (1965) has found that pea seedling root explants contain two distinct populations of cells which differ in their nutritional requirements for growth, and Snijman *et al.* (1977) have further found that variability in the growth response of cultured cells of *Nicotiana tabacum* pith explants can be related to differences in the endogenous levels of growth regulators.

As explant cells prepare for and undergo cell division, a radical change in the pattern of metabolism is initiated. A number of the major primary metabolic changes have been characterized in the Jerusalem artichoke system (see Yeoman and Forche, 1980, for review), and these can be summarized as follows. During callus induction, there are recognized two phases, namely the "wound response" and the "growth response." The former may be identified (Yeoman and Aitchison, 1973) when explants are cultured on an auxin-free medium and is characterized by a rapid increase in metabolic activity which, in all but a few isolated examples, does *not* lead to callus formation. The "growth response," on the other hand, does result in cell division and is dependent on an exogenous supply of auxin (MacLeod *et al.*, 1979). During the growth response there are changes in cell structure apparent under the light microscope (Yeoman and Street, 1977) and the electron microscope (Israel and Steward, 1966; Bagshaw *et al.*, 1969; Vasil, 1973). These structural changes, including increased numbers of mitochondria and polyribosomes, and the disappearance of storage products such as starch and phytoferritin, reflect concomitant changes in metabolic activity. Thus the wound response and the growth response are both characterized by distinct periods of RNA synthesis (MacLeod *et al.*, 1979), and the levels of tRNA and rRNA are regulated independently (MacLeod, unpublished data). In the Jerusalem artichoke tuber explant system, there are transient periods of accumulation of RNA at discrete stages, and a similar pattern emerges when the activities of enzymes associated with nucleic acid synthesis are examined. Ribonuclease activity peaks at cytokinesis (Yeoman *et al.*, 1978), and thymidine kinase, thymidine monophosphate kinase, and DNA polymerase show cell-cycle-dependent increases in activity (Harland *et al.*, 1973). The rate of respiration increases on the transfer of explanted tissues to a culture medium (Steward *et al.*, 1958; Evans, 1967), as do the synthesis and accumulation of total proteins (Nicholson and Flamm, 1965; Aitchison *et al.*, 1977). The activities of enzymes associated with respiration, such as hexokinase, glucose-6-phosphate dehydrogenase, and malic dehydrogenase, also increase before DNA replication (Yeoman and Aitchison, 1973; Aitchison and Yeoman, 1974a,b), and in at least one case, i.e., for glucose-6-phosphate dehydrogenase, this is due to net synthesis of the enzyme (Yeoman and Aitchison, 1976).

There is some evidence that differences in biosynthetic potential between cell cultures are influenced directly by the biosynthetic potential of the plant from which the culture was derived. There is an indication that plants which accumulate relatively high yields of specific secondary metabolites produce high-yielding tissue cultures, and vice versa. This might be expected, since the capacity for the biosynthesis of a secondary product is a genetically determined characteristic. Zenk *et al.* (1977) have measured and compared the levels of the indole alkaloid serpentine in cell cultures derived from high- and low-yielding plants of *Catharanthus roseus*, the Madagascar periwinkle. They found that cultures from high-yielding plants (which accumulated 0.93% dry weight of alkaloid compared with 0.17% in low-yield plants) produced greater quantities of serpentine (70 mg/liter of medium) than did cultures from the low-yielding intact plants (16 mg/liter). There was much variation in the alkaloid production of cultures derived from homozygous seeds, but the differences between the two types of cell line were distinct enough to suggest a significant difference in their biosynthetic potential. Kinnersley and Dougall (1980b) have similarly found that the nicotine contents of cell lines derived from low-alkaloid tobacco cultivars were significantly lower than in cultures derived from high-alkaloid plants, despite the fact that the cultivars were isogenic, except for the two loci, for alkaloid accumulation.

These examples illustrate how the genetic composition of the intact plant determines, broadly, the synthetic capabilities of cultured cells. There is also some evidence that the secondary metabolic characteristics of individual organs or tissues of a plant may directly influence the ability of cultured cells derived from that organ to produce secondary compounds. Kadkade (1982), for example, has found evidence that the contents of podophyllotoxin and other lignanes in cultures of *Podophyllum peltatum* were dependent on the plant part from which they were derived, as well as on the conditions used for callus initiation. Furthermore, Nagel and Reinhard (1975) have found some correlation between the types of volatile oil accumulated in *Ruta graveolens* callus and the organ from which the callus had been derived. However, in apparent contradiction to these results, there is evidence that, as explant cells "dedifferentiate," the pattern of secondary metabolic activity characteristic of a particular tissue or organ is temporarily lost, although the genetic basis for the biosynthetic potential appears to remain. Speake *et al.* (1964) have concluded from their study of alkaloid metabolism in cultured tobacco cells that there is no difference in nicotine productivity in callus cultures derived from root, stem, or leaf. Ibrahim *et al.* (1971) have shown that callus tissues of acyanic or cyanic species, whether initiated from tuber, cotyledon, or storage root, possess the complete machinery for *de novo* anthocyanin synthesis, if the culture conditions are appropriate. Furthermore, callus cultures initiated from various parts of *Hyoscyamus niger* seedlings were found by Dhoot and Henshaw (1977) to accumulate similar levels of alkaloids. Such results are consistent with a view of the cell being a totipotent entity, i.e., retaining, during development, a full genetic complement (with the exception of anucleate xylem and phloem sieve cells). If this concept of totipotency, i.e., *genetic* totipotency, is adhered to strictly, then most of the available evidence is not found to be in disagreement. However, there is now some doubt concerning the practical value of this term (see Henshaw *et al.*, 1982). It is becoming apparent that there are differences between cells of a nongenetic nature, which may, in some cases, act to restrict the flexibility of development implied by the term "totipotency." Let us briefly consider some of the characteristics which are carried over from explant to culture.

The ability of plant species to produce callus cultures depends, in a number of cases, on the part of the plant from which the explanted material was obtained. The most rapidly growing callus is usually obtained from the youngest, and physiologically most active, parts. King (1980) suggests that a number of morphological and presumably therefore also metabolic characteristics of cell lines are not determined by the culture conditions alone but may also be the result of an undefined critical event. This event may be more a function of the species and the nature of the explant than of the experimental conditions. For example, Henke et al. (1978) found that callus derived from roots and shoots of rice (Oryza sativa) required different auxin concentrations for induction and growth. Similarly, Meijer and Broughton (1981) found that root and hypocotyl callus of the legume Stylosanthes guyanensis showed different responses, in terms of callus development and root and shoot formation to the same hormone regimes. Radin et al. (1982) also found stable morphological differences between callus cultures obtained from inflorescences of Guayule (Parthenium argentatum). The difficulties of obtaining rapidly growing and friable cultures of cereal plants are well known, and the problems may in time be overcome as a result of studies on the epigenetic stability of phenotype rather than by any breakthrough in cell culture techniques (Vasil and Vasil, 1984a,b).

A clear example of how stable developmental patterns which obtain in the intact plant can be transferred to cell cultures is given by the transmission of juvenile and mature characterisics of Hedera helix (English ivy). The characteristics of juvenile and mature plants originate in and develop from the shoot meristem, and callus derived from stem tissue retains certain morphological and physiological features associated with the original, explanted material. Stoutemeyer and Britt (1965; 1969) have described how callus derived from explanted seedling stem grows more quickly and is composed of larger cells than callus from stems of adult ivy. Robbins and Hervey (1970) confirmed this result and described how seedling callus is more variable in nature and perhaps as a consequence better able to adapt to different culture conditions. Although there was always found to be a significant difference in the growth rate of callus from juvenile and adult plants, there was evident a certain variability in growth response of callus derived from different parts of an individual plant, so that proliferation in the fastest-growing adult callus cultures was often greater than that of the slowest-growing juvenile callus. The developmental capacity of the two types of callus was also found to differ: seedling callus produced fewer roots than did mature callus, when cultured on an inductive (high-salt) medium.

Pith tissue of tobacco has been found to exhibit a gradient in its tendency to undergo cytokinin habituation when cultured on an auxin-containing medium at 35°C (Meins *et al.*, 1980). It was found that pith explanted from below the eighth to the eleventh internode rarely habituates for cytokinin, whereas explants from above this position habituate rapidly. The size of the explant was found to be critical, and there is a pronounced interaction between size and position on the subsequent frequency of habituation, so that the threshold position for cytokinin habituation shifted upwards with decreasing explant size. Furthermore, the rate of habituation of tobacco pith was found to vary with the time of year (Meins and Lutz, 1980). Tissue which was explanted from plants during the spring habituated about seven times faster than that isolated during the winter. A possibly related phenomenon has been recorded in the Jerusalem artichoke system. The intensity of callus formation from tuber explants varies throughout the year, increasing in the spring (Blanarikova and Karacsonyi, 1978), and metabolic activity of the tubers shows seasonal variation. The "lag phase" before synchronous cell division after explantation remains relatively constant at 20–24 hr during the first 5 months of tuber storage; after 1 year's storage, the length of the lag phase increases sharply to reach a duration of about 60 hr (Robertson, 1966; Evans, 1967). The gradual decrease in metabolic activity on excision is associated with decreasing levels of food reserves and stored RNA during storage (Macleod *et al.*, 1979).

It is therefore apparent that explant characteristics with a nongenetic basis can affect the structural and metabolic behavior of cultured cells. In some cases, it is apparent that the development and/or biosynthetic potential of the cells has been limited to some extent, but perhaps not irreversibly so. In practical terms, therefore, the concept of plant cell totipotency may not be particularly useful except to describe the genetic complement of a cell. Nevertheless, it is known that there do exist genetic differences between cells of an explant which potentially could contribute to heterogeneity in cultured cells. Most evidence points to quantitative rather than qualitative differences in DNA. Within a tissue or organ there may be observed differences in ploidy level between cells, due to endoreduplication, whereby DNA replication proceeds in the absence of mitosis. This gives rise to polytene cells (Bennici *et al.*, 1968), and it is believed that, since such polyploid cells divide, along with diploid cells, to form callus, they may be a source of polyploidy in cell cultures (Bennici *et al.*, 1971; Sunderland, 1977; Bayliss, 1980; Swedlund and Vasil, 1985). Occasionally, explant cells may contain extra fragments of DNA, which confer stable and abnormal characteristics on cultured cells. The now classic example is that of crown gall, a tumorlike tissue which develops as a result of infection of plants by *Agrobacterium tumefaciens*. Virulent strains of the bacterium contain a tumor-inducing (Ti) plasmid (van Larabecke *et al.*, 1974), a small part of which becomes integrated into the plant genome where it is transcribed

(Schröder *et al.*, 1981). Once such transformation has been initiated, the cells are characterized both *in vivo* and *in vitro* by their rapid growth and ability to synthesize unusual amino acids (opines). On explantation from the host plant, the transformed cells can grow in culture in the absence of exogenously supplied growth regulators (Braun, 1956).

It is possible that genetic abnormalities other than those described may be carried over from the explant during callus formation, but if the alteration is not conducive to rapid cell division, the "cell line" will be overtaken in number by faster-dividing cells and so come to represent only a small proportion of the total cell population (see Section IV,B). Nevertheless, this represents a potential source of heterogeneity in cell cultures.

Certain enzymatic differences are known to occur between callus cultures derived from different tissues of the same plant, but the nature of the origins of these differences is not known. Arnison and Boll (1975) examined by electrophoretic analysis patterns of isoenzymes in extracts of cell suspension cultures derived from root, hypocotyl, and cotyledon of single seedlings of *Phaseolus vulgaris*. It was found that for some isoenzymes, notably peroxidases, polyphenol oxidases, esterases, and malate dehydrogenases, there were large differences in the activity of the enzymes derived from the different tissues. Sánchez de Jimenez and Fernandez (1983) measured the activities of enzymes related to nitrogen assimilation in homogenates of callus derived from leaves and radicles of *Bouvardia ternifolia* on different nitrogen sources. When supplied with glutamine as the sole nitrogen source, the activity of glutamate synthase increased in leaf callus but was more or less completely inhibited in root callus. These results suggest that, perhaps, not all tissues "dedifferentiate" to the same extent in response to callus induction. Haddon and Northcote (1976) similarly found differences in the extent of dedifferentiation in cultures from different tissues of bean (*Phaseolus vulgaris*). Callus cultures from a variety of somatic tissues all lost their morphogenetic potential after five to seven subcultures, but callus from anther tissue continued to form vascular tissue over a prolonged culture period.

III. METABOLIC CHANGES DURING THE CULTURE GROWTH CYCLE

The growth of callus and liquid cell suspension cultures, as measured by increases in, for example, fresh weight, dry weight, or cell number against time, takes the generalized form of a sigmoidal curve (Street, 1977; Aitchison *et al.*, 1977). Thus there can be recognized a period of little or no cell division (the lag phase) and a period of exponential cell division followed by a steady rate of division (the linear phase), which in turn is followed by a gradual cessation of cell division as nutrients become depleted (the stationary phase). Each of these stages is associated with characteristic biochemical and structural cellular features which can be considered to constitute a continuously changing cycle if the cells are subcultured onto fresh medium at the stationary phase. There also may occur metabolic changes in cultures which are maintained for a large number of such growth cycles, or in which the composition of the nutrient medium is manipulated experimentally. We will now consider some of these phenomena.

A. The Lag Phase

On transferring cultured cells to fresh nutrient medium, metabolic changes are initiated which act to prepare the cells for division. Major increases in the levels of NADPH, ATP, and energy charge occur, demonstrating the activation of the cells' energy-generating system (Shimizu *et al.*, 1977). The glycolytic pathway and the pentose phosphate pathway make appreciable contributions to carbohydrate oxidation, which provides reducing power in the form of NADPH for biosynthesis (Fowler, 1971), and invertase activity, associated with the onset of cell wall synthesis, increases (Copping and Street, 1972). Cell RNA content increases, and the maximum rate of [2-^{14}C]uracil incorporation has been observed in *Vinca rosea* cells to occur during this period (Kanamori *et al.*, 1979). The rates of total protein synthesis also increase during the lag phase after subculture, and this has been found to be due in part to a more efficient ribosomal translational capacity (Verma and Marcus, 1974). Bevan and Northcote (1981a) similarly found that subculture causes a rapid stimulation of polysome formation and an increase in the translatable levels of a small group of mRNAs in both *Phaseolus vulgaris* and *Glycine max* cultures, but the mechanism of this effect is uncertain. Cytokinins were found to cause a slight increase in polysome levels after subculture but had no effect on the levels of *particular* mRNAs, nor on the distribution of mRNAs between nontranslating and translating pools, nor on polysome levels in the absence of subculture. Auxins, on the other hand, may have a more important role. By starving the cultured cells of the auxins 2,4-D and NAA, and then subculturing them into media with or without auxins, it was found that the increase in mRNA for certain proteins was strongly dependent on these growth regulators (Bevan and Northcote, 1981b). Schröder *et al.* (1978), however, found no effect of dilution on the rate of incorporation of labeled precursors into protein for at least 25 hr, and it seems likely therefore that the synthetic activity of the lag phase will be determined to a large extent by the composition of the nutrient medium.

During the lag phase, the levels of secondary products found are primarily determined by the content of the subcultured cells; usually only low levels of new compounds are synthesized. Phenylalanine ammnonia-lyase (PAL), an enzyme associated with the anabolism from phenylalanine of a variety of phenolic secondary metabolites, has been found to be rapidly synthesized and to increase in activity during the lag phase of culture in a number of experimental systems, but this is a phenomenon associated with culture dilution, and both the synthesis and activity of the enzyme decrease toward the end of the lag phase (Matsumoto et al., 1973; Hahl-brock et al., 1978). Complex phenolic secondary products do not accumulate: the activity probably results only in the formation of cinnamic esters. One case in which synthesis and accumulation of a secondary product does, however, take place in the lag phase is reported by Noguchi and Sankawa (1982). Incorporation experiments with [1-^{14}C]acetate demonstrated the induction of the pigment germichrysone (an octaketide hydroanthracene) in subcultured cell cultures of Cassia torosa.

B. The Cell Division Phase

The cytological changes which accompany the exponential and linear phases of growth have been well described (see Yeoman and Street, 1977), and the details need not be repeated here. Once cell division has been induced, it proceeds rapidly and, if fresh nutrients are supplied, will continue. The increase in cell number usually keeps ahead of the increase in fresh weight, resulting in a sharp decrease in average cell size. Yeoman et al. (1965) found that, over the first 7 days of culture, Jerusalem artichoke explant underwent increases in cell number of over 1000% (at 25°C), and such a burst in cellular activity is associated with the increased synthesis of a range of primary metabolites. The nucleic acid content of, for example, cultured artichoke cells, which is composed of about 90% RNA, rises dramatically during the first 3 days of culture but drops on a per cell basis concomitant with the drop in cell size as division outpaces expansion (Davidson, 1971). Robertson (1966) has also demonstrated the occurrence of large increases in alcohol-insoluble nitrogen and oxygen uptake as cell proliferation proceeds. During the linear phase of growth the average cell size remains constant.

In callus tissue, rapid cell division and expansion is mostly confined to the periphery of the tissue (see Aitchison et al., 1977). Proliferation is usually located either in nodules or in sheet meristems at the surface and is the result primarily of periclinal division. As the rate of division slows down, the plane of division may alter, and meristematic activity is initiated deeper in the callus, as recognized by the formation of nodules. In cell suspension cultures in particular, there is little differentiation (either cytodifferentiation or cell organization) apparent in the exponential and linear phases, most cells resembling the meristematic cells of stem apices. There is evidence that the extent of greening in rapidly dividing cells decreases (Lindsey and Yeoman, 1983a), and Laetsch and Stetler (1965) have shown that linear-phase cells of tobacco contain only poorly developed chloroplasts.

Reflecting this process of "dedifferentiation" is the limitation imposed on secondary metabolic activity, accumulation usually being confined to the stationary phase (see Section III,C; for reviews see Yeoman *et al.*, 1980, 1982b). Phillips and Henshaw (1977) showed that, in cell cultures of *Acer pseudoplatanus*, the incorporation of radioactively labeled amino acids, which are common precursors of proteins and phenolics, was preferentially into protein. A similar result has been obtained by Lindsey and Yeoman (1983a) for labeled-phenylalanine incorporation into proteins and tropane alkaloids in cell cultures of *Datura innoxia*. By experimentally changing the rate of cell division or synthesis of primary metabolites, especially proteins, it is often possible to alter the levels of secondary metabolites produced (Lindsey, 1985).

By manipulating the auxin concentration supplied to callus of *Datura innoxia*, the growth rate of the cultures can be increased, and the tropane alkaloid content and chlorophyll content are observed to drop, along with an increase in friability of the callus (Lindsey and Yeoman, 1983a). A similar observation has been made by Hiraoka and Tabata (1974), who found that, by increasing the growth rate of tobacco callus by raising the concentration of 2,4-D in the medium, the levels of nicotine accumulated by the callus dropped dramatically. Phillips and Henshaw (1977) were able to inhibit phenolics accumulation in *Acer pseudoplatanus* cell suspension cultures with urea and auxin treatments, which stimulate protein accumulation and turnover, respectively. Alkaloid synthesis in *Ephedra gerardiana* callus cultures proceeds under experimental conditions in which protein synthesis is precluded by omitting auxin and nitrogen from the culture medium (Ramawat and Arya, 1979). Mizukami *et al.* (1977) have found that, by inhibiting protein and RNA synthesis using streptomycin sulphate, the accumulation of shikonin (1,4-naphthoquinone) derivatives can be stimulated in cultured *Lithospermum erythrorhizon* cells. A high nitrogen supply had the opposite effect. Cycloheximide inhibits protein synthesis and growth of callus and suspension cultures of *Nicotiana tabacum*, but alkaloid accumulation remains unaffected (Neumann and Mueller, 1971). Neumann and Mueller (1974) have further demonstrated that, in callus cultures of *Macleaya cordata*, actinomycin D promotes the biosynthesis of Papaveraceae alkaloids while inhibiting protein synthesis. Moreover, P. A. Aitchison (unpublished data) has found that growth-inhibitory concentrations of cycloheximide stimulated the incorporation of [^{14}C]phenylalanine and [^{14}C]valine into capsaicin in callus cultures of *Capsicum frutescens*. Results using such metabolic inhibitors are difficult to interpret but support the view that, in some cases at least, there is an inverse relationship between protein synthesis and the synthesis of certain secondary products. The restriction of growth by limiting the levels of essential nutrients has been found to have similar effects. The level of incorporation of radioactive amino acid precursors into capsaicin can be increased by reducing the supply of nitrogen and sucrose to cultures of Chili pepper callus (Yeoman *et al.*, 1980) or immobilized pepper cells (Lindsey, 1985). Nettleship and Slaytor (1974) have found that a variety of nutrient-limited media, but particularly a medium lacking inorganic phosphate, were conducive to the production of alkaloids and other secondary metabolites by callus cultures of *Peganum harmala*. The effect of phosphate in this context is especially interesting and has been further investigated by Knobloch and co-workers (Knobloch and Berlin, 1981; Knobloch *et al.*, 1981). The accumulation of cinnamoyl putrescines was greatly enhanced by phosphate limitation, while growth of the cultures was reduced under such conditions. They conclude that the levels of inorganic phosphate used routinely for cell culture are prohibitively high for certain types of secondary metabolism to proceed. Abe and Ohta (1983) have recently found a similar effect of phosphate on lunularic acid accumulation in cultures of *Marchantia polymorpha*. Furthermore, Lindsey (1985) has shown that, by reducing both the growth index and, simultaneously, the incorporation of L-[U-[^{14}C]phenylalanine into protein in immobilized Chili pepper cells by limiting the supply of nitrate, the incorporation of [^{14}C]phenylalanine into capsaicin is correspondingly increased.

These results therefore suggest that, in a number of cases, rapid rates of cell division are not conducive to a high productivity of secondary metabolites (Fig. 3). Nevertheless, there do seem to be some exceptions to this pattern. Zenk *et al.* (1975) have found that, by using a nutrient medium which was *suboptimal* for the growth of suspended cells of *Morinda citrifolia*, the rate of anthraquinone formation was at a maximum in the linear phase of growth. Moreover, the levels produced exceeded those produced by root tissue. However, nutrient media which allowed an *optimum* growth rate inhibited anthraquinone production by more than 80%, and completely "undifferentiated" cell suspension cultures, composed of about 60% single cells or pairs of cells, showed a complete absence of anthraquinones. Noguchi and Sankawa (1982) found that germichrysone production by cell suspension cultures of *Cassia torosa* peaked not only in the lag

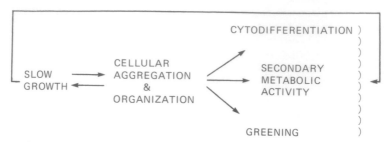

Fig. 3. A schematic representation of the relationship among the growth, differentiation, and secondary metabolic activity of cell cultures. (After Yeoman *et al.*, 1982a.)

phase, as described above, but also in the linear phase. It was not, however, indicated whether the culture conditions were optimum for growth.

As might be expected from a consideration of the above observations, the reduction of the rate of cell division, followed by an increase in average cell size, is associated with a number of changes in metabolic activity. Some of these will now be examined.

C. The Stationary Phase and Regeneration

The stationary phase in callus cultures is associated with a cessation of cell division. The rate of respiration declines, and RNA synthesis and content and the activities of key enzymes of the pentose phosphate pathway decrease (Khavkin and Varakina, 1981). In many cases, if the spent nutrient medium is not replaced, structural and biochemical differentiation may occur. This may involve the activity of meristematic regions within the body of callus cultures, leading to the regeneration of organs. White (1967) has suggested that this initiation of organization within the tissue is a response to gradients of characteristic steepness. All cells in the intact plant are surrounded by other cells, although the relative position of a cell with respect to others may vary during development. Thus the position of the cell is determined ultimately by the rate and pattern of division of it and the surrounding cells, and the degree and form of differentiation of that cell is characteristic of its position in the plant body. Along with its position, the concentration of a variety of physical and chemical factors will also change, and many have been implied in the regulation of differentiation and development. Important gradients include those of light, oxygen, carbon dioxide, growth regulators, temperature, and mechanical pressure, and it will be apparent that the extent and nature of the structural organization of cultured cells will in itself influence the metabolic behavior of those cells as a result of the interaction between the genome and environmental stimuli (see Section IV,C).

Particularly in the case of solanaceous species in culture, the stationary phase is associated with the initiation of structural organization of the cells, which in turn is often associated with increased secondary metabolite production. For example, Thomas and Street (1970) found that morphogenesis in cell suspension cultures of *Atropa belladonna* was favored late in the growth cycle and was increased by delaying subculture. It was further suggested that cellular organization was necessary for tropane alkaloid production. Nikolaeva and Vollosovich (1972) similarly found that the production of alkaloids by *Datura* spp. and indole alkaloids by *Rauwolfia serpentina* tissue cultures was restricted to those conditions in which the level of auxin permitted organogenesis. The production of the flavor components of onion (*Allium cepa*) were found to occur in tissue cultures to levels approaching those in fresh onion, but only in cultures in which roots had differentiated (Freeman *et al.*, 1974). Tabata and Hiraoka (1976) found that the nicotine content of tobacco (*Nicotiana tabacum*) callus decreased to trace amounts in cultures which had lost the ability to regenerate roots. Light regimes inducing tissue differentiation and lignification in *Citrus* tissue cultures also stimulated flavonoid production (Brunet and Ibrahim, 1973). Furthermore, Lindsey and Yeoman (1983a) found that organization and alkaloid accumulation increase together at the end of the growth phase in a number of solanaceous species, evidence which supports the view that cellular organization, if not a prerequisite, is in many cases conducive to the synthesis and accumulation of relatively high yields of secondary metabolites. The implication of these results is that experimentally induced structural differentiation is important for the full expression of metabolism (Yeoman *et al.*, 1982a; Fig. 3). Investigations into the metabolism of immobilized cells (Fig. 4) and cell aggregates are providing information on the role of multicellular organization in regulating metabolism (Lindsey and Yeoman, 1983a,b, 1984, 1985; Lindsey *et al.*, 1983; see Section IV,C).

Disorganized cell cultures not only usually accumulate lower levels of secondary metabolites, but often accumulate the compounds in different proportions to the whole plant (Boulanger *et al.*, 1973; Ibrahim *et al.*, 1981; Sejourne *et al.*, 1981), or may even accumulate novel compounds (reviewed by Nickell, 1980). However, the retention of the capability for secondary metabolite synthesis is not lost during culture, for regeneration restores the metabolic characteristics of the original explant. For example, Tabata *et al.* (1972) noted that roots regenerated from callus cultures of *Scopolia parviflora* accumulated the pattern of alkaloids characteristics of the intact plant, whereas the disorganized tissues did not. If "undifferentiated" callus (containing 0.009–0.01% dry weight of alkaloids) was transferred to a medium conducive to root development, alkaloids accumulated to a level of 0.08% dry weight. Hiraoka and Tabata (1974) cultured cellular aggregates derived from suspension cultures of *Datura innoxia* so that they underwent regeneration. Most plantlets so formed were diploid and the incidence of

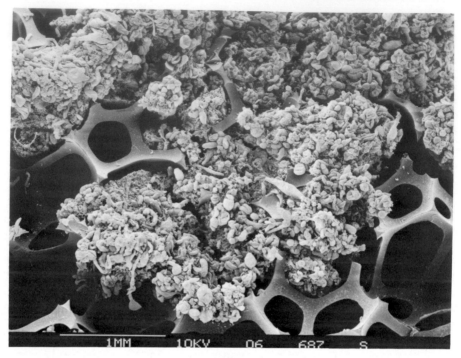

Fig. 4. Cells of *Daucus carota* immobilized in reticulate polyurethane foam.

aneuploidy was lower than in the suspended cells. During root differentiation and plant development there was an initiation of synthesis of the tropane alkaloid scopolamine and a progressive increase in alkaloid content. Consequently, the general pattern of alkaloid composition was restored to a normal state in most of the regenerated plants, including polyploids and aneuploids. Similar results were also obtained by Ikuta *et al.* (1974) for cultured cells and regenerated plants of species of the Papaveraceae.

The dramatic change in metabolism evident during the stationary phase therefore appears to be associated, in part at least, with the onset of structural organization which is limited, as shown particularly in the Solanaceae, by rapid and disorganized cell division favored by the culture conditions of the early part of the growth cycle. As the prevailing environmental conditions (i.e., the concentrations of nutrients, growth regulators, and metabolites) change, and cells may fail to separate after division, so the metabolic behavior of the cells is influenced. In some species, notably *Acer pseudoplatanus*, cell aggregation decreases during the stationary phase (King, 1980), but this species accumulated phenolics to a maximum level during the period of declining growth rate, indicating the importance of a changing growth rate, rather than organization per se, in affecting the expression of secondary metabolism (Davies, 1972; Phillips and Henshaw, 1977). We have already seen how secondary metabolic activity is often incompatible with rapid cell proliferation.

Both structural differentiation and biochemical differentiation in the stationary phase can be further described by concomitant changes in the activities of particular enzyme systems. Forrest (1969) studied polyphenol metabolism in tissue cultures of tea (Camellia sinensis) and found that the activity of polyphenol oxidase was inversely correlated with growth rate. PAL activity has been correlated with the onset of synthesis of polyphenols and structural differentiation. Haddon and Northcote (1975) found that PAL activity in homogenates of bean (Phaseolus vulgaris) callus rose to a maximum when the rate of differentiated nodule formation was greatest, and the activity of β-1,3-glucan synthetase also increased. Moreover, the decline in nodule formation over a prolonged period of culture was paralleled by a decrease in the activity of both enzymes. Brunet and Ibrahim (1973) similarly showed a correlation, in light-grown tissue, between high PAL activity and lignification and flavonoid production in Citrus peel tissue cultures. Total peroxidase activity and isoenzyme patterns change in a characteristic manner as tobacco callus differentiates (Mäder et al., 1975), and it was found that by experimentally inhibiting callus growth it was possible to induce an isoenzyme pattern very similar to that of differentiation. Further, Westcott and Henshaw (1976) demonstrated that in Sycamore cell suspension cultures, a three to fourfold increase in PAL activity could be correlated with the exhaustion of nitrogen from the medium in the latter half of the growth cycle, prior to an increase in the synthesis of tannins. Hahlbrock's laboratory has produced much information on the regulation of enzymes of phenylpropanoid metabolism during the growth cycle. They have discovered that the activity of two related groups of enzymes, including PAL, cinnamic acid 4-hydroxylase, and two isoenzymes of p-coumarate, CoA ligase (involved in general phenylpropanoid metabolism) on the one hand and S-adenosylmethionine : O-dihydric phenol M-O-methyltransferase (involved in lignin synthesis) on the other, are independently regulated (Ebel et al., 1974). Enzymes of the former, but not the latter, group increase in activity prior to the stationary phase, when, in the case of Glycine max and parsley (Petroselinum hortense) cell culture, phenylpropanoids are accumulated. This may be the result of increased mRNA accumulation (Hahlbrock et al., 1982).

It is therefore apparent that a decline in the rate of cell division permits, under appropriate conditions, both an accumulation of secondary metabolites and the structural organization of cells. Both phenomena seem to be interrelated, and the changes in secondary metabolic activity during the transformation from explant to cultured cell to regenerated plant vary in a way that reflects the changes in structural differentiation. The metabolic heterogeneity described has dealt with variation both throughout a population of cells and throughout the phases of the growth cycle. The variations with time, however, are not limited simply to metabolic changes through a growth cycle. It has often been reported that both the biosynthetic and the morphogenetic potentials of cultured cells vary (and usually are reported to decline) with successive subcultures, and such variability may be considered to provide the background over which growth phase–related heterogeneity occurs. For example, Partanen *et al.* (1955) found that cultured fern prothalli became tumorlike and lost their capacity for regeneration over a period of culture, and Murashige and Nakano (1965) discovered that single-cell clones of tobacco similarly gradually lost the ability to initiate roots and shoots. Wilson and Street (1975) found that new cultures of *Hevea brasiliensis* produced roots spontaneously but lost the ability to do so during serial subcultures, although a few scattered latex vessels were always found. Bean (*Phaseolus vulgaris*) callus appears to stop producing vascular tissue after prolonged culture (Haddon and Northcote, 1976), and Sharma and Chowdury (1977) found a similar decline in morphogenetic capacity. The potential for secondary metabolite production has, in some cases, been seen to decline along with the ability to structurally differentiate. Thomas and Street (1970) found that, in *Atropa belladonna* cell cultures, morphogenesis and alkaloid accumulation dependent upon root formation declined with serial propagation. The nicotine content of tobacco cultures also decreases to trace amounts in succeeding subcultures in association with a decline in root-regenerating activity (Tabata and Hiraoka, 1976), and Dhoot and Henshaw (1977) found that the degree of aggregation of cell suspension cultures of henbane (*Hyoscyamus niger*) declined, and the alkaloid composition changed and levels dropped after approximately the eighth passage of subculture.

In spite of this evidence, however, the inevitability of the loss of developmental potential is by no means certain. Chaturvedi and Mitra (1975), for example, have demonstrated a change, rather than loss, in the pattern of morphogenesis in *Citrus* callus in long-term culture. Under appropriate conditions, recently initiated cultures produced only shoots, but after a prolonged culture period differentiation took the form of embryoid formation, concomitant with a shift in callus appearance from pigmented, compact, and nodular tissue to a white, friable, and faster-growing type. In some cases it has been shown possible to restore a "lost" developmental potential by altering the conditions of culture (Harry *et al.*, 1977), and in other cases, there seems to be no loss at all in potential with prolonged culture (Vasil, 1985; Vasil and Vasil, 1984a,b). Corduan and Spix (1975) have shown that haploid callus of *Digitalis purpurea* is able to retain the ability to differentiate over 20 subcultures of 4 weeks each, and Cummings *et al.* (1976) similarly report that callus from *Avena sativa* embryos was still capable of regenerating plants though 13 subcultures, over a total period of 18 months. Tomita *et al.* (1970) have described how callus cultures of *Dioscorea tokoro* maintain steroidal sapogenin and sterol synthesis after 1 year in culture, and Tewes *et al.* (1982) have found that embryonic strains of *Digitalis* retained, undiminished, a morphogenetic capacity for more than 3 years. Colomas and Bulard (1977) have similarly found that callus of *Myrtillocactus geometrizans* accumulated betalains after 45 subcultures, and the variation throughout each growth cycle, with an increase in pigment production in the stationary phase, was maintained. In our laboratory, we have studied the stability of various characteristics of *Datura innoxia* callus (Lindsey, 1982), and it was found that, even after 93 weeks in culture, the callus was still capable of synthesizing tropane alkaloids and chlorophyll in undiminished quantities (Table I), and the morphology of the callus, which was fairly compact, was unaltered. Furthermore, like the *Myrtillocactus* cells, the callus continued to show variation in alkaloid and chlorophyll

TABLE I

The Alkaloid and Chlorophyll Content and Chromosome Number of Callus Derived from Diploid [2n] *Datura innoxia* Mill. over a Prolonged Culture Period[a]

Week of culture	Mean chromosome number per cell	Alkaloid content (mg/g dry wt)	Chlorophyll content (μg/g fresh wt)
0		0.85	49.5
2		0.85	62.0
4	58.2 ± 17.8 (5n)	0.85	49.3
6		0.85	40.8
20		0.50	33.7
23		0.50	37.0
27		0.50	47.9
30	82.4 ± 21.1 (7n)	0.50	40.1
33		0.50	37.2
42		1.00	41.5
47		0.85	48.7
51		0.50	42.9
62	70.9 + 28.3 (6n)		
67		0.85	43.8
87		0.75	49.3
89	88.7 ± 27.3 (7n)		
93		0.75	39.2

[a] The *approximate* ploidy level is given in parentheses. Alkaloid and chlorophyll contents were determined as described in Lindsey and Yeoman (1983a). Karyotype determinations were the mean for 30 cells per sample. Although the ploidy level increased with prolonged culture, the alkaloid and chlorophyll contents of the cells did not decline. Alkaloids were not released into the medium.

content during the growth cycle. The capacity for the synthesis of capsaicin by cell cultures of *Capsicum frutescens* is also preserved for at least 6 years. A comparison of old and newly initiated strains demonstrated no loss in the quantity of capsaicin produced (P. A. Aitchison and M. M. Yeoman, unpublished observations).

The apparent "loss" of biosynthetic and morphogenetic potential may be due, therefore, not to some permanent alteration to the cell's synthetic machinery (although genetic sources of heterogeneity do exist and are discussed in Section IV), but to an inability of particular culture conditions to allow the realization of that potential. We shall now consider some of the factors which influence the changes and variation in the developmental characteristics described above.

IV. THE ORIGINS OF HETEROGENEITY
IN CELL CULTURES

So far, we have described some of the features of heterogeneity: (1) those which occur within the explant and within a cell population at a single point in the growth cycle ("spatial" heterogeneity) and (2) those which vary with time throughout a growth cycle and successive growth cycles (which can be called "temporal" heterogeneity). We shall attempt now to define some of the factors which contribute to such variability once cultures are established, and conclude by attempting to bring together evidence and ideas on the regulation of metabolism *in vitro*.

A. Genetic Factors

We have seen (Section II,B) how genetic differences between cells of the explant may provide a source of heterogeneity in cultured cells. However, there is evidence that genetic changes occur during the culture period which may account, to some extent, for spatial heterogeneity, temporal heterogeneity, and the loss of developmental potential. The genetic alterations can be classed in one of two groups, namely point mutations and chromosomal aberrations, and we shall briefly consider these in turn.

1. Point Mutations

Although there is no doubt about the occurrence of variant cell lines in cultures, it is far from certain that such plasticity is to a significant extent the result of somatic mutations. There is little convincing evidence that mutations arise spontaneously which reduce or eliminate the biosynthetic potential of a culture (Henke, 1981), and before an alteration in potential can be described as being the result of a mutation, it first must satisfy a number of criteria: (1) it must be irreversible and involve an alteration in the primary structure of DNA, (2) it must be of rare occurrence, (3) it must occur randomly, and (4) it must be heritable. To our knowledge, in no instance have all these criteria successfully been applied to a cell culture producing a specific secondary product, and it is extremely difficult to delineate between a single genetic mutation and a metabolic change of epigenetic origin (Binns, 1981). Nevertheless, Akasu et al. (1976) suggest that, in callus cultures of Stephania cepharantha, which were found to be incapable of synthesizing the principal alkaloids of the original plant, cepharanthine and isotetrandrine, the enzyme controlling specific methylation and methyldioxy group formation were absent. Furthermore, Furuya et al. (1978) found that cell cultures of Papaver somniferum were incapable of synthesizing morphinane alkaloids from the precursor $(-)(R)$-reticuline and concluded that the phenol oxidation enzyme required for the transformation to salutaridine was absent. However, it is possible in both these examples that it was merely the expression of the gene that was lost (and presumably therefore could be regained under the appropriate experimental conditions) rather than the gene itself. In support of this view, results of clonal analysis of wild carrot cell cultures by Dougall et al. (1980) indicated that changes in the ability of the cells to accumulate anthocyanins involved no qualitative change in the DNA content, and Mok et al. (1976) similarly could not confirm that mutation in cultured carrot cells was a cause of altered pigmentation. Experimentally induced mutations, however, may change the pattern of pigment formation (Nishi et al., 1974).

2. Chromosomal Aberrations

Chromosomal aberrations arising during the growth cycle are potentially the largest source of genetic heterogeneity in cell cultures and have been well reviewed by Sunderland (1977) and Bayliss (1980). The main alterations to the karyotype have been categorized as (1) aberrations due to spindle abnormalities, resulting in changes in the amount of DNA and in the number of chromosomes, with the retention of the original chromosome number (e.g., polyploidy), or with the loss of individual chromosomes (aneuploidy); (2) aberrations due to chromosome breakage, resulting in changes in the amounts of DNA per chromosome, such as in translocations; and (3) aberrations by endoreduplication, leading to an increase in the circumference and DNA content of chromosomes without a change in chromosome number (polyteny).

Growth regulators, and notably 2,4-D, have been suspected as agents which induce aneuploidy in cell cultures (see the review by Bayliss, 1980), but Bayliss (1975) has indicated a greater importance of tissue organization (which may be determined in part by the hormone regime) in maintaining karyo-type stability. The frequency of structural changes appears to be influenced by the composition of the nutrient medium (Steffenson, 1961).

It might be expected that cells which were variously aberrant would have different biosynthetic capabilities. However, the emphasis that should be laid upon the effect of such genetic abnormalities in altering the capacity of cultured cells to produce designated secondary metabolites is uncertain. The accumulation of aneuploid or otherwise genetically altered cells has been considered to be a cause of a change in or even the loss of biosynthetic and morphogenetic potential in cell cultures (Torrey, 1967). Although Mok et al. (1976) could find no evidence for the role of somatic mutations in pigment variation in carrot cell cultures, it was considered that aneuploidy may have made some contribution. Similarly, Murashige and Nakano (1965) suggested that aneuploidy found in cultured tobacco cells may have blocked morphogenesis. In opposition to a generalized concept of a genetic basis for the loss of developmental potential, there are documented cases in which chromosomal aberrations had no seriously deleterious effects on developmental potential. Sacristan and Melchers (1969), for example, found that, despite severe genetic abnormalities, tobacco cells were still capable of regeneration. Hiraoka and Tabata (1974), furthermore, found that plants could be regenerated successfully from aneuploid and poly-ploid callus of Datura innoxia, and the alkaloid pattern and content of the plants was unaffected. Similarly, Lindsey (1982) found that, although Datura innoxia callus cultured over a prolonged period rapidly accumulated aberrations, the chlorophyll and alkaloid content was unaffected (Table I). Tabata et al. (1968) even found that autotetraploid tobacco plantlets regener-ated from pith callus accumulated slightly enhanced levels of nicotine, compared with the diploid plant. Sunderland (Strickland and Sunderland, 1972; Sunderland, 1977) has described cultures of Haplopappus gracilis Strain A1 in which 90% of the cells had irregular karyotypes. Nevertheless, these cells were capable of producing large amounts of anthocyanin pigments for several years. Another strain (A2) of the same species, which was genet-ically more abnormal, even produced anthocyanins in the dark.

These results suggest some interesting points:

1. Some chromosomal aberrations at least are not deleterious to the biosynthetic or morphogenetic potential, but may give rise to culture vari-ability. The extent of variability which can be directly attributed to chro-mosomal abnormalities, however, is difficult to ascertain.

2. The aberrations can, in some cases, be considered beneficial, for ex-

ample (a) in studying the regulation of a specific product, and (b) in providing variability for selection purposes, with the aim of establishing high-yielding cell lines (Ogino *et al.*, 1978; Tabata *et al.*, 1978).

Since the general pattern of an increasing yield of designated secondary metabolites throughout the growth cycle is more or less identical over successive culture periods (albeit the *absolute* accumulated yield might decrease with progressive subcultures), it seems unlikely that such variability is the direct result of genetic change. However, it is possible that there is some selective mechanism at work which may have either a genetic or an epigenetic basis, and this is now considered.

B. Selection Pressure

Usually, conditions for cell culture are chosen as being conducive to a maximum growth rate. These conditions, therefore, provide the best opportunity for the selective accumulation of cells which are capable of rapid proliferation, and so possess characters associated with rapid growth. Under such an experimental regime, then, genetic abnormalities which are most likely to arise during rapid cell division (see Bayliss, 1973) will be selected for. An example of cell line selection in culture has been given by Singh *et al.* (1975). The proportions of 4-chromosome and 16-chromosome cell lines of *Haplopappus gracilis* cultures were followed over 90 days. It was found that, although there were approximately equal numbers of the two cell lines at the beginning of the experimental period, there was a dramatic increase in the relative numbers of the 4-chromosome line. As early as about day 10, almost 80% of the observed metaphases were of 4-chromosome cells, compared with about 20% 8-chromosome cells. This example illustrates how the selection of a genetically distinct cell line can result in its predominance in a cell population. The selection pressure determining which line would dominate was in this case a competitive one: the cells with the shortest cycle time, a function in part of the amount of DNA to be replicated, obviously proliferated at a faster rate. Vanzulli *et al.* (1980) have similarly suggested the preferential proliferation of haploid nuclei derived from pollen embryoids may have been due in part to their shorter mitotic cycle.

It is conceivable that, for a species which accumulates a specific secondary product, a similar selection mechanism might dictate whether, under a particular set of culture conditions, that product will be accumulated. The determining factor again would be the cell cycle time: if accumulation was in some way linked to rapid cell division, then a rapidly dividing cell line would inevitably predominate, and the culture would concomitantly accumulate the secondary product in the stationary phase, as is often observed. This model seems plausible in principle but is, of course, highly simplistic, and there is no evidence to support it as it stands. Indeed, the evidence available tends to suggest an *inverse* relationship between secondary metabolite production and growth rate, although the isolation of constitutively producing mutant cell lines with short cell cycle times, if discovered, might prove to be valuable exceptions to this generalization. This model might, however, account for the generally low levels of secondary products found in cell cultures.

C. Environmental Factors

The expression of the developmental potential is a product of the interaction between the genome of the cell and its environment. For this reason it is necessary, when discussing heterogeneity in secondary metabolite production (or in any other character which has a biochemical basis), to consider how the regulation of genes can be modulated by external factors. The relative contribution of genetic and environmental or epigenetic factors in the production of a particular phenotype is almost impossible to quantify; suffice it to say, at present, that cell culture systems provide a valuable tool with which that phenotype can be modified.

Callus and suspended cell cultures are composed of cell aggregates, each of which may be composed of a greatly different number of cells, some of which will be dividing, some of which will be quiescent (Yeoman and Street, 1977). The environment of a cell may be considered to comprise an intercellular "microenvironment," namely the degree of association with other cells, and a "macroenvironment," which is more readily controlled by the experimenter and comprises the nutrient medium and physical environment in which the cells are cultured. The macroenvironment determines to a large extent the degree of aggregation and growth rate of the cultured cells and is continuously changing in composition during a growth cycle. The roles of the micro- and macroenvironments in regulating metabolism and development are obviously complex and have been considered in some detail elsewhere (e.g., Davidson *et al.*, 1976, Mantell and Smith, 1983), but we shall examine briefly some of the possibilities here.

1. The Macroenvironment

In order to grow, plant cells require a carefully constructed environment. The nutrient medium (see Chapter 3, this volume) is supplied either as a liquid or is solidified with agar, and contains a range of salts, vitamins, growth-regulating substances (usually auxins and cytokinins), and, since cultures tend to be heterotrophic, a carbon source (usually sucrose). Thus the medium is usually chemically defined, but occasionally some species appear to be dependent on undefined additives such as coconut milk or casein hydrolysate. Since the important constituent(s) of these latter substances are unknown and the efficacy of the substance can vary from batch to batch, the tendency is for research workers to use defined media as much as possible. The pH of the medium is usually adjusted initially to between 5.0 and 6.0, but tends to drift during the culture period according to the buffering capacity of the salts present. Temperature is usually maintained at about 25°C, but the light requirement can vary enormously between species (Seibert and Kadkade, 1980).

Manipulation of the chemical environment of cultured cells has most commonly involved altering the levels of growth regulators in the medium (see Everett et al., 1978). The literature on the effects of growth regulators on secondary metabolism is vast, and for recent reviews the reader is directed to Staba (1980) and Mantell and Smith (1983). In many cases, evidence for the involvement of auxins and cytokinins in the regulation of specific secondary metabolite synthesis and accumulation is conflicting, and it is correspondingly difficult to ascribe a particular regulatory role of one molecule with a particular biosynthetic pathway. For example, Furuya et al. (1971), using tobacco callus cultures, discovered that the presence of 2,4-D in the nutrient medium precluded the accumulation of tropane alkaloids, while IAA allowed the production of nicotine, anatabine, and anabasine. By transferring 2,4-D–grown callus to an IAA medium, and vice versa, it was shown that 2,4-D appeared to suppress nicotine accumulation while IAA permitted it. Tabata et al. (1971), however, using the same system, found that kinetin promoted nicotine accumulation in the absence of an auxin source (IAA or 2,4-D), but auxin strongly inhibited nicotine formation even in the presence of kinetin. Takahashi and Yamada (1973), on the other hand, found no difference in the effects of 2,4-D and IAA on nicotine accumulation. Perhaps significantly, though, they found an antithetical relationship between the rate of tobacco callus growth and alkaloid accumulation, and similar results were found by Konoshima et al. (1970) and Tabata and Hiraoka (1976). It therefore seems possible that is is in this way, i.e., by promoting conditions which are either favorable or unfavorable to culture growth, that auxins exert their effects on the production of alkaloids and other secondary metabolites (see Lindsey and Yeoman, 1983a). There is also evidence that 2,4-D promotes protein turnover in cultured cells and may limit secondary metabolism by restricting the availability of common precursors (Phillips and Henshaw, 1977). Cytokinins also affect secondary metabolite production in cell cultures, but since these growth substances are usually supplied in association with auxins, cytokinin-specific effects are not clearly apparent. Kinnersley and Dougall (1980a) suspect that the effect of high concentrations of cytokinins in inhibiting anthocyanin accumulation in carrot cell cultures is associated with effects on increasing cell aggregation.

The effect of nutrients on the growth and metabolism of cultured cells has been reviewed by Dougall (1980). As exemplified by the case of orthophosphate described earlier (Section III,B), the modulatory effects of nutrients may, like those of the aptly named growth regulators, be a function of their effects on culture growth rate. Mantell and Smith (1983) describe how nicotine biosynthesis in tobacco cultures is associated with the exhaustion from the nutrient medium of phosphate, nitrate, and sucrose. Similarly, in our laboratory it has been shown that enhanced yields of capsaicin by immobilized and callus cells of *Capsicum frutescens* and of steroidal glycoalkaloids by immobilized cells of *Solanum nigrum* are associated with, and can be induced by, low levels of phosphate, nitrate, and sucrose in the nutrient medium (Yeoman *et al.*, 1980; Lindsey and Yeoman, 1983b; Lindsey, 1985).

These results therefore imply that a variety of factors initially present in and subsequently added to the nutrient medium may exert effects on differentiation and metabolism via a common means, namely by affecting culture growth rate. Adverse physical factors such as high temperatures also increase the incidence of differentiation concomitant with decreased growth rate (Naik, 1965).

Another important factor which appears strongly to affect differentiation and secondary metabolism in cell cultures is the manner in which the nutrient medium is supplied to the cells, i.e., in a solid or liquid form. In general, callus cultures tend to accumulate high levels of secondary metabolites and exhibit greater structural differentiation than do suspension cultures (Yeoman *et al.*, 1980). King and Street (1977) suggest that the mitotic index may be greater in aggregates of suspended cells than in callus cultures. The reasons for this seem not to depend on whether the medium is solid or liquid per se, but rather on whether the cells are either in the form of a physically stationary callus or are dispersed and freely suspended in liquid. Zeleneva and Khavkin (1980) have found, for example, that in callus cells of maize (*Zea mays*), the activities of a number of enzymes were similar to those in the intact plant, but this was not so for freely suspended cells. It was concluded that the metabolism of suspended cells is probably unique and quite dissimilar to that of the whole plant, due to the peculiarity of their environment. The implications of this and related evidence will now be discussed.

2. The Microenvironment

The appearance of organized structures in suspension cultures is often associated with increased yields of secondary metabolites, as we have already seen, and such organization, though it may not be necessarily into morphologically recognizable structures, would seem to favor a fuller expression of metabolism than occurs in dispersed and rapidly growing cells (Yeoman et al., 1982a; Lindsey and Yeoman, 1983a). In the intact plant, the cytoplasms of almost all the living cells are connected by plasmodesmata. Such junctions between cells permit the so-called symplastic transport of nutrients, growth regulators, ions, etc., which is essential for the establishment of intercellular gradients and polarity. This seems to be a critical factor for differentiation and development. Indeed, it seems that multicellularity is a prerequisite for the normal expression of the genetic information. This concept is illustrated by the restoration of metabolism in regenerated organs and plantlets discussed previously, and is supported by work on immobilized cells and cell aggregates. Immobilized cells, which are usually derived from freely suspended cells, are cultured in such a way that they have certain similarities with both callus cultures and the organized (aggregated) structures present in stationary-phase suspension cultures of certain species. These similarities are that they allow (1) high cell–cell contact, (2) slow growth rate, and (3) the establishment of physical and chemical gradients, particularly if grown physically stationary (Lindsey and Yeoman, 1983b). Furthermore, there are an increasing number of examples in which both primary and secondary metabolism is more active in immobilized than in freely suspended cells. Brodelius and Nilsson (1980) have found that cells of Catharanthus roseus entrapped in a gel of calcium alginate exhibited an increased efficiency of ajmalicine synthesis from the precursors tryptamine and secologanin compared with suspension cultures. Cells of Capsicum frutescens immobilized on polyester matting, in a calcium alginate gel or reticulate polyurethane foam (Fig. 4), produce higher yields of capsaicin (microgram or rarely milligram levels) than do suspension cultures (usually only nanogram levels), and the same phenomenon is also associated with steroidal alkaloid production by Solanum nigrum cells immobilized on polyester matting or in calcium alginate (Yeoman et al., 1980; Lindsey and Yeoman, 1983b; Lindsey et al., 1983). Similarly, Bartlett (1982) has found that experimentally induced cell aggregates of Populus alba are significantly better at producing and glycosylating salicylic acid than are suspended cells. Kinnersley and Dougall (1980a) and Ozeki and Komamine (1981) have implicated aggregation as playing a role in the regulation of anthocyanin synthesis in carrot cultures, and in this case it was found to be possible to delineate between and define the biosynthetic capabilities of aggregates according to their size: in the larger aggregates (of diameter greater than 170 μM) anthocyanin accumulation was inhibited (Kinnersley and Dougall, 1980a).

The extent of aggregation (the microenvironment) is determined principally by the macroenvironment, although factors associated with the original explant may have some influence on the friability of the culture (King, 1980), and as we have seen, aggregate size itself may influence the metabolic behavior of the constituent cells. Further, the proportions of cells in division may vary between aggregates of different size, and there is little doubt that the rate of cell division has profound effects on metabolism. Also, in callus cultures, compact cell aggregates often accumulate higher levels of secondary metabolites than do more friable cultures (Lindsey and Yeoman, 1983a; Table II), and although there is no strict correlation, compact tissues are often composed of a greater proportion of nondividing cells than friable cultures (Grant and Fuller, 1968; King and Street, 1977).

The main point to emerge from this evidence, therefore, is that differences in the extent of organization (i.e., in aggregate size or the degree of compactness/friability) provide a source of biochemical heterogeneity by providing variation in the establishment of physical and chemical gradients and (perhaps as a consequence) variation in cytokinetic activity (Fig. 3). The relationship between structural organization, growth rate, and the regulation of the activity of metabolic pathways will now be considered in

TABLE II

The Habit and Alkaloid Content of Callus of a Variety of Solanaceous Species[a,b]

Species	Callus habit	Alkaloid content mg/g dry weight
Datura clorantha	Friable, colorless, actively growing	0.1
	Compact, green, slow growing	1.5
Datura innoxia	Friable, colorless, actively growing	<0.1
	Compact, green, slow growing	1.5
Datura stramonium	Pale green, actively growing	0.5
	Green–brown, slow growing	1.0
Solanum dulcamara	Friable, colorless, actively growing	<10
	Compact, green, structurally differentiated, very slow growing	30
Solanum nigrum	Friable, colorless, actively growing	<10
	Compact, green, slow growing	13

[a] After Lindsey and Yeoman (1983a).

[b] In general, compactly aggregated and slow-growing cultures accumulated the highest yields of alkaloids.

the light of evidence provided by the developmental phenomenon habituation.

D. The Nature of the Metabolic Switch

Habituation is a progressive change in metabolism which appears to have an epigenetic rather than a genetic basis (Meins and Binns, 1977), although Sacristan (1967) found that auxin-habituated tobacco cells were triploid, and its regulation may prove to be an interesting parallel to the regulation of secondary metabolic activity. It is characterized by the loss of dependence on either auxins ("auxin habituation") or cytokinins ("cytokinin habituation") as cells are maintained in culture. Habituated tissues can arise spontaneously, but the condition can also be induced by manipulation of the culture conditions, such as by temperature shock (Meins et al., 1980). In most cases reported, the biochemical aspects of the habituated state are correlated with a characteristic morphological condition of the cultured cells. Habituated callus is often more friable and faster growing than the nonhabituated type and may be less lignified and less green, and show less organogenesis (Hadačová et al., 1975; Chaturvedi and Chowdhury, 1977; Vyskot and Novak, 1977). This therefore implies that what may be described broadly as a "nondifferentiated" condition represented by disorganized, tumorlike proliferation can be stably maintained and controlled by a nongenetic mechanism. Meins and Binns (1979) suggest that control is by a positive-feedback mechanism, in which the synthesis of cytokinins (in the case of cytokinin habituation) is induced by, initially, cytokinins in the medium and then by cytokinins produced by the cells.

We have discussed in this article how secondary metabolites often begin to accumulate at a distinct period of the growth cycle, and since control seems unlikely to be the result of a genetic alteration, the epigenetic "trigger" requires definition, albeit somewhat speculatively. Only a limited amount of work has been performed on the enzymology of secondary metabolism, but Hahlbrock's work in particular has provided valuable information (Hahlbrock and Grisebach, 1979). Although there is a suggestion of regulation at the transcriptional level, there is little evidence that secondary metabolism is controlled, in general, at this point. Blue or ultraviolet light is necessary for the activation of PAL and perhaps other enzymes involved in the synthesis of phenolics, but factors other than light are undoubtedly necessary for the switching from primary to secondary metabolism, since illumination is usually constant during a culture period.

Much of the evidence presented in this article has demonstrated an inverse relationship between culture growth rate on the one hand and cellular organization and secondary metabolite formation on the other. In order to account for the roles of the macro- and microenvironmental effects on metabolism, therefore, we can consider two possible mechanisms, which are interrelated and may provide an insight into the switching from primary to secondary metabolism. The first describes a possible *kinetic* mechanism, in which the pathways of primary and secondary metabolism compete for common precursors, according to whether culture conditions permit rapid or slow growth. The evidence of Phillips and Henshaw (1977) and Lindsey and Yeoman (1983a) directly suggests an antagonism between two pathways. It is proposed that alternative pathways such as those of protein synthesis and phenolic or alkaloid synthesis will be active depending on whether one of the reactions is kinetically favorable. In a generalized situation in which pathways A and B have common precursors, if the rate of reaction A is greater than that of B, then a greater proportion of the precursors will be utilized by the more demanding reaction. The enzymes in pathway B would be relatively inactive until reaction A stops or slows down sufficiently so that precursors accumulate to such a level that reaction B can proceed. So, it can be visualized that in rapidly dividing cells, the pathways of protein synthesis are highly active, and amino acid precursors are utilized in this process in preference to any secondary pathway. This is obviously a highly simplistic model, and takes no account of differences in K_m values of the enzymes of the alternative pathways. Nevertheless, it would be expected that, under nutritional and hormonal conditions which are favorable to rapid cell proliferation, primary metabolic activity will be stably maintained; habituated tissues would be expected to express mainly primary metabolic activity, and the evidence described above supports this hypothesis. On the slowing down of growth, either, for example, as nutrients are naturally depleted or as the concentrations of growth regulators alter in the medium, or if the culture conditions are deliberately manipulated (Lindsey, 1985), the rate of protein synthesis will be reduced, causing a diversion of amino acids into secondary metabolism. It is suggested, in this hypothesis, that a *kinetic* separation of enzyme and substrate determines the activity of pathways, and, as in habituation, a stable metabolic condition is maintained by an epigenetic means.

The second possible mechanism to be discussed is, in a sense, an extension of the first, and suggests the importance of cellular organization in regulating metabolism. We have described much evidence which points to a strong correlation between organization and secondary metabolic activity, and this can be interpreted in terms of the establishment of regulatory gradients of a variety of physical and chemical factors which are known to influence metabolism. Although there is a lack of direct evidence from cell culture systems, it is also possible that cellular organization may affect metabolism by changing the ultrastructural organization of the cells. There is little doubt that the endomembrane system of the cell is involved in the processes of cell differentiation and secondary metabolite production, and there are examples of the possible involvement of endomembrane changes (both spatial and permeability changes) in the differentiation of vascular tissues (e.g., Northcote, 1968; Cronshaw, 1974; O'Brien, 1974). The arrangement of the cytoplasm, in these cases, reflects the pattern of structural differentiation of the cell. In cell cultures the association of changes in subcellular organization with changes in cell growth rate and organization is beyond doubt (e.g., Laetsch and Stetler, 1965; Yeoman and Street, 1977), and although such changes are not necessarily implied in the *control* of differentiation, it is possible that alterations in enzyme/substrate compartmentalization may be involved in the regulation of secondary metabolism. It is also apparent that cell aggregates of differing size and degree of compactness may have different growth rates and proportions of cells involved in active division. Thus, according to the "kinetic" mechanism described earlier, aggregates or calluses in which the majority of the cells are quiescent would be expected to accumulate secondary metabolites; as we have seen, there is evidence that this is the case, particularly in cultures of solanaceous species.

Differences in the secondary biosynthetic activity of cell lines provide a valuable system for the study of biochemical heterogeneity and the stability of synthetic capabilities of cells. In our laboratory (M. Aitken, K. Lindsey and M. M. Yeoman, unpublished observations) we have observed morphological differences between cell lines of *Capsicum frutescens* which produce high and low yields of capsaicin (Fig. 5). The high-yield character is stable over a number of growth cycles but is experimentally reversible. Like habituation, it is possible that the high-yield characteristic is the result of an overproduction mechanism, induced in response to an environmental "trigger" but perhaps influenced by the physiological nature of the explant. This possibility is under investigation.

V. CONCLUDING REMARKS

Despite the aspirations of the pioneering cell culturists of 25 or 30 years ago to produce physiologically homogeneous experimental systems, it is now apparent that such an ideal appears at present to be impossible to achieve. In this chapter we have attempted to give an introduction to some of the changes in metabolic activity which occur during the transition from intact plant to cell culture to regenerated plant, with special reference to

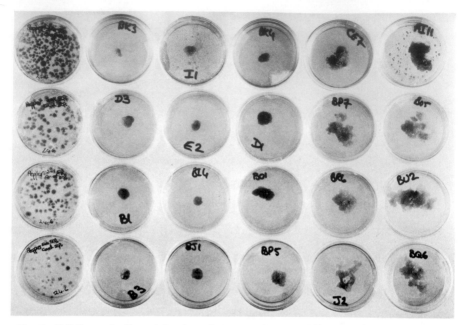

Fig. 5. Cell lines of callus of *Capsicum frutescens,* showing differences in morphology.

the dynamics of secondary metabolism. Metabolic activity is the product of the genetic information available, which may alter during cell division, and the physical and chemical environment, which certainly does change as culture growth proceeds. It is now becoming evident that the less well defined physiological status of the explant itself adds another factor to this equation.

From the practical viewpoint, metabolic heterogeneity can be exploited by selecting cell lines and by manipulating their environment for the large-scale production of valuable secondary products. Occasionally, as in the case of habituated cell lines and cell lines producing relatively high levels of specific secondary products, a particular metabolic condition, characterized by an ability to overproduce a specific compound, becomes temporarily fixed but can be experimentally reversed. This may be explained in terms of an epigenetic mechanism such as positive feedback control. But other factors such as the extent of organization of the cells and their growth characteristics, which may also become "fixed," may well be involved and require further investigation.

Thus, there is a structural and biochemical heterogeneity at every stage of the development of a cell culture, and the origins of the heterogeneity are gradually becoming clear. Studies on the molecular regulation of secondary metabolism using techniques of enzymology and labeled precursor feeding are badly needed and hopefully will act as useful models for the regulation of cell differentiation itself.

REFERENCES

Abe, S., and Ohta, Y. (1983). Lunularic acid in cell suspension cultures of *Marchantia polymorpha*. *Phytochemistry* **22**, 1917–1920.

Aitchison, P. A., and Yeoman, M. M. (1974a). The use of 6-MP to investigate the control of glucose-6-phosphate dehydrogenase levels in cultured artichoke tissue. *J. Exp. Bot.* **24**, 1069–1083.

Aitchison, P. A., and Yeoman, M. M. (1974b). Control of periodic enzyme synthesis in dividing plant cells. *In* "Cell Cycle Controls," (G. M. Padilla, I. L. Cameron, and A. Zimmerman, eds.) pp. 251–263. Academic Press, New York and London.

Aitchison, P. A., Macleod, A. J., and Yeoman, M. M. (1977). Growth patterns in tissue (callus) cultures. *In* "Plant Tissue and Cell Culture" (H. E. Street, ed.), Botanical Monograph Series Vol. 11, 2nd ed.), pp. 267–306. Blackwell, Oxford.

Akasu, M., Itokawa, H., and Fujita, M. (1976). Biosoclaurine alkaloids in callus cultures of *Stephania cepharantha*. *Phytochemistry* **15**, 471–473.

Arnison, G., and Boll, W. G. (1975). Isoenzymes in cell cultures of bush bean (*Phaseolus vulgaris* cv. Contender): Isoenzymatic differences between stock suspension cultures derived from a single seedling. *Can. J. Bot.* **53**, 261–271.

Bagshaw, V., Brown, R., and Yeoman, M. M. (1969). Changes in the mitochondrial complex accompanying callus growth. *Ann. Bot. (London)* **33**, 35–44.

Bartlett, D. J. (1982). The production of compounds of pharmaceutical interest using plant tissue cultures. D.Phil. Thesis, Oxford University, Oxford.

Bayliss, M. W. (1973). Origin of chromosome variation in cultured plant cells. *Nature (London)* **246**, 529–530.

Bayliss, M. W. (1975). The effects of growth *in vitro* on the chromosome complement of *Daucus carota* L. suspension cultures. *Chromosoma* **51**, 401–411.

Bayliss, M. W. (1980). Chromosomal variation in plant tissues in culture. *Int. Rev. Cytol. Suppl.* **11A**, 113–144.

Bennici, A., Buiatti, M., and D'Amato, F. (1968). Nuclear conditions in haploid *Pelargonium in vivo* and *in vitro*. *Chromosoma* **24**, 194–201.

Bennici, A., Buiatti, M., D'Amato, F., and Pagliali, M. (1971). Nuclear behaviour of *Haplopappus gracilis* callus grown *in vitro* on different culture media. *Colloq. Int. C.N.R.S.* **193**, 245–250.

Bevan, M., and Northcote, D. H. (1981a). Subculture-induced protein synthesis in tissue cultures of *Glycine max* and *Phaseolus vulgaris*. *Planta* **152**, 24–31.

Bevan, M., and Northcote, D. H. (1981b). Some rapid effects of synthetic auxins on mRNA levels in cultured plant cells. *Planta* **152**, 32–35.

Binns, A. N. (1981). Developmental variation in plant tissue culture. *Environ. Exp. Bot.* **21**, 325–32.

Blanarikova, V., and Karacsonyi, S. (1978). The isolation of tissue culture of *Populus alba* L. 'pyramidalis.' *Biol. Plant.* **20**, 14–18.

Boulanger, D., Bailey, B. K., and Steck, W. (1973). Formation of edulinine and furoquinoline alkaloids from quinoline derivatives by cell suspension cultures of *Ruta graveolens*. *Phytochemistry* **12**, 2399–2405.

Braun, A. C. (1956). The activation of two growth substance systems accompanying the conversion of normal to tumour cells in crown gall. *Cancer Res.* **16**, 53–56.

Brodelius, P., and Nilsson, K. (1980). The entrapment of plant cells in different matrices. *FEBS Lett.* **122**, 312–316.

Brunet, G., and Ibrahim, R. K. (1973). Tissue culture of *Citrus* peel and its potential for flavonoid synthesis. *Z. Pflanzenphysiol.* **69**, 152–162.

Chaturvedi, H. C., and Chowdhury, A. R. (1977). Habituation in root callus of *Citrus aurantifolia:* Free amino acid contents of normal and habituated callus. *Indian J. Exp. Biol.* **15**, 581–582.

Chaturvedi, H. C., and Mitra, G. C. (1975). A shift in morphogenetic pattern in *Citrus* callus tissue during prolonged culture. *Ann. Bot. (London)* **39**, 683–687.

Colomas, J., and Bulard, C. (1977). Comportement en culture *in vitro* des tissues de tige de *Myrtillocactus geometrizans* (Mart.) Cons. (T.) et biosynthese de betalains. *Bull. Soc. Bot. Fr.* **124**, 385–393.

Copping, L. G., and Street, H. E. (1972). Properties of the invertases of cultured Sycamore cells and changes in their activity during culture growth. *Physiol. Plant.* **26**, 346–354.

Corduan, G., and Spix, C. (1975). Haploid callus and regeneration of plants from anthers of *Digitalis purpurea* L. *Planta* **124**, 1–11.

Cronshaw, J. (1974). Phloem differentiation and development. *In* "Dynamic Aspects of Plant Ultrastructure" (A. W. Robards, ed.), pp. 391–413. McGraw-Hill, New York.

Cummings, D. P., Green, C. E., and Stuthman, D. P. (1976). Callus induction and plant regeneration in oats. *Crop Sci.* **16**, 465–470.

Davey, M. R., Fowler, M. W., and Street, H. E. (1971). Cell clones contrasted in growth, morphology and pigmentation isolated from a callus culture of *Atropa belladonna* v. Lutea. *Phytochemistry* **10**, 2559–2575.

Davidson, A. W. (1971). Effect of light on developing *Helianthus tuberosus* L. callus cultures. Ph.D. Thesis, Univ. of Edinburgh, Edinburgh.

Davidson, A. W., Aitchison, P. A., and Yeoman, M. M. (1976). Disorganized systems. *In* "Cell Division in Higher Plants" (M. M. Yeoman, ed.), pp. 407–438. Academic Press, New York and London.

Davies, M. E. (1972). Polyphenol synthesis in cell suspension cultures of Paul's Scarlet Rose. *Planta* **104**, 50–65.

Dhoot, G. K., and Henshaw, G. G. (1977). Organisation and alkaloid production in tissue cultures of *Hyoscyamus niger*. *Ann. Bot. (London)* **41**, 943–949.

Dougall, D. K. (1980). Nutrition and metabolism. *In* "Plant Tissue Culture as a Source of Biochemicals" (E. J. Staba, ed.), pp. 21–58. CRC Press, Boca Raton, Florida.

Dougall, D. K., Johnson, J. M., and Whitten, G. H. (1980). A clonal analysis of anthocyanin accumulation by cell cultures of wild carrot. *Planta* **149**, 292–297.

Ebel, J., Schaller-Hekeler, B., Knobloch, K-H., Wellman, E., Grisebach, H., and Hahlbrock, K. (1974). Coordinated changes in enzyme activities of phenylpropanoid metabolism during the growth of soybean cell suspension cultures. *Biochem. Biophys. Acta* **362**, 417–424.

Evans, P. K. (1967). Studies on cell division during early callus development in tissue isolated from Jerusalem artichoke tubers. Ph.D. Thesis, Univ. of Edinburgh, Edinburgh.

Everett, N. P., Wang, T. L., and Street, H. E. (1978). Hormone regulation of cell growth and development *in vitro*. *In* "Frontiers of Plant Tissue Culture 1978" (T. A. Thorpe, ed.) pp. 307–316. Univ. of Calgary Press, Calgary, Alberta, Canada.

Forrest, G. I. (1969). Studies on the polyphenol metabolism of tissue cultures derived from the tea plant (*Camellia sinensis* L.). *Biochem. J.* **113**, 765–772.

Fowler, M. W. (1971). Studies on the growth in culture of plant cells. XIV. Carbohydrate oxidation during the growth of *Acer pseudoplatanus* L. cells in suspension culture. *J. Exp. Bot.* **22**, 715–724.

Freeman, G. G., Whenham, R. J., Mackenzie, I. A., and Davey, M. R. (1974). Flavour component in tissue cultures of onion (*Allium cepa*). *Plant Sci. Lett.* **3**, 121–125.

Furuya, T., Kojima, H., and Syono, K. (1971). Regulation of nicotine biosynthesis by auxins in tobacco callus tissues. *Phytochemistry* **10**, 1529–1532.

Furuya, T., Nakano, M., and Yoshikawa, T. (1978). Biotransformation of (RS)-reticuline and morphinan alkaloids by cell cultures of *Papaver somniferum*. *Phytochemistry* **17**, 891–893.

Grant, M. E., and Fuller, K. W. (1968). Tissue culture of root cells of *Vicia faba*. *J. Exp. Bot.* **19**, 667–680.

Hadacóvá, V., Kaminek, M., and Lustinec, J. (1975). Glucose-6-phosphate dehydrogenase in tobacco callus strains differing in their growth and their requirement for auxin and cytokinin. *Biol. Plant.* **17**, 448–451.

Haddon, L., and Northcote, D. H. (1976). The effect of growth conditions and origin of tissue on the ploidy and morphogenetic potential of tissue cultures of bean (*Phaseolus vulgaris* L.). *J. Exp. Bot.* **27**, 1031–1051.

Hahlbrock, K., and Grisebach, H. (1979). Enzymic controls in the biosynthesis of lignin and flavonoids. *Annu. Rev. Plant Physiol.* **30**, 105–130.

Hahlbrock, K., Betz, B., Gardiner, S. E., Kreutzaler, F., Matern, V., Ragg, H., Schafer, E., and Schroder, J. (1978). Enzyme induction in cultured cells. In "Frontier of Plant Tissue Culture 1978" (T. A. Thorpe, ed.) pp. 317–324. Univ. of Calgary Press, Calgary, Alberta, Canada.

Hahlbrock, K., Kreuzaler, F., Ragg, H., Fautz, E., and Kuhn, D. N. (1982). Regulation of flavonoid and phytoalexin accumulation through mRNA and enzyme induction in cultured plant cells. *Hoppe-Seyler's Z. Physiol. Chem.* **363**, 121–122.

Harland, J., Jackson, J. F., and Yeoman, M. M. (1973). Changes in some enzymes involved in DNA biosynthesis following induction of division in cultured plant cells. *J. Cell Sci.* **13**, 121–138.

Harry, E., Mestre, J-C., and Guignard, J-L. (1977). La modification des capacités d'expression des potentialités embryogenes d'une souche de cal de carotte (*Daucus carota* L.) en fonction du milieu. *Bull. Soc. Bot. Fr.* **124**, 395–405.

Henke, R. R. (1981). Selection of biochemical mutants in plant cell cultures: Some considerations. *Environ. Exp. Bot.* **21**, 347–357.

Henke, R. R., Mansur, A., and Constantin, M. J. (1978). Organogenesis and plantlet formation from organ- and seedling-derived calli of rice (*Oryza sativa*). *Physiol. Plant* **44**, 11–14.

Henshaw, G. G., O'Hara, J. F., and Webb, K. H. (1982). Morphogenetic studies in plant tissue cultures. In "Differentiation *in Vitro*" (M. M. Yeoman and D. E. S. Truman, eds.) Brit. Soc. Cell Biol. Symp. Vol. 4, pp. 231–252, Cambridge Univ. Press, London and New York.

Hiraoka, N., and Tabata, M. (1974). Alkaloid production by plants regenerated from cultured cells of *Datura innoxia*. *Phytochemistry* **13**, 1671–1675.

Ibrahim, R. K., Thakur, M. L., and Permanand, B. (1971). Formation of anthocyanins in callus tissue cultures. *Lloydia* **34**, 175–182.

Ikuta, A., Syono, K., and Furuya, T. (1974). Alkaloids of callus tissues and redifferentiated plantlets in the Papaveraceae. *Phytochemistry* **13**, 2175–2179.

Israel, H. W., and Steward, F. C. (1966). The fine structure of quiescent and growing carrot cells: Its relation to growth induction. *Ann. Bot. (London)* **30**, 63–79.

Kadkade, P. G. (1982). Growth and podophyllotoxin production in callus tissues of *Podophyllum peltatum*. *Plant Sci. Lett.* **25**, 107–115.

Kanamori, I., Ashihara, H., and Komanine, A. (1979). Changes in the activities of the pentose phosphate pathway and pyrimidine nucleotide biosynthesis during the growth of *Vinca rosea* cells in suspension culture. *Z. Pflanzenphysiol.* **93**, 437–448.

Khavkin, E. E., and Varakina, N. N. (1981). Respiration of maize cells in batch suspension culture as compared to the intact root tip and coleoptile. *Z. Pflanzenphysiol.* **104**, 419–429.

King, P. J. (1980). Cell proliferation and growth in suspension cultures. *Int. Rev. Cytol. Suppl.* **11A**, 25–53.

King, P. J., and Street, H. E. (1977). Growth pattern in cell cultures. In "Plant Cell and Tissue Culture" (H. E. Street, ed.) Botanical Monograph Series Vol. 11, 2nd ed., pp. 307–387. Blackwell, Oxford.

Kinnersley, A. M., and Dougall, D. K. (1980a). Increase in anthocyanin yield from wild-carrot cell cultures by a selection system based on cell-aggregate size. *Planta* **149**, 200–204.

Kinnersley, A. M., and Dougall, D. K. (1980b). Correlation between the nicotine content of tobacco plants and callus cultures. *Planta* **149**, 205–206.

Knobloch, K-H., and Berlin, J. (1981). Phosphate-mediated regulation of cinnamoyl biosynthesis in cell suspension cultures of *Nicotiana tabacum*. *Planta Med.* **42**, 167–172.

Knobloch, K-H., Beutnagel, G., and Berlin, J. (1981). Influence of accumulated phosphate on culture growth and formation of cinnamoyl putrescines in medium-induced cell suspension cultures of *Nicotiana tabacum*. *Planta* **153**, 582–585.

Konoshima, M., Tabata, M., Yamamoto, H., and Hiraoka, N. (1970). Growth and alkaloid production of *Datura* tissue cultures. *Yakugaku Zasshi* **90**, 370–377.

Kurz, W. G. W., Chatson, K. B., Constabel, F., Kutney, J. P., Choi, L. S. L., Kolodziejczyk, P., Sleigh, S. K., Stuart, K. L., and Worth, B. R. (1981). Alkaloid production in *Catharanthus roseus* cell cultures VIII: Characterization of the PRL #200 cell line. *Planta Med.* **42**, 22–31.

Kutney, J. P., Choi, L. S. L., Kolodziejczyk, P., Sleigh, S. K., Stuart, K. L., Worth, B. R., Kurz, W. G. W., Chatson, K. B., and Constabel, F. (1980). Alkaloid production in *Catharanthus roseus* cell cultures: Isolation and characterisation of alkaloids from one cell line. *Phytochemistry* **19**, 2589–2595.

Laetsch, W. M., and Stetler, D. A. (1965). Chloroplast structure and function in cultured tobacco tissue. *Am. J. Bot.* **52**, 798–804.

Lindsey, K. (1982). Studies on the growth and metabolism of plant cells cultured on fixed-bed reactors. Ph.D. Thesis, Univ. of Edinburgh, Edinburgh.

Lindsey, K. (1985). The manipulation, by nutrient limitation, of the biosynthetic activity of immobilized cells of *Capsicum frutescens* Mill. cv. annuum. *Planta* **165**, 126–133.

Lindsey, K., and Yeoman, M. M. (1983a). The relationship between growth rate, differentiation and alkaloid accumulation in cell cultures. *J. Exp. Bot.* **34**, 1055–1065.

Lindsey, K., and Yeoman, M. M. (1983b). Novel experimental systems for studying the production of secondary metabolites by plant tissue cultures. In "Plant Biotechnology," SEB Seminar Series Vol. 18 (S. H. Mantell and H. Smith, eds.), pp. 39–66. Cambridge Univ. Press, London and New York.

Lindsey, K., and Yeoman, M. M. (1984). The synthetic potential of immobilized cells of *Capsicum frutescens* Mill. cv. annuum. *Planta* **162**, 495–501.

Lindsey, K., and Yeoman, M. M. (1985). Immobilised plant cells. In "Plant Cell Culture Technology" (M. M. Yeoman, ed.), Botanical Monograph Series, pp. 226–263. Blackwell, Oxford.

Lindsey, K., Yeoman, M. M., Black, G. M., and Mavituna, F. (1983). A novel method for the immobilisation and culture of plant cells. *FEBS Lett.* **155**, 143–149.

Luckner, M. (1980). Expression and control of secondary metabolism. In "Secondary Plant Products," Encyclopedia of Plant Physiology Vol. 8" (E. A. Bell and B. V. Charlwood, eds.), pp. 23–63. Springer-Verlag, Berlin, Heidelberg, New York.

Macleod, A. J., Mills, E. D., and Yeoman, M. M. (1979). Seasonal variation in the pattern of RNA metabolism of tuber tissue in response to excision and culture. *Protoplasma* **98**, 343–354.

Mäder, M., Münch, P., and Bopp, M. (1975). Regulation of peroxidase patterns during shoot differentiation in callus cultures of *Nicotiana tabacum*. *Planta* **123**, 257–265.

Mantell, S. M., and Smith, H. (1983). Cultural factors that influence secondary metabolite accumulations in plant cell and tissue cultures. In "Plant Biotechnology" (S. H. Mantell and H. Smith, eds.), SEB Seminar Series Vol. 18, pp. 75–108. Cambridge Univ. Press, London and New York.

Matsumoto, T., Nishida, K., Noguchi, M., and Tamaki, E. (1973). Some factors affecting the anthocyanin formation by *Populus* cells in suspension cultures. *Agric. Biol. Chem.* **37**, 561–567.

Meijer, E. G. M., and Broughton, W. J. (1981). Regeneration of whole plants from hypocotyl-, root- and leaf-derived tissue cultures of the pasture legume *Stylosanthes guyanensis*. *Physiol. Plant* **52**, 280–284.

Meins, F., and Binns, A. N. (1977). Epigenetic variation of cultured somatic cells: Evidence for gradual changes in the requirement for factors promoting cell division. *Proc. Natl. Acad. Sci. U.S.A.* **74**, 2928–2932.

Meins, F., and Binns, A. N. (1979). Cell determination in plant development. *BioScience* **29**, 221–225.

Meins, F., and Lutz, J. (1980). The function of cytokinin habituation in primary pith explants of tobacco. *Planta* **149**, 402–407.

Meins, F., Lutz, J., and Foster, R. (1980). Factors influencing the incidence of habituation for cytokinin of tobacco pith tissue in culture. *Planta* **150**, 264–268.

Mizukami, H., Konoshima, M., and Tabata, M. (1977). Effect of nutritional factors on shikonin derivative formation in *Lithospermum* callus cultures. *Phytochemistry* **16**, 1183–1186.

Mok, M. C., Gabelman, W. H., and Skoog, F. (1976). Carotenoid synthesis in tissue cultures of *Daucus carota* L. *J. Am. Soc. Hortic. Sci.* **101**, 442–449.

Murashige, T., and Nakano, R. (1965). Morphogenetic behaviour of tobacco tissue cultures and implication of plant senescence. *Am. J. Bot.* **52**, 819–827.

Nagel, M., and Reinhard, E. (1975). Das Aetherische Oel der Calluskulturen von *Ruta graveolus* L. Die Zuzammensetzung des Oeles. *Planta Med.* **27**, 151–158.

Naik, G. G. (1965). Studies on the effects of temperature on the growth of plant tissue cultures. M.Sc. Thesis, Univ of Edinburgh, Edinburgh.

Nettleship, L., and Slaytor, M. (1974). Adaptation of *Peganum harmala* callus to alkaloid production. *J. Exp. Bot.* **25**, 1114–1123.

Neumann, D., and Mueller, E. (1971). Beitrage zur Physiologie der Alkaloide: V. Alkaloidbildung in Kallus—und Suspensionkulturen von *Nicotiana tabacum*. *Biochem. Physiol. Pflanz.* **162**, 503–513.

Neumann, D., and Mueller, E. (1974). Formation of alkaloids in callus cultures of *Macleaya*. *Biochem. Physiol. Pflanz.* **165**, 271–282.

Nicholson, M. O., and Flamm, W. G. (1965). Properties and significance of free and bound ribosomes from cultured tobacco cells. *Biochim. Biophys. Acta* **108**, 266–274.

Nickell, L. G (1980) Products. In "Plant Tissue Culture as a Source of Biochemicals" (E. J. Staba, ed.), pp. 256–269. CRC Press, Boca Raton, Florida.

Nikolaeva, L. A., and Vollosovich, A. G. (1972). Effect of auxins on the biosynthesis of tropane and indole alkaloids. *Rastit. Resur.* **8**, 188–192.

Nishi, A., Yoshida, A., Mori, M., and Sugano, N. (1974). Isolation of variant carrot cell lines with altered pigmentation. *Phytochemistry* **13**, 1653–1656.

Noguchi, H., and Sankawa, U. (1982). Formation of germichrysone by tissue cultures of *Cassia torosa*: Induction of secondary metabolism in the lag phase. *Phytochemistry* **21**, 319–323.

Northcote, D. H. (1968). The organisation of the endoplasmic reticulum, the Golgi bodies and microtubules during cell division and subsequent growth. In "Plant Cell Organelles" (J. B. Pridham, ed.), pp. 179–197. Academic Press, London.

O'Brien, T. P. (1974). Primary vascular tissues. In "Dynamics of Plant Ultrastructure" (A. W. Robards, ed.), pp. 414–440. McGraw-Hill, New York.

Ogino, T., Hiraoka, N., and Tabata, M. (1978). Selection of high nicotine-producing cell lines of tobacco callus by single cell cloning. Phytochemistry 17, 1907–1910.

Ozeki, Y., and Komamine, A. (1981). Induction of anthocyanin synthesis in relation to embryogenesis in carrot (Daucus carota cv. Kurodagosun) suspension culture: Correlation of metabolic differentiation with morphological differentiation. Physiol. Plant. 53, 570–577.

Partanen, C. R., Sussex, I. M., and Steeves, T. A. (1955). Nuclear behaviour in relation to abnormal growth in fern prothalli. Am. J. Bot. 42, 245–256.

Phillips, R., and Henshaw, G. G. (1977). The regulation of synthesis of phenolics in stationary phase cell cultures of Acer pseudoplatanus L. J. Exp. Bot. 28, 785–794.

Radin, D. N., Behl, H. M., Proksch, P., and Rodriguez, E. (1982). Rubber and other hydrocarbons produced in tissue cultures of guayule (Parthenium argentatum). Plant Sci. Lett. 26, 301–310.

Ramawat, K. G., and Arya, H. C. (1979). Effect of amino acids on ephedrine production in Ephedra gerardiana callus cultures. Phytochemistry 18, 484–485.

Robbins, W. J., and Hervey, A. (1970). Tissue culture of callus from seedling and adult stages of Hedera helix. Am. J. Bot. 57, 452–457.

Robertson, A. I. (1966). Metabolic changes during callus development in tissue isolated from Jerusalem artichoke tubers. Ph.D. Thesis, Univ. of Edinburgh, Edinburgh.

Sacristan, M. D. (1967). Auxin-Autotrophie und Chromosomenzahl: Untersuchungen an alten, spontan habituierten und crowngall Kalluskulturen von Nicotiana tabacum aus dem Laboratorium Gautherets. Mol. Gen. Genet. 99, 311–321.

Sacristan, M. D., and Melchers, G. (1969). The caryological analysis of plants regenerated from tumourous and other callus cultures of tobacco. Mol. Gen. Genet. 105, 317–333.

Sánchez de Jimenez, E., and Fernandez, L. (1983). Biochemical parameters to assess cell differentiation of Bouvardia ternifolia Schlecht callus. Planta 158, 377–383.

Schröder, J., Betz, B., and Hahlbrook, K. (1978). Messenger RNA-controlled increase in phenylalanine ammonia-lyase activity in parsley: Light-independent induction by dilution of cell suspension cultures into water. Plant Physiol. 60, 440–445.

Schröder, J., Schröder, G., Huisman, H., Schilperoort, R. A., and Schell, J. (1981). The mRNA for lysopine dehydrogenase in plant tumour cells is complementary to a Ti plasmid fragment. FEBS Lett. 129, 166–168.

Seibert, M., and Kadkade, P. G. (1980). Environmental factors: Light. In "Plant Tissue Culture as a Source of Biochemicals (E. J. Staba, ed.) pp. 123–142. CRC Press, Boca Raton, Florida.

Sejourne, M., Viel, C., Bruneton, J., Rideau, M., and Chenieux, J. C. (1981). Growth and furoquinoline alkaloid production in cultured cells of Choisya ternata. Phytochemistry 20, 353–355.

Sharma, D. R., and Chowdury, J. R. (1977). Effects of different media on cultured anthers of Datura innoxia Mill. and comparative morphogenetic potentiality of haploid and diploid tissues. Indian J. Exp. Biol. 15, 616–618.

Shimizu, T., Clifton, A., Komamine, A., and Fowler, M. W. (1977). Changes in metabolite levels during growth of Acer pseudoplatanus (Sycamore) cells in batch suspension culture. Physiol. Plant 40, 125–129.

Singh, B. D., Harvey, B. L., Kao, K. N., and Miller, R. A. (1975). Karyotypic changes and selection pressure in Haplopappus gracilis suspension cultures. Can. J. Genet. Cytol. 17, 109–116.

Snijman, D. A., Noel, A. R. A., Bornman, C. H., and Abbott, J. G. (1977). Nicotiana tabacum callus studies. II. Variability in cultures. Z. Pflanzenphysiol. 82, 367–370.

Speake, T., McCloskey, P., Smith, W., Scott, T., and Hussey, H. (1964). Isolation of nicotine from cell cultures of Nicotiana tabacum. Nature (London) 201, 614–615.

Staba, E. J. (ed.) (1980). "Plant Tissue Culture as a Source of Biochemicals." CRC Press, Boca Raton, Florida.

Steffenson, D. M. (1961). Chromosome structure with special reference to the role of metal ions. *Int. Rev. Cytol.* **12**, 163–197.

Steward, F. C., Bidwell, R. G. S., and Yemm, E. W. (1958). Nitrogen metabolism, respiration and growth of cultured plant tissue. *J. Exp. Bot.* **9**, 11–51.

Stickland, R. G., and Sunderland, N. (1972). Production of anthocyanins, flavonols and chlorogenic acids by cultured callus tissues of *Haplopappus gracilis*. *Ann. Bot. (London)* **36**, 443–457.

Stoutemeyer, V. T., and Britt, O. K. (1965). The behaviour of tissue cultures from English and Algerian ivy in different growth phases. *Am. J. Bot.* **52**, 805–810.

Stoutemeyer, V. T., and Britt, O. K. (1969). Growth and habituation in tissue cultures of English ivy, *Hedera helix*. *Am. J. Bot.* **56**, 222–226.

Street, H. E. (1977). Differentiation in cell and tissue cultures—regulation at the molecular level. *In* "Regulation of Developmental Processes in Plants" (H. R. Schutte and D. Gross, eds.), pp. 192–218. VEB Gustav Fischer Verlag, Jena.

Sunderland, N. (1977). Nuclear cytology. *In* "Plant Tissue and Cell Culture" (H. E. Street, ed.), pp. 177–205. Botanical Monograph Series Vol. 11, Blackwell, Oxford.

Swedlund, B., and Vasil, I. K. (1985). Cytogenetic characterization of embryogenic callus and regenerated plants of *Pennisetum americanum* (L.) K. Schum. *Theor. Appl. Genet.* **69**,

Tabata, M., and Hiraoka, N. (1976). Variation of alkaloid production in *Nicotiana rustica* callus cultures. *Physiol. Plant.* **38**, 19–23

Tabata, M., Yamamoto, H., and Hiraoka, N. (1968). Chromosome constitution and nicotine formation of mature plants derived from cultured pith of tobacco (var. "Wisconsin 38"). *Jpn. J. Genet.* **43**, 319–322.

Tabata, M., Ogino, T., Yoshioka, K., Yoshikawa, N., and Hiraoka, N. (1978). Selection of cell lines with higher yield of secondary products. *In* "Frontiers of Plant Tissue Culture 1978" (T. A. Thorpe, ed.), pp. 213–222. Univ. of Calgary Press, Calgary, Alberta, Canada.

Tabata, M., Yamamoto, H., Hiraoka, N., and Konoshima, M. (1972). Organisation and alkaloid production in tissue cultures of *Scopolia parviflora*. *Phytochemistry* **11**, 949–955.

Tabata, M., Ogino, T., Yoshioka, K., Yoshikawa, N., and Hiraoka, N. (1978). Selection of cell lines with higher yield of secondary products. *In* "Frontiers of Plant Tissue Culture 1978" (T. A. Thorpe, ed.), pp. 213–222. Univ. of Calgary Press, Calgary, Alberta, Canada.

Takahashi, M., and Yamada, Y. (1973). Regulation of nicotine production by auxins in tobacco cultured cells *in vitro*. *Agric. Biol. Chem.* **37**, 1755–1757.

Tewes, A., Wappler, A., Deschke, E-M., Garve, R., and Nover, L. (1982). Morphogenesis and embryogenesis in long-term cultures of *Digitalis*. *Z. Pflanzenphysiol.* **106**, 311–324.

Thomas, E., and Street, H. E. (1970). Organogenesis in cell suspension cultures of *Atropa belladonna* L. and *Atropa belladonna* cultivar *lutea* DOLL. *Ann. Bot. (London)* **34**, 657–669.

Tomita, Y., Uomori, A., and Minato, H. (1970). Steroidal sapogenins and sterols in tissue cultures of *Dioscorea tokoro*. *Phytochemistry* **9**, 111–114.

Torrey, J. G. (1965). Cytological evidence of cell selection by plant tissue. *In* "Plant Tissue Culture" (P. R. White and A. R. Grove, eds.), pp. 473–484. McCutchan, Berkeley, California.

Torrey, J. G. (1967). Morphogenesis in relation to chromosomal constitution in long-term plant tissue cultures. *Physiol. Plant.* **20**, 265–275.

van Larabecke, N., Engler, G., Holsters, M., van den Elsacker, S., Zaeneu, I., Schilperoort, R. A., and Schell, J. (1974). Large plasmid in *Agrobacterium tumefaciens* essential for crown gall-inducing ability. *Nature (London)* **252**, 169–170.

Vanzulli, L., Magnien, E., and Olivi, L. (1980). Caryological stability of *Datura innoxia* calli analysed by cytophotometry for 22 hormonal combinations. *Plant Sci. Lett.* **17**, 181–192.

Vasil, I. K. (1973). Morphological, histochemical and ultrastructural effects of plant growth substances *in vitro*. *Biochem. Physiol. Pflanz.* **164**, 58–71.

Vasil, I. K. (1985). Somatic embryogenesis and its consequences in the Gramineae. *In* "Propagation of Higher Plants through Tissue Culture" (R. Henke, K. Hughes, and A. Hollaender, eds.), pp. 31–48. Plenum, New York.

Vasil, V., and Vasil, I. K. (1984a). Induction and maintenance of embryogenic callus cultures of Gramineae. *In* "Cell Culture and Somatic Cell Genetics of Plants," Vol. 1, Laboratory Procedures and Their Applications (I. K. Vasil, ed.), pp. 36–42. Academic Press, New York.

Vasil, V., and Vasil, I. K. (1984b). Isolation and maintenance of embryogenic cell suspension cultures of Gramineae. *In* "Cell Culture and Somatic Cell Genetics of Plants," Vol. 1, Laboratory Procedures and Their Applications (I. K. Vasil, ed.), pp. 152–158. Academic Press, New York.

Verma, D. P. S., and Marcus, A. (1974). Activation of protein synthesis upon dilution of an *Arachis* cell culture from the stationary phase. *Plant Physiol.* **53**, 83–87.

Vyskot, B., and Novak, F. J. (1977). Habituation and organogenesis in callus cultures of chlorophyll mutants of *Nicotiana tabacum* L. *Z. Pflanzenphysiol.* **81**, 34–42.

Westcott, R. J., and Henshaw, G. G. (1976). Phenolic synthesis and phenylalanine ammonia-lyase activity in suspension cultures of *Acer pseudoplatanus* L. *Planta* **131**, 67–73.

White, P. R. (1967). Promises and challenges of tissue culture for biology and for mankind. *In* "Plant Cell, Tissue and Organ Cultures," pp. 12–19. U.G.C. Centre of Advanced Study in Botany, Delhi, India.

Wilson, H. M., and Street, H. E. (1975). The growth, anatomy and morphogenetic potential of callus and cell suspension cultures of *Hevea brasiliensis*. *Ann. Bot. (London)* **39**, 671–682.

Yeoman, M. M., and Aitchison, P. A. (1973). Changes in enzyme activities during the division cycle of cultured plant cells. *In* "The Cell Cycle in Development and Differentiation" (M. Balls and F. S. Billet, eds.) Br. Soc. Dev. Biol. Symp. 1, pp. 185–201. Cambridge Univ. Press, London and New York.

Yeoman, M. M.,and Aitchison, P. A. (1976). Molecular events of the cell cycle: A preparation for division. *In* "Cell Division in Higher Plants" (M. M. Yeoman, ed.), pp. 111–113. Academic Press, New York and London.

Yeoman, M. M., and Davidson, A. W. (1971). Effect of light on cell division in developing callus cultures. *Ann. Bot. (London)* **35**, 1085–1100.

Yeoman, M. M., and Forche, E. (1980). Cell proliferation and growth in callus cultures. *Int. Rev. Cytol. Suppl.* **11A**, 1–24.

Yeoman, M. M., and Street, H. E. (1977). General cytology of cultured cells. *In* "Plant Cell and Tissue Culture" (H. E. Street, ed.), Botanical Monograph Series Vol. 11, 2nd ed. pp. 137–176. Blackwell, Oxford.

Yeoman, M. M., Dyer, A. F., and Robertson, A. I. (1965). Growth and differentiation of plant tissue cultures. I. Changes accompanying the growth of explants from *Helianthus tuberosus* tubers. *Ann. Bot. (London)* **29**, 265–276.

Yeoman, M. M., Aitchison, P. A., and Macleod, A. J. (1978). *In* "Regulation of Enzyme Synthesis and Activity in Higher Plants" (H. Smith, ed.), pp. 63–81. Academic Press, New York.

Yeoman, M. M., Miedzybrodzka, M. B., Lindsey, K., and McLauchlan, W. R. (1980). The synthetic potential of cultural plant cells. *In* "Plant Cell Cultures: Results and Perspectives" (F. Sala, B. Parisi, R. Cella, and O. Cifferi, eds.), pp. 327–343. Elsevier-North Holland, Amsterdam.

Yeoman, M. M., Lindsey, K., and Hall, R. D. (1982a). Differentiation as a prerequisite for the production of secondary metabolites. *In* "Proceedings of the Plant Cell Culture Conference," pp. 1–7. Oyez Scientific and Technical Services Ltd., London.

Yeoman, M. M., Lindsey, K., Miedzybrodzka, M. B., and McLauchlan, W. R. (1982b). Accumulation of secondary products as a facet of differentiation in plant cell and tissue cultures. *In* "Differentiation *in Vitro*" (M. M. Yeoman and D. E. S. Truman, eds.), Brit. Soc. Cell Biol. Symp. Vol. 4, pp. 65–82.

Zeleneva, I. V., and Khavkin, E. E. (1980). Rearrangement of enzyme patterns in maize callus and suspension cultures: Is it relevant to the changes in the growing cells of the intact plant? *Planta* **148**, 108–115.

Zenk, M. H., El-Shagi, H., and Schulte, U. (1975). Anthraquinone production by cell suspension cultures of *Morinda citrofolia*. *Planta Med. Suppl.* 79–101.

Zenk, M. H., El Shagi, H., Arens, H., Stockigt, J., Weiler, E. W., and Deus, B. (1977). Formation of the indole alkaloids serpentine and ajmalicine in cell suspension cultures of *Catharanthus roseus*. *In* "Plant Tissue Culture and Its Biotechnological Application" (W. Barz, E. Reinhard, and M. H. Zenk, eds.), pp. 27–43. Springer-Verlag, Berlin and New York.

CHAPTER 3

The Mass Culture of Plant Cells

A. H. Scragg
M. W. Fowler

Wolfson Institute of Biotechnology
University of Sheffield
Sheffield, England

I. INTRODUCTION

A. Preamble

The concept of using plant cell cultures to produce natural products commercially was first discussed in detail by Routier and Nickell in 1956. The same communication also described their own attempts to culture plant cells on a large scale, subsequently reported on further by Tulecke and Nickell in 1959. Over the years a wide range of process systems and format have been explored and proposed for a plant cell culture industrial process. No matter what the format, all have a common denominator, the need to produce quite large amounts of biomass effectively and rapidly, typically as a key first stage in the process. Since the early work of Nickell's group, major progress has been made in the large-scale culture of plant cells and it is toward a review of that progress and an assessment of the present state of the art that this chapter is directed.

B. Perspective

In spite of major advances in synthetic organic chemistry, the plant kingdom still contributes significantly in both quantity and product range to the specialty chemicals used by a number of industries, including foods, pharmaceuticals, and agrochemicals. Examples of these products have been extensively discussed in a previous review (Fowler, 1983). With increased knowledge and sophistication in the handling of plant cell cultures has arisen a situation where quite a large number of these products are synthesized in cell cultures, albeit in many cases at rather low levels. Nonetheless, the list is quite impressive in scale and range (see Table I, and also Nickell, 1980), and there are examples where specific cultures may synthesize desirable products at levels equivalent to or possibly higher than in the plant from which they were extracted. It is against this rapidly developing backcloth of potential product targets that we have assessed progress in large-scale culture.

Large-scale cultures of plant cells have a number of potentially advantageous features when compared with traditional methods of plantation growth.

These include:

1. Independence from various environmental factors, including climate, pests, and geographical and seasonal constraints.
2. Defined production systems, with production as and when required.
3. More consistent product quality and yield.
4. Reduction in land use for "cash crops."
5. Freedom from political interference.

One of the most attractive advantages is perhaps the defined production system in which one is dealing with an essentially homogeneous population of cells and in which a defined and controlled large-scale culture can be used to smooth out production variations observed with plantation crops.

TABLE I

Products Synthesized Using Cell Cultures

Industry	Product	Plant species	Use
Pharmaceuticals	Ajmalicine	*Catharanthus roseus*	Circulatory problems
	Berberine	*Coptis japonica*	Antimicrobial
	Codeine	*Papaver somniferum*	Sedative
	Quinine	*Cinchona ledgeriana*	Antimalarial
	Camptothecin	*Camptotheca acuminata*	Antitumor
	Digoxin	*Digitalis lanata*	Cardiatonic
	Diosgenin	*Dioscorea deltoidea*	Antifertility agents
	Ubiquinone 10	*Nicotiana tabacum*	Cardiac drug
	Shikonin	*Lihtospermum erythrorhizon*	Antimicrobial
	Quinidine	*Cinchona ledgeriana*	Cardiac drug
Food	Anthocyanins	*Daucus carota*	Pigment
	Quinine	*Cinchona ledgeriana*	Bittering agent
	Spearmint	*Mentha*	Flavor
Agriculture	Pyrethrin	*Chrysanthemum*	Insecticide
Cosmetics	Jasmine	*Jasminum*	Perfume

The acceptance of a plant cell culture system for producing various chemicals on an industrial scale depends upon the economics of the process, which, in turn, are dependent on the production side on (1) the ability to grow the particular cells rapidly in large quantities, and (2) the absolute need to obtain a high yield of product. Apart from a few specialized cases such as tobacco biomass, the products of value produced by plant cell cultures are "secondary metabolites" that are normally present at very low levels. However, with the development of new nutrient media, and techniques for the isolation and selection of cell lines, there have occurred enhancements in yield of a number of secondary products to values above those found in whole, or parts of whole, plants (Table II). This, in turn, has done much to encourage the development of mass culture of plant cells for the production of secondary products. That such a goal is a feasible target has recently been demonstrated with the announcement by Mitsui Petrochemical Industries Ltd. (Tokyo) of a process for the production of shikonin, a dye and pharmaceutical (Curtin, 1983).

TABLE II

Yields of Natural Products Obtained in Plant Cell Culture Compared with Whole Plants

Chemical	Plant species	Cell culture yield	Whole plant yield	Reference
Nicotine	*Nicotiana tabacum*	2–5	1	Tabata *et al.* (1978)
Crude saponins	*Panax ginseng*	21	4.5	Furuya *et al.* (1972)
Diosgenin	*Dioscorea deltoidea*	2.0	2.0	Kaul *et al.* (1969)
Shikonins	*Lithospermum erythrorhizon*	12	1.5	Tabata *et al.* (1976)
Rosmarinic acid	*Coleus blumei*	15	3.0	Razzaque *et al.* (1977)
Jatrorrhizine	*Berberis stolonifera*	7	1	Hinz *et al.* (1981)
Anthraquinones	*Morinda citrifolia*	18	2.2	Zenk *et al.* (1975)
Shikimic acid	*Galium mollugo*	10	—	Amrhein
Ubiquinone	*Nicotiana tabacum*	0.2	0.003	Matsumoto *et al.* (1981)
Serpentine	*Catharanthus roseus*	0.8	0.5	Zenk *et al.* (1977b)
Ajmalicine	*Catharanthus roseus*	1.0	0.3	Zenk *et al.* (1977b)
Caffeine	*Coffea arabica*	1.6	1.6	Frischkrecht *et al.* (1977)
Glutathione	*Nicotiana tabacum*	1.0	0.1	Fowler (1982a,b)
Visnagin	*Ammi visnaga*	0.31	0.1	Kaul and Staba (1967)

II. MASS CULTURE: HISTORICAL

The large-scale cultivation of plant cells began over 25 years ago when, in 1959, Tulecke and Nickell described the growth of *Ginkgo,* holly, *Lolium,* and rose cell suspensions in a simple sparged 20-liter carboy. Wang and Staba (1963) used a dual carboy system that was both stirred and sparged for the cultivation of 3 liters of spearmint cells. The carboy system was adapted to a large roller bottle by Lamport (1964) growing 2–5 liters of *Acer pseudoplatanus,* and by Short *et al.* (1969) for 4- to 5-liter cultures of *A. pseudoplatanus* cells. Graebe and Novelli (1966) used large (1- to 6-liter) sparged flasks where cell aggregates were broken up by periodic use of a magnetic stirrer to grow *Zea mays* cultures. In 1970, Velicky and Martin described a similar system in which the configuration of the fermenter was an inverted flask using a magnetic stirrer and air sparging to grow cultures of *Phaseolus vulgaris*. All these systems were developed specifically for the growth of plant cell suspensions. However, as early as 1962, Byrne and Koch reported the growth of carrot cells in a commercial stirred tank reactor (CSTR) (7.5 and 15 liters). These early designs are described in the review by Martin (1980). The initial work on mass culture of plant cell suspensions demonstrated clearly that morphological and physiological differences between plant cells and microorganisms had a significant effect on their mass culture and on the bioreactor conformation used for the process.

III. EFFECT OF STRUCTURE AND PHYSIOLOGY OF PLANT CELLS ON THEIR LARGE-SCALE CULTIVATION

Some of the important differences between plant and microbial cells are shown in Table III. One of the most striking features of plant cells is their enormous size when compared with microorganisms. Plant cells are 20–40 μ in diameter and can be 100–200 μ in length, which gives them a volume of up to 100,000 times that of most microorganisms. Plant cell suspensions

TABLE III

The Differences between Microbial and Plant Cells

Characteristics	Microbial cell	Plant cell
Size	$2\mu m^3$	$>10^5\mu m^3$
Individual cells	Often	Normally in clumps
Growth of individual cells into colonies	Yes	Not often
Doubling time	>1 hr	Days
Inoculation density	Small	5–20% of total culture
Chromosome number	Haploid/diploid	Mixture of haploid, diploid, polyploid, aneuploid
Shear stress	Insensitive	Sensitive
Variability within a single culture	Stable	Can vary greatly
Development	Can form spores, pseudomycelium	Organogenesis, embryogenesis
Aeration	High 1–2 vvm	Low 0.2–0.3 vvm
Product formation	Often into medium	Into vacuole

can vary greatly within a population in both size and shape, and these latter can also change markedly during the growth cycle (Liau and Boll, 1971; Kato et al., 1978).

Individual cells are rarely found in plant cell suspensions, the cells normally forming small groups, each containing from 2 to 10 or more cells. Such groups are probably derived from the nonseparation of the cells after division. The presence of these groups or clumps poses two problems. First, turbidometric methods of cell number estimation are impractical. Second, they may also impact upon the mixing and rheology of the system depending upon the number of cells per clump, the clump morphology, and the number of clumps per unit culture volume. The formation of groups of cells or clumps can be affected by the media composition, method of subculture, or environmental conditions. Alteration in environmental conditions can lead to the development of large cell clumps 2 mm in diameter and containing up to 200 cells (Henshaw et al., 1966). This, in turn, leads to problems with mixing and sampling. A second reason for clumping is the development of cell "stickiness" due to a coating of extracellular polysaccharides on the cell walls. Tobacco cells have been shown to be particularly prone to this form of clumping, becoming sticky in late lag phase of growth (Fowler, 1982a). Takayama et al., (1977) have shown that reduction in calcium ion concentration reduces the degree of cell clumping. However, this may also affect other aspects of cell physiology and furthermore appears to be only a short-term palliative. Aggregation due to polysaccharides may be reduced by pulsed or phased addition of the carbon supply during the exponential phase of growth (Smart, 1984). This reduces the level of polysaccharide excretion by the cells, especially in carbon overload situations, and so lessens the degree of stickiness.

A number of workers have used different approaches to obtain "fine" cell suspensions, that is, suspensions composed largely of free cells or small clumps. The use of cell wall degrading enzymes has been reported (Street et al., 1971), as has the separation of individual cells by fine mesh grids. In both cases, however, and with further culture growth, cell clumping recurred. A novel method of producing fine suspensions of plant cells has been developed in our institute (Morris and Fowler, 1981; Morris et al., 1982). Plant cells are immobilized in calcium alginate beads and then placed in a fluidized bed reactor. Production of fine cells occurs as daughter cells grow out of the beads and separate or break off into the surrounding medium due to abrasion. This method has the advantage that, where "cell conditioning" is a prerequisite for growth, released cells are conditioned in the growth medium and therefore can survive at low densities. Finely dispersed cell suspensions, containing small groups of embryogenic cells, have also been established in several species of the Gramineae (Vasil, 1985; Vasil and Vasil, 1984).

While microorganisms have a relatively flexible murein or chitin sheath, plant cells have a rigid cellulose cell wall. As a consequence of this and their large size, plant cells are very susceptible to shear. This specific property is of crucial importance in relation to approaches adopted for large-scale culture systems. In classical fermentation reactors, shear produced by the impeller is used to break up gas bubbles from the aeration stream to obtain a high mass transfer of oxygen into solution. A balance has therefore to be found in achieving controlled dissolved oxygen levels at shear levels acceptable for work with plant cells. This has led to the exploration of a wide variety of bioreactor conformations (see Section VIII).

A major portion of the plant cell is taken up by a central vacuole that gives the cell an unusually high water content (>90%). The vacuole is also the site of metabolite storage, since plant cells, unlike microorganisms, do not generally excrete secondary products into the surrounding environment (Neumann et al., 1983). As a consequence of their large size and high water content, plant cells have a low overall metabolic activity and low growth rate. The doubling time for plant cells is typically within the range of 20–100 hr (Noguchi et al., 1977) when compared with as little as 20 min for bacteria (Table IV).

Another significant difference between plant cells and microorganisms lies in the observation that, in order to divide, both single cells or cell populations at low density require substances not required by large cell masses. This implies that plant cells release into their surrounding environment key metabolites that assist the growth and maintain the viability of single cells. This phenomenon is known as "conditioning." Such a phenomenon makes the use of standard microbiological techniques, such as plate replication, very difficult with plant cells.

IV. MIXING AND AERATION

A fundamental requirement of any mass culture system is adequate mixing in order to achieve homogenous conditions within the bioreactor required for consistent nutrient supply. It is also key in achieving representative sampling.

Microbial mass growth is normally carried out in stirred-tank reactors where development has concentrated on the maximization of oxygen supply (the least soluble nutrient). The impeller system in a stirred tank, often a flat blade turbine, is not only used for mixing but is also used to break up the aeration bubbles in order to obtain a high mass transfer of oxygen. At

TABLE IV

The Growth Rates of Plant Cell Suspensions

Plant type	Td (days)	u	Vessel	Reference
Phaseolus vulgaris	63	0.011	3- to 6-liter conical vessel, stirred and sparged	Velicky and Martin (1970)
Nicotiana tabacum	1.65	0.42	600 liter STR	Yasuda *et al.* (1972)
N. tabacum	1.0–1.12	0.62–0.69	15- and 1500-liter bubble tanks	Kato *et al.* (1976a,b)
N. tabacum	0.72	0.96	15-, 30-, and 230-liter STR	Kato *et al.* (1977)
N. tabacum	0.63	1.09	20,000-liter (15,000 w/v) STR	Noguchi *et al.* (1977)
Morinda citrifolia	3.5	0.198	10-liter airlift	Wagner and Vogelmann (1977)
M. citrifolia	1.44	0.48	10-liter airlift	Vogelmann *et al.* (1978)
N. tabacum	1.98	0.35	20,000-liter (6340 w/v) STR	Hashimoto *et al.* (1982)
Catharanthus roseus	1.73	0.4	10-liter airlift	Smart and Fowler (1984)
Bush bean, tobacco, soy bean, *Datura innoxia*, rice	0.9–1.0	0.77–0.693	Shake flasks	Compiled in Noguchi *et al.* (1977)

even moderate revolutions the impellers used in CSTRs produce a high shear at their tips.

Plant cells are large and hence sediment rapidly in the absence of stirring; some form of mixing in bioreactors is therefore essential. However, and as described earlier, because of their large size and rigid cell walls, plant cells are particularly sensitive to shear and are often damaged and broken in small-scale suspension as well as in conventional stirred-tank bioreactors (Kato *et al.*, 1972). In particular, shear produced in shake flasks has been shown to affect the growth of tobacco cells (Matsumoto *et al.*, 1981) and *Cudramia tricuspidata* cell suspensions (Tanaka, 1981). In a comparative study of different reactor types, Wagner and Vogelmann (1977) found that turbines not only resulted in major cell lysis but that product yield, in this case anthroquinone from *Morinda citrifolia*, was also reduced by up to 60%. Cell breakage was confirmed by microscopic observation and also appeared to be associated with an increase in culture pH resulting from the release of cell contents. Judging from changes in medium pH, the M. citrifolia cells became more sensitive to shear toward the end of the growth phase. Beta vulgaris cell suspensions have been shown to be sensitive to mechanical stirring even at the low rotational speed of 28 rpm (Wilson, 1978). Other cell suspensions shown to be sensitive to stirring include Nicotiana tabacum and Daucus carota (Fowler, 1982a; Smart, 1984) and spinach (Dalton, 1978). All reports so far have indicated that cell viability is rapidly lost at high impeller speeds. The exception to this, with Catharanthus roseus cells, has now been noted (Fowler, 1982a). However, stirred-tank bioreactors have been used to grow plant cells (Byrne and Koch, 1962; Kato et al.; 1977), and it is clear that the effect of mechanical shearing varies from cell line to cell line and with the age of the culture.

Little is known about the rheology of plant cell suspensions and its effect on mixing. It has, however, been shown that viscosity increases with growth (Kato et al., 1978). Vogelmann et al. (1978) suggest that plant cell suspensions are complex in their rheological behavior and may exhibit non-Newtonian behavior of the Bingham type. Morinda citrifolia suspension cultures showed shear thinning, yield stress, and thixotrophy, at above 20 g/liter (dry weight) (Fig. 1).

In slight contrast, Tanaka (1982) has recently shown that cell lines of Cudraimia tricuspidata, Catharanthus roseus, and Agrostemma githago exhibit pseudoplastic rheological behavior at high concentrations (10–18 g/liter). This non-Newtonian behavior of plant cells could cause problems in culture scale-up (Blanch and Bhavaraju, 1976). Plant cells in suspension are analogous to fungal floc cultures described by Roels et al. (1974) in which measurement of rheology has also caused problems. In the work of Roels et al. (1974), problems in measuring the viscosity of dispersed particles were overcome by using an impeller system rather than the normal cup and bob method. Until more data are available, it is impossible to determine whether plant cell suspensions are in general Bingham or pseudoplastic (Oswald) in nature and, as a consequence, the effect that this may have on viscosity during the progress of a large-scale cell culture.

Although plant cells are large, they have a relatively low metabolic activity, and therefore should have a lower oxygen requirement than microorganisms. There have been comparatively few studies of the effects of aeration, $K_L a$ (volumetric mass transfer coefficient) values, and dissolved oxygen levels on plant cell growth and productivity. Muir et al. (1958) were the first to point out that plant cell suspensions do require oxygen. It has since been shown by many workers that the oxygen demand is, however, low (a maximum of about 1.8 mMO_2 per liter per hour). The most comprehensive study on the effect of $K_L a$ on growth and productivity in plant cell

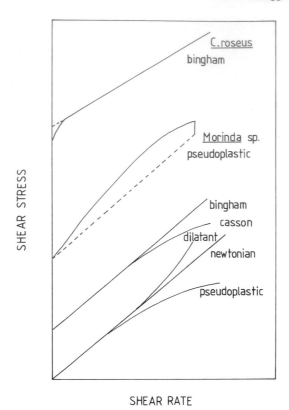

SHEAR RATE

Fig. 1. Plant suspension culture rheology.

suspensions has been that of Kato *et al.* (1975) working with tobacco cells. While growth was found to be limited with $K_L a$ values below 5/hr, a linear relationship was observed between $K_L a$ and final biomass yield over the range of 5–10/hr. Above $K_L a$ values of 10/hr. no increase in biomass yield was obtained. In a similar study with *Catharanthus roseus* cells, growth was observed to be severely limited below a $K_L a$ of about 5/hr, followed by a linear relationship between 5–/10hr, above which a decline in final biomass yield occurred (Smart and Fowler, 1981; Fig. 2). Tanaka (1981) has also shown an inhibitory effect of high $K_L a$ values on 15-day-old cell cultures of *Cudramia tricuspidata*. The lowest value of $K_L a$ used was 25/hr, so that no information relating to the possibility of a linear response between $K_L a$ and cell mass at lower $K_L a$ values to confirm or disprove the previous work was presented. Pareilleux and Vinas (1983) have recently shown an effect of aeration on the growth of *Catharanthus roseus* and in particular on product yield. A marked effect of the rate of aeration on initial growth rate has also been observed (Fowler, 1982a). Increasing the aera-

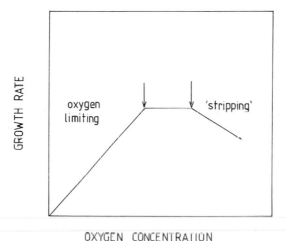

Fig. 2. The effect of $K_L a$ on the growth rate of plant cell cultures.

tion rate from 1.0 to 2.01 per minute led to an extended lag phase as well as suppression of the initial growth rate for *C. roseus*.

Another consequence of high aeration rates is the possible stripping of volatiles or reduction in other gases such as carbon dioxide. Reduced levels of carbon dioxide have previously been shown to reduce the growth of rose cells (Nesius and Fletcher, 1973), and in a detailed study Hegarty, N. J. Smart, and M. W. Fowler (unpublished results) have shown that carbon dioxide levels can influence the lag phase and growth rate of *C. roseus* cell suspensions in 7-liter airlift bioreactors.

V. FOAMING, SURFACE ADHESION, AND WALL GROWTH

Foaming is a common feature of all fermentations, and in the case of plant cells, foaming often causes the sticking of biomass to the surfaces of gas–liquid interfaces. The problem has been mentioned often, but there are a few studies of the factors involved. A number of workers have used antifoams (Wang and Staba, 1963; Wilson *et al.*, 1971; Kato *et al.*, 1976a; Tulecke and Nickell, 1960) that have proved to be nontoxic for a range of cell suspensions. However, Martin (1980) has observed a reduction in growth rate in *Ipomoea* cells when using polypropylene glycol as an antifoam (see also Fowler, 1982a). Yasuda *et al.* (1972) report the reduction of foaming and sticking by reduction in the bubble size and gas flow rate, and similarly Kato et al. (1977) reduced foaming and bulking by reduction in stirrer speed.

A particular problem of plant cell cultures occurs as the culture approaches the end of log phase, when cell surfaces can become sticky and surface adhesion increases. This causes a cell mass to build up around probes and at the top of the culture, forming a "dry" crust known as a meringue (Fowler, 1982a). Cell stickiness appears to be caused by extracellular polysaccharides and its production is encouraged by high carbon levels. This is perhaps analogous to the production of extracellular polysaccharides by *Pullaria*, in which nitrogen limits growth while considerable amounts of carbon remain. This problem can be reduced by batch feeding of the plant cell suspensions with a carbon source (Smart, 1984). Takayama et al. (1977) were also able to reduce cell adhesion to the bioreactor by reduction in the Ca^{2+} levels in the medium.

VI. EFFECT OF pH AND TEMPERATURE

While microbial cells usually prefer a constant pH during their growth and production phases, plant cells do not grow well at fixed pH values. In general the optimal pH range for the growth of plant cell suspensions is in the pH range 5–7. Inoculum pH is usually in the region of 5.8, which then falls rapidly to about 5.0, rising gradually over the next few days to around 6–7. Undefined media containing relatively large amounts of ingredients, such as coconut milk, may be fairly well buffered, but defined media can change rapidly in pH as various ions are released or taken up by the cells. Plant cells exhibit considerable ability to alter their ionic environment, and major ion fluxes have been demonstrated (Raven and Smith, 1974). In accord with this, Matsumoto et al. (1972), working with suspension cultures of *Populus, Nicotiana glutinosa*, and *N. tabacum*, and Takayama et al. (1977), investigating *Agrosemma githago* cultures, both showed that in spite of a wide range of initial pH values, after 3–5 days growth the cultures all had pH values in the range 5–7.1. Results from work in which an attempt was made to control pH are equivocal. Martin and Rose (1975) grew *Ipomoea* suspension cultures under controlled pH regimes, the data from which suggests that the influence of pH on growth was through its effect on the utilization of ammonia and nitrate rather than from any general effect on cell physiology.

Studies on a limited range of plant cell suspensions suggest that the optimum temperature for growth is between 25 and 30°C but that individual species could vary considerably within this range. Kato et al. (1976b) carried out a study of the growth of tobacco cell suspensions over a range of from 17.5 to 37°C. The optimum for both biomass and growth rate was 28°C, and the cells showed morphological damage above 34°C. A similar result was found by Rose and Martin (1975) for *Ipomoea* suspensions. Extreme optima of 20–21°C for cell suspensions of *Solium*, and 31–32°C for *Rosa* have been shown by Tulecke and Nickell (1960). There is no data on the effect of temperature on secondary metabolite synthesis by plant cell suspensions except for changes in the fatty acid composition of plant cell cultures (McCarthy and Stumpf, 1980).

VII. MEDIUM COMPOSITION

The media formulations used for plant cell cultures are in general less complex and expensive than those used for culturing animal cells. Selecting the best medium, however, is done more through "empirical feel" than through science. By the nature of cell culture technology the formulation normally selected first is that which supports the best growth. For secondary product synthesis, a second or "production" medium is usually required (Zenk *et al.*, 1977a; Curtin, 1983). Typically a plant cell medium contains major inorganic salts such as nitrogen, phosphate, magnesium, and calcium; minor trace elements such as manganese and cobalt; vitamins; a carbon source; a nitrogen source; and plant growth promotors (see Chapter 4, this volume). In some cases, in order to initiate growth it is necessary to add undefined components such as coconut milk. In practice five or six standard formulations are used with the key differences being in the type and quantity of growth promotor used. These compounds, sometimes known as plant hormones, are in the main cytokinins and auxins and can have profound effects on plant cells in terms of growth or product formation. Their mode of action at the molecular level is unknown, and therefore finding the best combination is often a sophisticated version of trial and error. Common auxins used are 2,4-dichlorophenoxyacetic acid (2,4-D), naphthaleneacetic acid (NAA), indole-3-acetic acid (IAA), and indolebutyric acid (IBA). The cytokinins used are the naturally occurring zeatin, or synthetics such as kinetin. Of the auxins, 2,4-D is widely used, but it is now recognized (Zenk *et al.*, 1975; Smart, 1984) that although it encourages growth it depresses secondary product formation. Another possible problem with 2,4-D on an industrial scale is that it is used as a herbicide, and suggestions have been made that 2,4-D may have genotoxic potential, although recent work has failed to show any such effects (Linnainmaa, 1983).

Traditionally plant cell suspensions have been grown using sucrose as the principal carbon source (2-10%), but alternative carbon sources have been investigated (Maretzki, et al., 1974; Misawa et al., 1980; Fowler, 1978, 1982b). Carbon sources subjected to investigation include glucose, fructose, maltose, galactose, lactose, and nonrefined sources such as whey, molasses, and starches. With our own cell lines of Catharanthus roseus, Nicotiana tabacum, and other species, glucose appears to be a most effective growth source. Growth on many other carbon sources, in particular the unrefined sources, often requires "conditioning" and even then the response is poor. There are strong indications that not only the type of carbon source but the concentration may affect both growth and product formation. For instance, Zenk et al., (1977b) detected increases in alkaloid yield in C. roseus cultures when the sucrose level was increased above 3%. The nature and level of the nitrogen source has also been shown to have an effect on biomass yield (Noguchi et al., 1977), and plant cells have shown an ability to use a wide range of nitrogen sources including urea, nitrate, glutamine, glutamate, casein hydrolysate, and amino acid mixtures. Nitrate and ammonia are the most usual nitrogen sources and are often used together. Few attempts have been made to optimize media for biomass production except for the work with N. tabacum suspension cultures by Kato et al. (1977).

VIII. BIOREACTOR SIZE AND DESIGN

The early bioreactors used for the mass culture of plant cells were little more than converted carboys or inverted flasks with mixing achieved by sparging air or through a magnetic stirrer. These bioreactors were only for use at laboratory scales of operation. Early on, commercially available stirred-tank bioreactors were used for plant cell growth (Byrne and Koch, 1962; Kaul and Staba, 1967), before the differences between plant cells and microorganisms were fully appreciated. Despite these problems a number of workers, and in particular scientists at the Japanese Salt and Tobacco Monopoly, have successfully used stirred-tank bioreactors for the growth of plant cells in considerable volumes. The progress of plant cell culture in terms of volumes grown and bioreactors used is shown in Table V. Following the early work of Nickell and co-workers (see earlier), Kato et al. (1972) used a 15-liter bioreactor for the growth of tobacco cells, and later for the same species (Kato, et al., 1977) vessels of 15, 20, and 230 liters were used. Yasuda et al. (1972) again with tobacco used 130- and 600-liter stirred-tank bioreactors. In 1974 Hahlbrock et al. reported the growth of Petroselinum,

TABLE V

Development of Large-Scale Plant Cell Culture

Year	Volume (liters)	Vessel type	Plant species	Reference
1959	10	Glass carboy sparged	Ginkgo, Lolium, Rose, Holly	Tulecke and Nickell (1959)
1960	30, 134	Sparged, stainless steel tank	Ginkgo, Lolium, Rose, Holly	Tulecke and Nickell (1960)
1963	7.5, 15	STR (New Brunswick)	Daucus carota	Byrne and Koch (1962)
1971	7.5, 150	Stirred jar or stainless steel	Morning glory	Puhaz and Martin (1971)
1972	130, 600	STR	Nicotiana tabacum	Yasuda et al. (1972)
1974	300	STR	Glycine max, Petroselinum	Hahlbrock et al. (1974)
1975–1976	65–1500	Bubble tanks	Nicotiana tabacum	Kato et al. (1975; 1976a,b)
1977	20,000 (15,500)	STR	Nicotiana tabacum	Noguchi et al. (1977)
1977	10	Airlift	Morinda citrifolia	Wagner and Vogelmann (1977)
1981	100	Airlift	Catharanthus roseus	Smart and Fowler (1984)

and *Glycine max* in volumes of 300 liters, again using a stirred-tank bioreactor. From 1975 to 1977 bioreactor volumes increased rapidly. Kato *et al.* (1975; 1976a,b) reported the growth of tobacco cells in 65- to 1500-liter bubble tanks, and Noguchi *et al.* (1977) described the growth of tobacco cells in a 20,000-liter stirred-tank bioreactor (working volume 15,500 liters). The use of alternative designs for plant cell cultures started in 1977 with Wagner and Vogelmann (1977) using a 10-liter airlift bioreactor for the growth of *Morinda citrifolia*, and Fowler (1981) using a 100-liter airlift bioreactor with *Catharanthus roseus*.

During this scale-up work it quickly became apparent that, although conventional stirred-tank bioreactors could be used to grow some plant cell lines, they were by no means successful with a large number of cell lines (e.g., *Beta vulgaris*). It also became clear that plant cells do not require the same levels of oxygen supply compared with microorganisms, due to their lower metabolic activity (the maximum requirement for plant cells was about 1.8 mM O_2 per liter per hour compared with 10–15 mM O_2 per liter per hour for yeasts, and 5–90 mM O_2 per liter per hour for bacteria). The problem resolves into the need to achieve good mixing of nutrient and biomass against a low shear and aeration regime. The only systematic analysis of growth and product formation in a range of bioreactor designs so far carried out is that of Wagner and Vogelmann (1977).

Using *Morinda citrifolia* suspension cultures they investigated the yields of biomass and anthraquinones in four vessel types: a normal stirred tank with flat blade impeller, a stirred tank with a perforated disk impeller, a draft tube bioreactor with a kaplan turbine, and an airlift bioreactor with a draft tube. Although the final dry weight values achieved were very similar for the different designs, the production of anthraquinones was much enhanced in the airlift bioreactor. More recently Furuya (1982), has investigated a number of different impeller designs, such as disk and anchor. In this case the angled disk proved to be the best design for the growth of *Panax ginseng* suspension cultures. As mixing is proportional to the impeller diameter cubed and shear to the diameter squared, an increase in the size of the impeller will achieve the same mixing with a considerable reduction in impeller speed. It is this general principle that should be applied to the cultivation of plant cell suspensions in stirred-tank bioreactors.

Airlift bioreactors provide a significant alternative approach to conventional stirred reactors. Over the years various types have been designed, constructed, and used for a wide range of experiments and applications (Hatch, 1975; Table VI). The first airlift bioreactor was patented by LeFrancois *et al.* in 1955 and has gained favor because of its savings in power costs, "reasonable" oxygen transfer rates, and lower shear due to hydrostatic agitation. There are two main types of airlift bioreactor: one type with an outer column and an inner cylinder referred to as the draft tube, the other type with an external downflow section. As plant cells become more sensitive to shear toward the end of log phase, and secondary product formation often occurs at or after cessation of growth, the lower shear conditions in the airlift reactor may explain the better production of anthraquinone by *Morinda citrifolia* cells grown in this reactor in the systematic study of Wagner and Vogelmann (1977). Zenk *et al.* (1977b) used a 30-liter draft tube airlift bioreactor containing 22 liters of medium to grow cultures of *Catharanthus roseus*. Both groups reported good growth with the formation of the alkaloids serpentine and ajmalicine, after growth had slowed or stopped. This pattern of alkaloid synthesis is, however, in marked contrast to that observed by Zenk *et al.* (1977a) with shake flask cultures in which alkaloid synthesis was observed to accompany cell growth and division. The reason for this difference has yet to be satisfactorily explained. The use of both internal (7-liter), and external (100-liter) airlift bioreactors for the growth of *C. roseus* suspension cultures has also been described (Fowler, 1981). The detailed design of an external 10-liter airlift bioreactor that gave a growth rate almost double that obtained by growing cells in shake flasks has recently been published (Smart and Fowler, 1984). From the above it is apparent that the airlift configuration of fermenter design shows great

TABLE VI

Types of Bioreactors

Vessel System	Capacity (liters)	Cell line	References
Glass carboy, sparged	10	*Ginkgo*, holly, *Lolium*, Rose	Tulecke and Nickell (1959)
Stainless steel tank, sparged	30, 134	*Ginkgo*, holly, *Lolium*, Rose	Tulecke and Nickell (1960)
STR (New Brunswick)	7.5, 15	*Daucus carota*	Byrne and Koch (1962)
Glass carboy, sparged	3	Spearmint	Wang and Staba (1963)
Roller bottles	10	*Acer pseudoplatanus*	Lamport (1964)
Sparged, carboy	10	*Hyoscyamus niger*	Metz and Lang (1966)
Sparged, carboy	6	*Zea mays*	Graebe and Novelli (1966)
STR (New Brunswick)	10	*Ammi visnaga*	Kaul and Staba (1967)
Roller bottles	4.5	*Acer pseudoplatanus*	Short et al. (1969)
Inverted flask, stirred and aerated	5	*Phaseolus vulgaris*	Veliky and Martin (1970)
Stirred, round-bottomed glass vessel	5	*Acer pseudoplatanus*	Wilson et al. (1971)
Bubble column	1.8	*Glycine max, Triticum monococcum*	Kurz (1971)
Stirred jar, or stainless steel vessel	7.5, 150	Morning glory	Puhaz and Martin (1971)
Jar fermenter	30	*Nicotiana tabacum*	Kato et al. (1972)
STR	130–600	*Nicotiana tabacum*	Yasuda et al. (1972)
STR	10–12, 30, 300	*Glycine max, Petroselinum*	Hahlbrock et al. (1974)
STR	15	*Nicotiana tabacum*	Kato et al. (1975)
Bubble column	65-1500	*Nicotiana tabacum*	Kato et al. (1975, 1976a,b)
STR	7.5	*Ipomoea*	Martin and Rose (1975)
STR	15, 30, 230	*Nicotiana tabacum*	Kato et al. (1977)
STR	20,000 (15,000)	*Nicotiana tabacum*	Noguchi et al. (1977)
Variety including airlift, STR	10	*Morinda citrifolia*	Wagner and Vogelmann (1977)
Airlift	20	*Catharanthus roseus*	Zenk et al. (1977b)
Airlift	30	*Catharanthus roseus*	Dollar (1977)
STR	1.7	*Spinacia oleracea*	Dalton (1978)
STR	2	*Phaseolus vulgaris*	Bertola and Klis (1979)
Mixture, STR, and airlift	1–5	*Cudrania triscuspidata*	Tanaka (1981)

(continued)

TABLE VI *(Continued)*

Vessel System	Capacity (liters)	Cell line	References
Airlift	10, 100	*Catharanthus roseus*	Fowler (1981)
Airlift	10	*Triptergium wildordi*	Townsley *et al.* (1983)
STR	20,000 (6500)	*Nicotiana tabacum*	Hashimoto *et al.* (1982)
STR	4	*Catharanthus roseus*	Pareilleux and Vinas (1983)
Rotating drum	1–4	*Vinca rosea*	Tanaka (1983)

promise in the growth of plant cells, although not without reservation. In a comprehensive study Tanaka (1981) has suggested that above biomass densities of 20 gm (dry weight) per liter the airlift reactor is not a suitable system. He found, as also reported by Wilson (1978), that at high cell–biomass densities air bubbles did not separate from the culture, with the result that normal circulation became impeded and unstirred regions developed in the culture, leading to a multiphasic system. An interesting recent development has been reported by Townsley *et al.* (1983). In this work a 10-liter airlift bioreactor was constructed for the growth of *Triptergium wildordii* cell suspensions in which the major volume of air required to mix the cells is recycled air. Sufficient new air is bled in to meet the respiratory demands of the cells. Such an approach lends itself to close control of the oxygen supply and in this particular case resulted in more effective product formation (tripdiolide).

IX. PROCESS SYSTEMS

Plant cells have now been grown in batch, semicontinuous (fed-batch), and continuous culture. Batch culture has long been the traditional way of growing plant cells in liquid culture and has been extensively reviewed in recent years (see Street 1977, and articles in Thorpe, 1978).

Semicontinuous and continuous culture systems have been much less well investigated in plant cell cultures than batch growth. Almost all the semicontinuous culture work has been essentially batch culture, in which large volumes of culture are removed from the culture vessel and replaced with the same amount of fresh nutrient (see Kato *et al.*, 1977). In a recent comprehensive study Dougall and co-workers (Dougall *et al.*, 1983a,b) have used semicontinuous culture systems to establish key parameters in balancing growth, biomass productivity, and anthocyanin production in carrot cultures. They found that the effects of limiting nutrients, dilution rate, culture pH, and temperature on the yield constant for biomass production were often markedly different from the effects on anthocyanin production. For example, in one combination of factors, increasing culture pH from 4.5 to 5.5 decreased the biomass yield constant by 20% while that for anthocyanin synthesis decreased by 90%. The data of Dougall *et al.* (1983a,b) illustrate two keypoints that very much affect mass culture: first, the obvious complexity of interacting chemical and physical factors on biomass and product synthesis and, second, the fact that, for anthocyanins at least, there is a distinct lack of positive coupling between biomass yield and product synthesis. The data do, however, indicate the possible rewards of a multifactorial approach using semicontinuous culture.

Attempts have been made to grow plant cells in continuous (turbidostat and chemostat) culture for many years, and a reasonable body of literature has begun to accumulate. The area has been reviewed in recent years by Fowler (1977) and Wilson (1980). Much of the early work in this area was taken up with considerations of whether or not plant cells in chemostats conform to classical Michaelis–Menten kinetics derived from microbial systems (Wilson *et al.*, 1971; Fowler and Clifton, 1974; Wilson, 1980). While for certain substrates plant cell cultures may conform to Michaelis–Menten kinetics, Dougall and Weyrauch (1980) pointed out that the position may not be so straightforward in all cases and that for some substrates kinetic models developed by Nyholm (1976) may be more appropriate. Further indications of the nonideality of continuous culture systems with plant cells come from studies by Sahai and Shuler (1982). These workers have shown that deviation from classic kinetic models may also be caused by such features as wall growth, bad mixing, and consequent development of heterogeneous zones in the bioreactor. Quite obviously, these are all factors that in turn could affect scale-up predictions for mass cell culture systems.

An important facet of continuous culture operation is whether or not it can be used as a process system for natural product synthesis. Continuous culture operation has a number of obvious advantages in this mode, including the possibility of continuous and controlled product synthesis together with high vessel usage and consequent saving in labor and process costs. Unfortunately in plant cells, as with microbial systems, secondary product synthesis appears to be uncoupled from cell growth and division. In consequence, the continuous culture is probably not going to be a major primary instrument of natural product synthesis. This is well illustrated by recent work of Tal and Goldberg (1982), who grew cells of *Dioscorea deltoidea* in a 2-liter fermenter under chemostat conditions with the carbon source as limiting nutrient. In these conditions, while the cells synthesized small amounts of the steroid diosgenin, the levels were low compared with those achieved in batch cultures. While this is no different a situation to that observed with many microbial systems, it should be said that the range of conditions explored with the chemostat was restricted and involved only manipulation of the carbon source. It is possible from the work of Dougall *et al.* (1983a,b) on semicontinuous culture that modification of other related factors may have changed this situation. That this discussion is so nebulous emphasizes the need for detailed investigation in this area. Although at this point in time it would appear that continuous culture is not suitable for product synthesis, it does have immediate application in the continuous and rapid production of biomass. In this mode it would form the first part of a two-stage process, the second stage being geared to product synthesis, and away from biomass production. Such an approach has already been used in the Mitsui shikonin process (Curtin, 1983) as well as being proposed as the first stage of a process for the biotransformation of digitoxin to digoxin (Reinhard and Alfermann, 1980).

X. ECONOMICS

The advantages of mass plant cell culture over plantations for the production of secondary products have been outlined in the introduction. The opportunity for the industrial exploitation of plant cells will depend on the ability to produce the plant product at a price equal to or preferably lower than the field-produced product. Plant cells will never be able to compete with microorganisms for the production of primary metabolites such as amino acids and citric acid or for biomass (SCP), although SCP from plant cell culture has been suggested (Ciferri *et al.*, 1980). Plant cell culture can only compete in the area of specific products of high value made only by higher plants, with the possible exception of tobacco biomass.

To be able to determine whether a particular plant cell process is commercially viable is difficult at present, because we lack sufficient data in the financial and technical areas. Japanese workers (see Section VIII) have over the last 10 years shown that large volumes (up to 15,500 liters) of plant cells can be cultivated. Hashimoto *et al.* (1982) have shown that large-scale continuous culture is also possible. A 20,000-liter bioreactor was used, containing 6500 liters of culture; the cell mass was 16.5 g/liter, the dilution rate 0.35/day, and productivity 5.82 g/liter/day. The culture was run asceptically for 66 days. These results show that industrial scale production is possible. At the present state of the art two independent analyses (Goldstein *et al.* 1980; Zenk *et al.*, 1977b) have estimated the running costs of plant cell culture products at about $500 per kilogram of product. Goldstein *et al.* (1980) have produced the most comprehensive cost analysis to date,

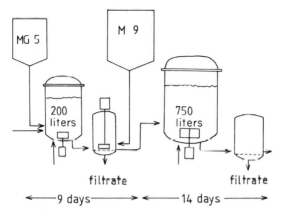

Fig. 3. The Mitsui process for the production of shikonin from plant cell cultures. (From Curtin, 1983).

where the cost of a secondary product was $551/kg under the following conditions: the yield of product was 0.05% of the fresh weight, the biomass concentration was 200 g/liter fresh weight, the doubling time was 1 day (μ = 0.693), and the market was 10^4 kg per annum. The capital cost of such a plant was $1625/kg, with a 5-year payback time. Although the yields of product and biomass are feasible, and the doubling time of 1 day may be possible, current ideas of non-growth-associated product formation complicate this calculation and markets of 10^4 kg per annum seem unlikely. The high-value, low-volume products considered for plant cell culture require production at levels of from 2 to 100 kg per annum. An example is the recently announced production of shikonin, a dye and pharmaceutical, by Mitsui Petrochemical Industries Limited. The requirement for shikonin is about 150 kg per annum, and Mitsui hopes to be able to produce 65 kg per annum, using the system described in Fig. 3. Smart (1984) quotes a production figure of $750/kg for the alkaloid serpentine based on a 30,000-liter bioreactor, 30-g/liter biomass levels attainable in 21 days, and a 1% product yield (dry weight). This value seems more reasonable, especially when compared with the wholesale price of ajmalicine, although the size of the bioreactors is very large when compared with the Mitsui shikonin process (Curtin, 1983).

XI. CONCLUSION

There are now sufficient examples to leave no doubt that it will be possible to process plant cells on a large scale commercially. The precise form that the process system will take is, however, still open to question. Much work is required in the fundamental aspects of plant cell physiology, reactor design and performance, and scale-up parameters. Plant cells from different species behave sufficiently differently in culture to suggest that no one process system will be applicable to all cell lines. Simplicity and versatility may therefore have to be the watchwords of those involved in the development of mass culture systems for plant cells.

REFERENCES

Blanch, H. W., and Bhavaraju, S. M. (1976). Non-Newtonian fermentation broths: Rheology and mass transfer. *Biotechnol. Bioeng.* **18,** 745–770.

Byrne, A. F. and Koch, M. B. (1962). Food production by submerged culture of plant tissue cells. *Science* **135,** 215–216.

Ciferri, O., Sala, F., Spadoni, M. A., Carnovale, E., and DiCorato, A. (1980). Plant cell culture as a source of single cell protein. *In* "Plant Cell Cultures: Results and Perspectives" (F. Sala, B. Parisi, R. Cella, and O. Ciferri, (eds.), pp. 411–417. Elsevier/North-Holland, Amsterdam.

Curtin, M. E. (1983). Harvesting profitable products from plant tissue culture. *Biotechnology* 649–657.

Dalton, C. C. (1978). The culture of plant cells in fermenters. Helioisynthase et Aquaculture seminar de Martiques, *C.N.R.S.* pp. 1–11.

Dollar, G. (1977). Influence of the medium on the production of serpentine by suspension cultures of *Catharanthus roseus* (L) G. Don. *In* "Plant Tissue Culture and Its Biotechnological Application" (W. Barz, E. Reinhard, and M. H. Zenk, eds.), pp. 107–117. Springer-Verlag, Berlin and New York.

Dougall, D. K., and Weyrauch, K. W. (1980). Growth and anthocyanin production by carrot suspension cultures grown under chemostat conditions with phosphate as the limiting nutrient. *Biotechnol. Bioeng.* **22,** 337–352.

Dougall, D. K., LaBrake, S., and Whitten, G. H. (1983a). The effects of limiting nutrients, dilution rate, culture pH, and temperature on the yield constant and anthocyanin accumulation of carrot cells grown in semicontinuous chemostat cultures. *Biotechnol. Bioeng.* **25,** 569–579.

Dougall, D. K., LaBrake, S., and Whitten, G. H. (1983b). Growth and anthocyanin accumulation rates of carrot suspension cultures grown with excess nutrients after semicontinuous culture with different limiting nutrients at several dilution rates, pH, and temperatures. *Biotechnol. Bioeng.* **25,** 581–594.

Fowler, M. W. (1977). Growth of cell cultures under chemostat conditions. *In* "Plant Tissue Culture and Its Biotechnological Application" (W. Barz, E. Reinhard, and M. H. Zenk, eds.), pp. 253–265. Springer-Verlag, Berlin and New York.

Fowler, M. W. (1978). Regulation of carbohydrate metabolism in cell suspension cultures. *In* "Frontiers of Plant Tissue Culture, 1978" (T. A. Thorpe, ed.) pp. 443–452. Univ. of Calgary Press, Calgary, Alberta, Canada.

Fowler, M. W. (1981). Plant cell biotechnology to produce desirable substances. *Chem. Ind. (London)* **7,** 229–233.

Fowler, M. W. (1982a). The large-scale cultivation of plant cells. *Prog. Ind. Microbiol.* **17,** 209–229.

Fowler, M. W. (1982b). Substrate utilisation by plant cell cultures. *J. Chem. Technol. Biotechnol.* **32,** 338–346.

Fowler, M. W. (1983). Commercial applications and economic aspects of mass plant cell culture. In "Plant Biotechnology, (S. H. Mantell and H. Smith, eds.), pp. 3–37. Cambridge Univ. Press, London and New York.

Fowler, M. W., and Clifton, A. (1974). Activities of enzymes of carbohydrate metabolism in cells of *Acer pseudoplatanus* L. maintained in continuous (chemostat) culture. *Eur. J. Biochem.* **45**, 445–450.

Frischinecht, P. M. (1977). Tissue culture of *Coffea arabica:* Growth and Caffeine formation. *Planta Med.* **31**, 344–350.

Furuya, T., and Iishii, T. (1972). The manufacture of *Panax:* Plant tissue culture containing crude saponins and crude sapogenins which are identical with those of natural *Panax* roots. Japanese Patent No. 48-31917.

Goldstein, W. E., Ingle, M. B., and Lasure, L. (1980). Product cost analysis. In "Plant Tissue Culture as a Source of Biochemicals" (E. J. Staba, ed.), pp. 191–234. CRC Press, Boca Raton, Florida.

Graebe, J. F. and Novelli, G. D. (1966). A practical method for large-scale plant tissue culture. *Exp. Cell. Res.* **41**, 509–520.

Hahlbrock, K., Ebel, J., and Oaks, A. (1974). Determination of specific growth stages of plant cell cultures by monitoring conductivity changes in the medium. *Planta* **118**, 75–84.

Hashimoto, T., Azechi, S., Sugita, S., and Suzuki, K. (1982). Large scale production of tobacco cells by continuous cultivation. In "Plant Tissue Culture 1982" (A. Fujiwara, ed.), pp. 403–404. Maruzen, Tokyo.

Hatch, R. T. (1975). Fermenter design. In "Single Cell Protein 11," pp. 46–68. MIT Press, Cambridge, Massachusetts.

Henshaw, G. G., Jha, K. K., Mehta, A. R., Shakeshaft, D. J., and Street, H. E. (1966). Studies on the growth in culture of plant cells. *J. Exp. Bot.* **17**, 362–377.

Hinz, H., and Zenk, M. H. (1981). Production of protoberberine alkaloids by cell suspensions of cultures of *Berbein* species. *Naturwissenschaften* **68**, 620–621.

Kato, K., Shiozawa, Y., Yamada, A., Nishida, K., and Noguchi, M. (1972). A jar fermenter culture of *Nicotiana tabacum* L. cell suspensions. *Agric. Biol. Chem.* **36**, 899–902.

Kato, A., Shimizu, Y., and Nagai, S. (1975). Effect of initial K_La on the growth of tobacco cells in batch culture. *J. Ferment. Technol.* **53**, 744–751.

Kato, A., Kawazoe, S., Iizima, M. and Shimizu, Y. (1976a). Continuous culture of tobacco cells. *J. Ferment. Technol.* **54**, 82–87.

Kato, A., Hashimoto, Y. and Soh, Y. (1976b). Effect of temperature on the growth of tobacco cells. *J. Ferment. Technol.* **54**, 754–757.

Kato, A., Fukasawa, A., Shimizu, Y., Soh, Y. and Nagai, S. (1977). Requirements of PO_4^{3-}, NO_3^-, SO_4^{2-}, K^+, and Ca^{2+} for the growth of tobacco cells in suspension culture. *J. Ferment. Technol.* **55**, 207–212.

Kato, K., Kawazoe, S., and Soh, Y. (1978). Viscosity of the broth of tobacco cells in suspension culture. *J. Ferment. Technol.* **56**, 224–228.

Kaul, B., and Staba, E. J. (1967). *Ammi visnaga* L. tissue cultures, multi-litre suspension growth and examination for furanochromes. *Planta Med.* **15**, 145–156.

Kaul, B., Stohs, S. J., and Staba, E. J. (1969). *Dioscorea* tissue culture. III. Influence of various factors on diosgenin production by *Dioscorea deltoidea* callus and suspension culture. *Lloydia* **32**, 347.

Kurz, W. G. W. (1971). A chemostat for growing higher plant cells in single cell suspension culture. *Exp. Cell Res.* **64**, 476–479.

Lamport, D. T. A. (1964). Cell suspensions of higher plants: isolation and growth energetics. *Exp. Cell Res.* **33**, 195–206.

LeFrancois, L., Mariller, C. G., and Mejane, J. V. (1955). Effectionements aux procedes de Cultures Pongiques et de Fermentations Industrielles, Brrevet d'Invention, France. Patent No. 1, 102-200.

Liau, D. F., and Boll, W. G. (1971). Growth and patterns of growth and division in cell suspension cultures of bush bean (*Phaseolus vulgaris* cv. Contender) *Can. J. Bot.* **49,** 1131–1139.

Linnainmaa, K. (1983). Sister chromatid exchanges among workers occupationally exposed to phenoxyacid herbicides 2,4-D and MCPA. *Teratog. Carcinog. Mutagen.* **3,** 269–279.

Maretzki, A., Thom, M., and Nickell, L. G. (1974) Utilization and metabolism of carbohydrates in cell and callus cultures. *In* "Tissue Culture and Plant Science" (H. E. Street, ed.), pp. 239–261. Academic Press, New York and London.

Martin, S. M. (1980). Mass culture systems for plant cell suspensions. *In* "Plant Tissue Culture as a Source of Biochemicals" (E. J. Staba, ed.), pp. 149–166. CRC Press, Boca Raton, Florida.

Martin, S. M., and Rose, D. (1975). Growth of plant cell (*Ipomoea*) suspension cultures at controlled pH levels. *Can. J. Bot.* **54,** 1264–1270.

McCarthy, J. J., and Stumpf, P. K. (1980). The effect of different temperatures on fatty acid synthesis and polyunsaturation in cell suspension cultures. *Planta* **147,** 389–395.

Matsumoto, T., Kanno, N., Ikeda, T., Obi, Y., Kisaki, T., and Noguchi, M. (1981). Selection of cultured tobacco cell strains producing high levels of Ubiquinone 10 by a cell cloning technique. *Agric. Biol. Chem.* **45,** 1627–1633.

Metz, H. and Lang, H. (1966). Verfahen zur Zuechtung von differenezierteum Wurzelgewebe. Ausleigeschrift, B.D. Patentant, Nr. 1,216,009.

Misawa, M. (1980). Industrial and government research. *In* "Plant Tissue Culture as a Source of Biochemicals" (E. J. Staba, ed.), pp. 167–190. CRC Press, Boca Raton, Florida.

Morris, P., and Fowler, M. W. (1981). A new method for the production of fine plant cell suspension cultures. *Plant Cell Tissue Organ Cult.* **1,** 15–24.

Morris, P., Smart, N. J., and Fowler, M. W. (1983). A fluidised bed vessel for the culture of immobilised plant cells and its application for the continuous production of fine cell suspensions. *Plant Cell Tissue Organ Cult.* **2,** 207–216.

Muir, W. H., Hildebrandt, A. C., and Riker, A. J. (1958). The preparation, isolation and growth in culture of single cells from higher plants. *Am. J. Bot.* **45,** 517–588.

Nesius, K. K., and Fletcher, J. S. (1973). Carbon dioxide and pH requirements of nonphotosynthetic tissue culture cells. *Physiol. Plant.* **28,** 259–263.

Neumann, D., Krauss, G., Hieke, M. and Groger, D. (1983). Indole alkaloid formation and storage in cell suspension cultures of *Catharanthus roseus*. *Planta Med.* **48,** 20–23.

Nickell, L. G. (1980). Products. *In* "Plant Tissue Culture as a Source of Biochemicals" (E. J. Staba, ed.), pp. 256–269. CRC Press, Boca Raton, Florida.

Noguchi, M., Matsumoto, T., Hirata, Y., Yamamoto, K., Katsuyama, A., Kato, A., Azechi, S., and Katoh, K. (1977). Improvement of growth rates of plant cell cultures. *In* "Plant Tissue Culture and Its Biotechnological Applications" (W. Barz, E. Reinhard, and M. H. Zenk, eds.), pp. 85–94, Springer-Verlag, Berlin and New York.

Nyholm, N. (1976). A mathematical model for microbial growth under limitation by conservative substrates. *Biotechnol. Bioeng.* **18,** 1043–1056.

Pareilleux, A., and Vinas, R. (1983). Influence of the aeration rate on the growth yield in suspension cultures of *Catharanthus roseus* (L) G. Don. *J. Ferment. Technol.* **61,** 429–322.

Puhanz, Z., and Martin, S. M. (1971). The industrial potential of plant cell culture. *In* "Progress in Industrial Microbiology" (D. J. Hockenhull, ed.), pp. 14–39. Gordon & Breach, New York.

Raven, J. A., and Smith, F. A. (1974). Significance of hydrogen ion transport in plant cells. *Can. J. Bot.* **52,** 1035–1047.

Razzaque, A., and Ellis, B. E. (1977). Rosmarinic acid production in *Coleus* cell cultures. *Planta* **137,** 287.

Reinhard, E., and Alfermann, A. W. (1980). Biotransformation by plant cell cultures. *In*

"Advances in Biochemical Engineering" (A. Fiechter, ed.) Vol. 16, Plant Cell Culture, I, , pp. 49–84. Springer-Verlag, Berlin and New York.

Roels, J. A., de Berg, J., and Voncken, R. M. (1974). The rheology of mycelial broths. *Biotechnol. Bioeng.* **16**, 181–208.

Routier, J. B., and Nickell, L. G. (1956). Cultivation of Plant Tissue. U.S. Patent 2,747,334.

Sahai, O. P., and Shuler, M. L. (1982). On the nonideality of chemostat operation using plant cell suspension cultures. *Can. J. Bot.* **60**, 692–700.

Short, K. C., Brown, E. G., and Street, H. E. (1969). Studies on the growth in culture of plant cells. V. Large scale culture of *Acer pseudoplatanus* L. cell suspensions. *J. Exp. Bot.* **20**, 572–590.

Smart, N. J. (1984). Plant cell technology as a route to natural products. *Lab. Prac.* **2**, 36.

Smart, N. J., and Fowler, M. W. (1981). Effect of aeration on large-scale cultures of plant cells. *Biotechnol. Lett.* **3**, 171–176.

Smart, N. J., and Fowler, M. W. (1984). An airlift column bioreactor suitable for large-scale cultivation of plant cell suspensions. *J. Exp. Bot.* **35** (in press).

Street, H. E. (ed.) (1977). "Plant Tissue and Cell Culture." Blackwood, Oxford.

Street, H. E., King, P. J., and Mansfield, K. J. (1971). Growth control in plant cell suspensions. *In* "Les Cultures de Tissues de Plantes," Colloques Intern. Strasbourg, pp. 17–40. CNRS, Paris.

Tabata, M., and Hiraoka, N. (1976). Variation of alkaloid production in *Nicotiana rustica* callus cultures. *Physiol. Plant.* **38**, 19–23.

Takayama, S., Misawa, M., and Misato, T. (1977). Effect of cultural conditions on the growth of *Agrostemma githago* cells in suspension culture and the concomitant production of an antiplant virus substance. *Physiol. Plant.* **41**, 313–320.

Tal, B., and Goldberg, I. (1982). Growth and diosgenin production by *Dioscorea deltoidea* cells in batch and continuous cultures. *Planta Med.* **44**, 107–110.

Tanaka, H. (1981). Technological problems in cultivation of plant cells at high density. *Biotechnol. Bioeng.* **23**, 1203–1218.

Tanaka, H. (1982). Oxygen transfer in broths of plant cells at high density. *Biotechnol. Bioeng.* **24**, 425–442.

Tanaka, H. (1983). Rotating drum fermenter for plant cell suspension cultures. *Biotechnol. Bioeng.* **25**, 2359–2370.

Thorpe, T. A. (ed.) (1978). "Frontiers of Plant Tissue Culture 1978." Univ. of Calgary Press, Calgary, Alberta, Canada.

Townsley, P. M., Webster, F., Kutney, J. P. Salisbury, P., Hewitt, G., Kawawura, N., Choi, L., and Kurihara, T. (1983). The recycling air-lift transfer fermenter for plant cells. *Biotechnol. Lett.* **5**, 13–18.

Tulecke, W., and Nickell, L. G. (1959). Production of large amounts of plant tissue by submerged culture. *Science* **130**, 863–864.

Tulecke, W., and Nickell, L. G. (1960). Methods, problems, and results of growing plant cells under submerged conditions. *Trans. N.Y. Acad. Sci.* **22**, 196–204.

Velicky, I., and Martin, S. M. (1970). A fermenter for plant cell suspension cultures. *Can. J. Microbiol.* **16**, 223–226.

Vogelmann, H., Bischof, A., Pape, D., and Wagner, F. (1978). Some aspects on mass cultivation. *In* "Production of Natural Compounds by Cell Culture Methods," (A. W. Alfermann and E. Reinhard, eds.), pp. 130–146. Gesellschaft für Strahlen-und Umweltforschung mbH, Bereich ProjekHragerschaflen, Munich.

Wagner, F., and Bogelmann, H. (1977). Cultivation of plant tissue culture in bioreactors and formation of secondary metabolites. *In* "Plant Tissue Culture and Its Biotechnological Applications" (W. Borg, E. Reinhard, and M. H. Zenk, eds.), pp. 245–252, Springer-Verlag, Berlin and New York.

Wang, C. J. and Staba, E. (1963). Peppermint and spearmint tissue culture. II. Dual-carboy culture of spearmint tissues. *J. Pharm. Sci.* **52**, 1058–1062.

Wilson, G. (1978). Growth and product formation in large-scale and continuous culture systems. *In* "Frontiers of Plant Tissue Culture 1978" (T. A. Thorpe, ed.), pp. 169–177. Univ. of Calgary Press, Calgary, Alberta, Canada.

Wilson, G. (1980). Continuous culture of plant cells using the chemostat principle. *In* "Advances in Biochemical Engineering" (A. Fiechter, ed.), Vol. 16, Plant Cultures, I, pp. 1–25. Springer-Verlag, Berlin and New York.

Wilson, S. B., King, P. J., and Street, H. E. (1971). Studies on the growth of plant cells. XII. A versatile system for the large-scale batch or continuous culture of plant suspensions. *J. Exp. Bot.* **21**, 177–207.

Yasuda, S., Satoh, K., Ishii, T., and Furuya, T. (1972). Studies on the cultural conditions of plant cell suspension culture. *Ferment. Technol. Today Proc. Int. Ferment. Symp., 4th, 1972,* p. 697.

Zenk, M. H., El-Shagi, H., and Schulte, U. (1975). Anthraquinone production by cell suspension cultures of *Morinda citrifolia. Plant Med. Suppl.* **79**, 101.

Zenk, M. H., El-Shagi, H., and Ulbrich, B. (1977a). Production of rosmarinic acid by cell-suspension cultures of *Coleus blumei. Naturwissenschaften* **64**, 585.

Zenk, M. H., El-Shagi, H., Arens, H., Stockigt, J., Weiler, E. W., and Deus, B. (1977b). Formation of the indole alkaloids serpentine and ajmalicine in cell suspension cultures of *Catharanthus roseus. In* "Plant Tissue Culture and Its Biotechnological Application" (W. Barz, E. Reinhard, and M. H. Zenk, eds.), pp. 27–43. Springer-Verlag, Berlin and New York.

CHAPTER 4

Nutrition of Plant Tissue Cultures

Peggy Ozias-Akins
Indra K. Vasil

Department of Botany
University of Florida
Gainesville, Florida

I. INTRODUCTION

There are three essential sources of nutrition for plants growing in nature. The mineral nutrients are obtained, along with water, from the soil through the root system. Atmospheric carbon dioxide is used in the process of photosynthesis to provide carbon as a source of basic energy. Lastly, the plant body, particularly its meristematic regions and young organs such as leaves, using fixed carbon and minerals, synthesizes all of the vitamins and various plant growth substances that are critical and essential for normal growth and development of the plant.

The requirements of plant tissues grown *in vitro* are similar in general to

129

those of intact plants growing in nature. However, in a vast majority of cases only isolated plant tissues or at best small plant organs are cultured instead of whole plants. These isolated tissues and organs lack the capacity to synthesize their own supply of carbohydrates (for growth of photosynthetic green cell and tissue cultures, see Chapter 6, this volume), most vitamins, and plant growth substances. Accordingly, all the substances needed by intact plants in nature must be provided artificially to cultured tissues.

The successful establishment and growth of plant cells *in vitro* generally is determined by the nature of the explant and the composition of the nutrient medium (White, 1951).[1] The earliest attempts to culture plant tissues and organs *in vitro* utilized the simple inorganic solutions of Knop (1865, 1884) and Hoagland and Snyder (1933), which were used widely for the growth of whole plants. The growth obtained in these early experiments generally was poor. The difficulties encountered in obtaining sustained growth of plant tissues *in vitro* can be attributed to inadequate nutrition and poor choice of plant material. With the successful growth of excised root tips and undifferentiated tissues of cambial origin (Kotte, 1922a,b; Robbins, 1922a,b; White, 1932, 1933a,b, 1934; Gautheret, 1937, 1938, 1939; Nobécourt, 1937, 1938, 1939, 1940), the remaining unresolved problem was the development of a nutrient medium that would allow optimal growth *in vitro*.

II. HISTORICAL PERSPECTIVE

Many of the early investigators used Knop's (1865) mineral solution, generally at half its normal concentration (Haberlandt, 1902; Gautheret, 1939; Nobécourt, 1939), and often supplemented it with Berthelot's (1934) microelements. Robbins (1922a) used the mineral solution recommended by Pfeffer. White (1932) found both of these solutions to be unsatisfactory for use over a wide range of pH values. He thus developed a medium based on the nutrient solution used by Uspenski and Uspenskaia (1925) for the growth of algae, and the microelements of Trelease and Trelease (1933). White's medium (White, 1943, 1963), in its various modified forms, was used extensively until about 1965 for the growth of excised roots, embryos, callus tissues, floral organs, etc. A systematic study of mineral requirements of plant tissues and organs in culture resulted in media formulations

[1]Several chapters in Volume 1 of this treatise have discussed at length the response in culture of a wide variety of explants.

that improved growth over that of Gautheret's or White's nutrient solutions (Hildebrandt et al., 1946; Burkholder and Nickell, 1949; Heller, 1953; Hildebrandt, 1962; Vasil and Hildebrandt, 1966). A common feature of all these media was the greatly elevated levels of mineral salts.

A major effort was made by Murashige and Skoog (1962) to optimize the growth of tobacco pith callus in vitro. They found that addition of an aqueous extract of tobacco leaves to White's medium resulted in a four- to fivefold increase in growth rate. It was shown that this improvement in growth was caused largely by the inorganic constituents of the leaf extract. Very similar results were obtained by adding to White's medium either the ash of tobacco leaf extracts or substantially higher levels of ammonium, nitrate, phosphate, and potassium salts. On the basis of these results a chemically defined medium was developed (MS), which also included chelated iron, myo-inositol, and a mixture of four vitamins.

While Murashige and Skoog (1962) determined the mineral nutrient re quirements of tobacco pith tissue in culture, a later study in Skoog's laboratory (Linsmaier and Skoog, 1965) examined the organic growth factor requirements and found that only one of the four vitamins, thiamine, was necessary, although at a slightly higher level than provided in the MS medium.

Vasil and Hildebrandt (1966) compared the growth of callus tissues of a wide variety of plant species grown in MS medium and other high-salt formulations and showed that optimal increases in fresh weight, dry weight, and chlorophyll development occurred in the MS medium. Several other media were later developed by substantially increasing the level of inorganic salts. Some of these are in common use today (Gamborg et al., 1968; Schenck and Hildebrandt, 1972), and in specific instances may be superior to the MS medium.

The high-salt media, which are excellent for supporting callus growth and morphogenesis, have not proved very suitable for the growth of excised roots, anthers, and other floral organs. Thus, White's medium is still considered to be one of the best for the culture of excised roots (Street, 1967) and early meiotic anthers (Vasil, 1973, 1977). The Nitsch (1951) formulation is best for the culture of excised floral organs, and the Nitsch and Nitsch (1969) medium is widely used for obtaining haploid tissues or embryoids from cultured anthers.

Most other media described in the literature are modifications of the formulations developed by White, Gautheret, Heller, and Murashige and Skoog. There is, however, an unfortunate and common tendency to give new names to media even with minor modifications. This has caused considerable confusion in the literature and should be avoided. The compositions of seven commonly employed tissue culture media are given in Table I.

TABLE I

Inorganic and Organic Components of Major Plant Tissue Culture Media (mg/liter)[a]

	White[1]	Heller[2]	Nitsch[3]	N_6[4]	MS[5]	B_5[6]	KM[7,b]
Inorganic							
KNO_3	80	—	950	2830	1900	2500	1900
$Ca(NO_3)_2 \cdot 4H_2O$	300	—	—	—	—	—	—
$NaNO_3$	—	600	—	—	—	—	—
NH_4NO_3	—	—	720	—	1650	—	600
$(NH_4)_2SO_4$	—	—	—	463	—	134	—
$MgSO_4 \cdot 7H_2O$	720	250	185	185	370	250	300
Na_2SO_4	200	—	—	—	—	—	—
$CaCl_2$	—	—	166	—	—	—	—
$CaCl_2 \cdot 2H_2O$	—	75	—	166	440	150	600
KCl	65	750	—	—	—	—	300
KH_2PO_4	—	—	68	400	170	—	170
$NaH_2PO_4 \cdot H_2O$	19[c]	125	—	—	—	150	—
$FeSO_4 \cdot 7H_2O$	—	—	27.8	27.8	27.8	—	—
$Na_2\text{-}EDTA \cdot 2H_2O$[d]	—	—	37.3	37.3	37.3	—	—
Sequestrene 330Fe	—	—	—	—	—	28	28
$Fe_2(SO_4)_3$	2.5	—	—	—	—	—	—
$FeCl_3 \cdot 6H_2O$	—	1	—	—	—	—	—
$MnSO_4 \cdot H_2O$	—	—	—	—	—	10	10
$MnSO_4 \cdot 4H_2O$	7	0.1	25	4.4	22.3	—	—
H_3BO_3	1.5	1	10	1.6	6.2	3	3
$ZnSO_4 \cdot 7H_2O$	3	1	10	1.5	8.6[e]	2	2
$NaMoO_4 \cdot 2H_2O$	—	—	0.25	—	0.25	0.25	0.25
$CuSO_4$	—	—	—	—	—	0.025	—
$CuSO_4 \cdot 5H_2O$	—	0.03	0.025	—	0.025	—	0.025
KI	0.75	0.01	—	0.8	0.83	0.75	0.75
$CoCl_2 \cdot 6H_2O$	—	—	—	—	0.025	0.025	0.025
$AlCl_3$	—	0.03	—	—	—	—	—
$NiCl_2 \cdot 6H_2O$	—	0.03	—	—	—	—	—
Organic							
Myo-inositol	—	—	100	—	100	100	100
Nicotinic acid	0.5	—	5	0.5	0.5	1	1
Glycine	3	—	2	2	2	—	—
Pyridoxine·HCl	0.1	—	0.5	0.5	0.5	1	1
Thiamine·HCl	0.1	—	0.5	1	0.1	10	1
Folic acid	—	—	0.5	—	—	—	0.4
Biotin	—	—	0.05	—	—	—	0.01
pH	5.5	5.5	5.5	5.8	5.7–5.8	5.5	5.6

[a] (1) White (1943, 1963); (2) Heller (1953, 1965); (3) Nitsch and Nitsch (1969); (4) Chu (1978); (5) Murashige and Skoog (1962); (6) Gamborg *et al.* (1968); (7) Kao and Michayluk (1975).

[b] For low-density culture of protoplasts, KM8p medium included 8 additional vitamins, 4 organic acids, 10 sugars and sugar alcohols, casein hydrolysate, and coconut water.

[c] Amount presumed by Singh and Krikorian (1981) to be what White (1943) intended.

[d] $Na_2\text{-}EDTA \cdot 2H_2O$ used to give equimolar amounts of Fe and EDTA (see discussion by Singh and Krikorian, 1980).

[e] Original paper included 8.6 mg/liter $ZnSO_4 \cdot 4H_2O$.

III. INORGANIC NUTRIENTS

The same essential elements that support growth of intact plants are necessary for sustained growth and development of plant tissues *in vitro*. The mineral nutrients required for whole plants have been divided by Clarkson and Hanson (1980) into two major groups: (1) those elements (N, P, S) that are covalently bonded in carbon compounds and are vital constituents of the macromolecules, DNA, RNA, and protein, and (2) all other elements (K, Na, Mg, Ca, Mn, Fe, Cu, Zn, Mo, B, Cl) that participate in a variety of often overlapping functions including regulation of osmotic and electrical gradients, protein conformation, and oxidation–reduction reactions of metalloproteins.

A. Macroelements

Deficiencies of the macroelements N, P, S, K, Mg, and Ca are more easily manifested in tissue culture when cells are cultured in liquid rather than on agar media since impurities present in agar are considerable (Heller, 1953). Sodium and chlorine, although considered nonessential but perhaps stimulatory at certain concentrations (Heller, 1953; Murashige and Skoog, 1962), are arbitrarily included in media as salts of other elements.

Of all the mineral nutrients, the form of nitrogen (oxidized or reduced, organic or inorganic) probably is responsible for the most pronounced effects on growth and differentiation of cultured tissues. Nitrogen generally is supplied in the form of NH_4^+ along with NO_3^-. Ammonium ion as the sole nitrogen source is usually unsuitable, probably because under such situations the pH of the medium has a tendency to fall below 5 during culture. This drop in pH, which may restrict the availability of nitrogen, has been demonstrated adequately with soybean (Gamborg *et al.*, 1968), rice (Ohira *et al.*, 1973), and carrot (Wetherell and Dougall, 1976) suspension cultures. The inclusion of 20–40 mM KNO_3 can prevent the extreme fluctuations of pH. In addition, NH_4^+ can serve as the sole nitrogen source when the medium is provided with an organic acid such as malate, succinate, citrate, or fumarate (Gamborg and Shyluk, 1970). The function of the organic acid in this situation, however, has not been elucidated. Although NH_4^+ is rarely adequate as the sole nitrogen source, a small amount of it may be essential for good growth. Such was the case with soybean suspension cultures in which KNO_3 alone did not support satisfactory growth, and the addition of 2 mM NH_4^+ greatly increased growth (Gamborg *et al.*, 1968; Bayley *et al.*, 1972). Wheat suspensions grew equally well in the presence or absence of NH_4^+ (Bayley *et al.*, 1972). In both soybean and wheat suspensions NH_4^+ was preferentially utilized by the cells and had disappeared from the medium by the third day, at which time NO_3^- utilization increased.

Growth as well as differentiation can be controlled by the various media components. When differentiation is the objective, attention is usually focused on growth regulators; however, mineral nutrition also may play an important role in the process. For example, the form of nitrogen available to cultured tissues has been demonstrated to quantitatively affect somatic embryogenesis in certain systems. Along with NO_3^-, reduced nitrogen supplied as NH_4^+ or certain amino acids, particularly glutamine, stimulates prolific somatic embryo formation in carrot (Halperin and Wetherell, 1965; Wetherell and Dougall, 1976; Kamada and Harada, 1979), alfalfa (Walker and Sato, 1981; Stuart and Strickland, 1984), *Atropa belladonna* (Thomas and Street, 1972), and *Digitalis lanata* (Kuberski *et al.*, 1984). In certain cases NH_4^+ may be detrimental to survival and growth of plant cells. Protoplasts of potato (Shepard and Totten, 1977), *Salpiglossis* (Boyes *et al.*, 1980), and three species of Asteraceae (Okamura *et al.*, 1984) are apparently sensitive to NH_4^+ and show greatly reduced division frequencies in its presence.

Although phosphorus and sulfur are usually supplied as phosphates and sulfates, certain other inorganic and organic forms are usable. Phosphate is the primary buffering constituent in tissue culture media. Heller's medium differs from White's mostly in the increased levels of phosphorus and potassium. Murashige and Skoog (1962) found that phosphorus levels greater than 2 mM were often inhibitory to growth of tobacco pith tissues; therefore, they selected 1.25 mM as the near-optimal level. This concentration was apparently suboptimal for *Haplopappus gracilis* suspension cultures, which showed an increase of 50% in growth rate when the phosphorus concentration was doubled (Eriksson, 1965). Certain organic forms of sulfur may be assimilated by plant tissues. Hart and Filner (1969) found that L-cysteine, L-methionine, and glutathione were satisfactory sources of sulfur for tobacco tissues. It is worthwhile to note that the sucrose had to be purified with ion-exchange resin to remove contaminating sulfur. Growth of *Rumex* virus tumor tissues in culture was supported only by cysteine as a substitute for sulfate (Nickell and Burkholder, 1950).

Potassium at a concentration of 1 mM is sufficient for growth of carrot suspension cultures; however, a much higher level (20 mM) is necessary for full expression of embryogenic potential (Brown *et al.*, 1976). In fact, much of the effect of NH_4^+ on embryogenesis may be dependent on the K^+ concentration (Tazawa and Reinert, 1969).

An antagonism between calcium and magnesium has been demonstrated, and it was found that an increase in the concentration of one element increased the requirement for the other (Heller, 1965; Ohira *et al.*, 1973).

B. Microelements

The microelements usually included in plant tissue culture media are Fe, Mn, B, Zn, Mo, Cu, I, and Co. Of all the microelements, Fe deficiency reduced growth of rice suspensions most dramatically; Zn, Cu, B, Mn, and Mo deficiencies also had an inhibitory effect that was most pronounced with Zn and least detectable with Mo (Ohira et al., 1975). Heller (1953) also observed deficiencies of Fe, Zn, Cu, B, and Mn in cultured carrot tissues. He did not include Mo as a micronutrient, and Murashige and Skoog (1962) did not observe any differences in growth of tobacco tissues with varying Mo concentrations. Molybdenum is a component of nitrate reductase and is essential for whole-plant nutrition; thus it generally is included in plant tissue culture media. Iron, the availability of which is reduced at high pH due to precipitation, is supplied in chelated form either as commercially available Sequestrene 330Fe or as an EDTA complex that can be prepared easily (Murashige and Skoog, 1962; Vasil and Hildebrandt, 1966). Most media contain about 0.1 mM Fe. Murashige and Skoog (1962) included 27.8 mg/liter FeSO$_4$·7H$_2$O (0.1 mM) and 37.3 mg/liter Na$_2$-EDTA (0.11 mM) per liter of medium. As pointed out by Singh and Krikorian (1980), these are not equimolar amounts of Fe and EDTA, and the excess EDTA present can chelate other metals in the complete medium. The error probably arose when the water of hydration of Na$_2$-EDTA·2H$_2$O was not taken into account. The amount of Na$_2$-EDTA should be adjusted to 33.6 mg/liter.

Two additional microelements not proven essential to whole-plant nutrition are nevertheless included in most tissue culture media, namely Co and I. White (1938) and Hildebrandt et al. (1946) showed that I was clearly beneficial for growth of excised tomato roots and tobacco tissues, respectively. Cobalt was included by Murashige and Skoog (1962) because of its demonstrated effects on plant metabolism (Salisbury, 1959).

Although Al and Ni are not essential for growth, they were included in Heller's (1953) micronutrient solution because they slightly stimulated growth of callus tisssues. In contrast, White (1938) determined that Al was inhibitory to growth of excised tomato roots. Recently Ni, a constituent of the metalloenzyme urease, has been demonstrated—after rigorous purification of the nutrient media—to be an essential trace element for soybean plants (Eskew et al., 1983).

Microelement toxicities also have been observed. Suspension cultures of *Haplopappus gracilis* at low inoculum densities failed to grow in media containing the suggested micronutrient concentrations of Murashige and Skoog (1962) with KI omitted. However, good growth was obtained when the concentration of microelements was reduced to 10% of the original level (Eriksson, 1965). Potassium iodide was omitted because of its toxicity to dark-grown cultures.

IV. ORGANIC NUTRIENTS

Three groups of organic nutrients are required by virtually all tissues cultured *in vitro*, i.e., carbohydrates, plant growth regulators, and vitamins. In addition, numerous complex natural extracts and liquid endosperms have been included in culture media.

A. Carbohydrates

The disaccharide sucrose, at a concentration of 2–3%, is the most commonly used carbohydrate in plant tissue culture media. Other disaccharides that may be utilized depending upon species are maltose, lactose, cellobiose, melibiose, and trehalose (Hildebrandt and Riker, 1949; Nickell and Burkholder, 1950; Nickell and Maretzki, 1970; Nash and Boll, 1975; Verma and Dougall, 1977). The trisaccharide raffinose, tetrasaccharide stachyose, and polysaccharide starch may be metabolized by some tissues (Hildebrandt and Riker, 1949; Nickell and Maretzki, 1970; Verma and Dougall, 1977). Monosaccharides glucose and fructose are sometimes adequate substitutes for sucrose. Other monosaccharides that are suitable only for certain species or clonal variants/mutants within a species are galactose (Hildebrandt and Riker, 1949; Sievert and Hildebrandt, 1965; Kochba *et al.*, 1982), mannose (Verma and Dougall, 1977), mannitol (Wolter and Skoog, 1966), sorbitol (Chong and Taper, 1974), ribose (Nickell and Maretzki, 1970), and glycerol (Scala and Semersky, 1971; Grout *et al.*, 1976; Chaleff and Parsons, 1978; Ben-Hayyim and Neumann, 1983). Arabinose, xylose, and rhamnose will not support growth of cultured plant tissues.

As with nitrogen, the form of carbon may influence differentiation quantitatively. Although *Helianthus tuberosus* tuber slices proliferate on appropriate media containing sucrose, substitution of maltose for this disaccharide reduces the number of tracheids formed by about 60% (Minocha and Halperin, 1974). Embryo formation from *Petunia* anthers (Raquin, 1983) and *Digitalis lanata* suspension cultures (Kuberski *et al.*, 1984) reportedly is enhanced by the use of maltose rather than sucrose in the media. Embryo formation in hormone autotrophic *Citrus* nucellar callus can be stimulated by omission of sucrose during one subculture period (Kochba and Button, 1974) or by replacing sucrose with galactose, lactose (Kochba *et al.*, 1982), or glycerol (Ben-Hayyim and Neumann, 1983). Carbohydrate concentration also may have a pronounced effect on growth and morphogenesis (Hildebrandt and Riker, 1953; Negrutiu and Jacobs, 1978; Lu *et al.*, 1982).

B. Plant Growth Regulators

In addition to mineral salts, carbohydrates, and vitamins, most tissue cultures require an exogenous supply of auxin or auxin and cytokinin. Tissues capable of sustained growth in the absence of growth regulators, e.g., habituated or crown gall tissues, synthesize sufficient quantities of auxin and cytokinin to provide for their own needs. The classic work on growth and morphogenesis *in vitro*, as controlled by auxin–cytokinin ratios, was done by Skoog and Miller (1957) with stem pith tissue of *Nicotiana tabacum*.

Cytokinins are 6-substituted purine compounds. The synthetic cytokinins, kinetin (6-furfurylamino purine) and BAP (6-benzylamino purine), are very active in tissue cultures, as are the naturally occurring zeatin and $N^6 -$ (δ^2-isopentenyl)adenine (2iP) and their respective ribosides. Exogenous cytokinins are not always essential, and many tissues grow indefinitely with only a supply of auxin.

The natural auxin, indole-3-acetic acid (IAA), although quite labile, may provide a more stable supply of auxin when conjugated with certain amino acids (Hangarter and Good, 1981). The synthetic auxins, α-naphthaleneacetic acid (NAA) and 2,4-dichlorophenoxyacetic acid (2,4-D), are used more commonly in tissue cultures. Depending on the plant species to be cultured, the most potent auxin, 2,4-D, may be used in the range of 0.1–10.0 mg/liter, although for palms the extraordinarily high concentration of 100 mg/liter may be necessary. In addition to inducing unorganized cell proliferation, 2,4-D is very effective in the induction of somatic embryogenesis (Kohlenbach, 1978), particularly in the Gramineae (Vasil, 1985). Other synthetic auxins employed frequently in tissue cultures are 2,4,5-trichlorophenoxyacetic acid (2,4,5-T), 2-methyl-4-chlorophenoxyacetic acid (MCPA), 2-methoxy-3,6-dichlorobenzoic acid (dicamba), and 4-amino-3,5,6-trichloropicolinic acid (picloram).

Three other classes of growth regulators that may exert some control on morphogenesis are gibberellins (GAs), abscisic acid (ABA), and ethylene. Abscisic acid often positively influences the maturation of somatic embryos by acting in some way to reduce the spectrum and frequency of abnormalities observed in developing embryoids (Ammirato, 1974). The effects of ABA, GA, and zeatin on embryogenesis and organogenesis have been examined by several investigators (Ammirato, 1977; Thorpe, 1980). Ethylene, since it is a gas, has been largely ignored in plant tissue culture research, although a few studies of its influence on cytodifferentiation (Phillips, 1980) and organogenesis (Huxter *et al.*, 1981) have been carried out. It is not unlikely, however, that ethylene may have considerable influence on the growth of plant tissues *in vitro*, especially in the closed environment of the culture vessels commonly used.

C. Vitamins

Although myo-inositol is a carbohydrate, it usually is considered along with the vitamins since it cannot serve as a sole carbon source. Inositol is seldom essential; however, growth rate of tobacco tissue on a medium without this compound was only 50% of the control during the first passage and declined further during the second passage (Linsmaier and Skoog, 1965). Growth of Paul's Scarlet rose suspension cultures was reduced by 90% in the absence of inositol (Nesius *et al.*, 1972). Since inositol has been proved beneficial in a number of cases, it is included in most media at 100 mg/liter (Table I). Nitsch (1974) used 5 gm/liter of inositol to stimulate androgenesis in cultures of isolated pollen grains.

Among the vitamins, only thiamine seems to be universally required (White, 1951). In many cases, pyridoxine, nicotinic acid, and/or the amino acid glycine are necessary in small amounts. Murashige and Skoog (1962) included these four substances in their medium, which is often modified by increasing the content of thiamine to 0.4 mg/liter as suggested by Linsmaier and Skoog (1965). B5 medium (Gamborg *et al.*, 1968) also contains these four supplements, although only a requirement for thiamine was demonstrated with soybean suspension cultures. Ohira *et al.* (1973) retained only thiamine in their modified B5 medium for rice suspension cultures. Eriksson (1965) determined that thiamine, pyridoxine, and nicotinic acid were essential for growth of *Haplopappus gracilis* suspensions. Other vitamins sometimes considered stimulatory are folic acid and biotin. Kao and Michayluk (1975) included numerous additional vitamins (and several other addenda such as amino acids, organic acids, sugars and sugar alcohols, nucleic acid bases, and complex natural substances) for low-density culture of protoplasts of *Vicia hajastana*. Specific requirement for most of the substances included in the medium was not demonstrated.

D. Other Supplements

A wide variety of complex natural extracts have been used to supplement tissue culture media when completely defined media did not give the desired results. Early investigations employing coconut water, other fluid endosperms, casein hydrolysate, yeast extract, malt extract, and tomato and orange juice have been reviewed by Gautheret (1955) and Hildebrandt (1962).

While the beneficial effects of complex natural extracts and liquid endosperms are more pronounced in low-salt media due to the contribution of inorganic (Ziebur *et al.*, 1950; Nolan and Nolan, 1972; Raghavan, 1977) as well as organic constituents, their primary role in the more recently developed high-salt media is the supplementation of carbohydrates, plant growth regulators, and vitamins. In addition, they provide amino acids.

Coconut water was first used by van Overbeek *et al.* (1941) for the successful culture of very young embryos of *Datura*, and subsequently was employed by Steward and associates for embryogenic suspension cultures of carrot (Steward *et al.*, 1964). The composition of coconut milk has been investigated extensively (Steward and Shantz, 1959; Pollard *et al.*, 1961; Tulecke *et al.*, 1961), but the analyses have been complicated by the variability in age of coconuts from which the liquid endosperm was obtained. It is clear, however, that the endosperm contains a number of amino acids, malic acid, several vitamins, sugars and sugar alcohols, growth regulators, and other unidentified substances, none of which alone is totally responsible for its growth-promoting qualities. More recently, cytokinins, gibberellins, and auxins have been identified in coconut water and malt extract (van Staden and Drewes, 1975a,b; Dix and van Staden, 1982). It is very likely that other liquid endosperms and natural extracts such as fruit juices may contain similar substances.

Another complex substance often used in tissue culture media is casein hydrolysate. The activity of components comprising it varies with the tissue being cultured. In tobacco callus the promotion of bud formation by casein hydrolysate in the presence of kinetin could be accounted for by tyrosine (Skoog and Miller, 1957). Similarly, the activity of yeast extract can be replaced largely by thiamine (White, 1951). Potato extract, a highly variable and complex mixture, has been widely used by Chinese scientists for the culture of cereal anthers and the production of haploid plants in many species.

V. OTHER FACTORS

A. Method of Sterilization

A number of components in tissue culture media may be decomposed or altered or may react with other components when exposed to heat. Therefore, care must be taken when solutions are autoclaved. It is too time-consuming to filter-sterilize all media, but filter sterilization is a crucial step when certain factors are under consideration or comparison. Stehsel and Caplin (1969) investigated the effect of autoclaved fructose on growth of carrot root tissues. Filter-sterilized sucrose, glucose, and fructose were comparable in their capacity to support growth; however, all media became somewhat inhibitory after autoclaving. Autoclaved fructose media completely inhibited growth. Fructose autoclaved separately and then added to filter-sterilized medium produced about 50% inhibition. Nash and Boll (1975) further investigated the growth inhibition of autoclaved fructose with Paul's Scarlet rose suspension cultures by autoclaving fructose together with various other components in the medium. The greatest inhibition was observed when fructose and $MgSO_4$ were autoclaved together.

A number of amino acids as well as growth regulators are heat labile; these include glutamine, IAA, GA, and ABA. Some degradation of thiamine at high temperature apparently occurs, and the effect may be detectable when levels are limiting (Linsmaier and Skoog, 1965); however, thiamine is usually present at sufficient concentrations for small losses to be insignificant.

B. Hydrogen Ion Concentration

The pH of most tissue culture media is adjusted to 5.5–5.8 (Table I) before sterilization since drift toward this range occurs during culture. Some investigators have deviated from these routinely employed pH values with interesting results. *Nicotiana tabacum* pollen embryogenesis in filter-sterilized medium was significantly enhanced at pH 6.8 over pH 5.8 (Rashid and Reinert, 1983).

Hydrogen ion concentration in media changes during growth of plant tissues. These fluctuations have been monitored closely in suspension cultures, and it has been shown that pH usually declines during the first day of culture but then subsequently increases (Gamborg et al., 1968; Bayley et al., 1972; Karlsson and Vasil, 1985). Most tissue culture media are poorly buffered, and pH fluctuations that occur may be detrimental to long-term survival and to growth of cells at either low density (Caboche, 1980) or as single cells (Koop et al., 1983). Media buffered with 1 mM 2-(N-morpholino)ethanesulfonic acid (MES) or 1% PB74 were far superior to unbuffered media for growth of single cells of *Datura innoxia* in the pH range of 5.0–6.0 (Koop et al., 1983). MES-buffered medium was also beneficial for sustained growth of protoplast-derived colonies of *Santalum album* (Rao and Ozias-Akins, 1985).

The pH may fall several tenths of a unit when the medium is autoclaved, and this relatively small drop in pH may alter morphogenetic processes under certain conditions, particularly during culture of thin cell layers in liquid media (Cousson and Tran Thanh Van, 1981). Thin cell layers from inflorescences of *Nicotiana tabacum* produced flowers on agar medium but not under the same conditions in liquid medium. Flower formation on liquid media could be achieved by controlling the pH, changing the equimolar concentration of IBA and kinetin, and including glass beads of a certain diameter. The authors speculated that the ill-defined effect of glass bead diameter on morphogenesis could have been due to a stabilization of the pH of the medium and Na^+ and Ca^{2+} released by dissolution of the glass. With embryogenic carrot suspensions, Wetherell and Dougall (1976) adjusted the culture pH every 8 hr and thereby determined the optimal pH for growth and embryogenesis. Good growth occurred between pH 5–6, but there was a relatively narrow pH optimum for embryogenesis at pH 5.4.

C. Physical Form of Medium

Murashige (1974) and Thorpe (1980) recommended that care should be taken when deciding upon the physical form of the tissue culture medium, since growth rate and pattern of morphogenesis may be affected. Tissues cultured in stationary liquid media may sink to the bottom of the culture vessel where the diffusion of gases is reduced. This problem of aeration may be circumvented by adding substances such as sucrose or Ficoll to the liquid medium that allow the tissues to remain afloat (Kao, 1981). Rate of embryo formation from *Nicotiana tabacum* anthers was reduced on agar medium relative to liquid medium due to inhibitors present in the agar. These inhibitors could be adsorbed by activated charcoal (Kohlenbach and Wernicke, 1978). Liquid media were employed in Heller's early nutritional studies because it was recognized that agar contained significant macroelement impurities. Debergh (1983) determined that Difco "Bacto" agar contributed appreciable quantities of Ca, K, Na, Mg, and Mn to the media. A comparison of different agar brands and agar quality revealed differences in solidity of the gel formed at equivalent concentrations (Debergh, 1983; Kevers et al., 1984). Different brands of agar and agarose were tested with protoplasts of four plant species, three of which grew well in liquid media (Lörz et al., 1983). Plating efficiencies of protoplasts in agar media were dramatically reduced; however, this inhibition was not observed with agarose.

Unidentified inhibitors or oxidized phenolic compounds released from certain tissues or inadvertently included in agar media may prevent growth or morphogenesis. Addition of 1% activated charcoal to anther culture media was found to substantially enhance haploid plant formation in *Nicotiana tabacum* (Anagnostakis, 1974). Positive effects of activated charcoal on differentiation have been observed in numerous other systems

(Fridborg *et al.*, 1978; Kohlenbach and Wernicke, 1978; Johansson, 1983; Ho and Vasil, 1983). Activated charcoal has been shown to adsorb phenolic compounds and other inhibitory substances (Kohlenbach and Wernicke, 1978); however, one must be aware that it also may adsorb growth regulators (Weatherhead *et al.*, 1978; Johansson, 1983). Other substances that may enhance callus initiation or differentiation due to adsorption of phenolic compounds or inhibition of phenol oxidases have occasionally been incorporated into plant tissue culture media, e.g., polyvinylpyrrolidone (PVP), cysteine (Babbar and Gupta, 1982), dithiothreitol (Katterman *et al.*, 1977), and ascorbic acid.

REFERENCES

Ammirato, P. V. (1974). The effects of abscisic acid on the development of somatic embryos from cells of caraway (*Carum carvi* L.). *Bot. Gaz. (Chicago)* **135**, 328–337.
Ammirato, P. V. (1977). Hormonal control of somatic embryo development from cultured cells of caraway: Interactions of abscisic acid, zeatin, and gibberellic acid. *Plant Physiol.* **59**, 579–586.
Anagnostakis, S. L. (1974). Haploid plants from anthers of tobacco—enhancement with charcoal. *Planta* **115**, 281–283.
Babbar, S. B., and Gupta, S. C. (1982). Promotory effect of polyvinylpyrrolidone and L-cysteine-HCl on pollen plantlet production in anther cultures of *Datura metel*. Z. *Pflanzenphysiol.* **106**, 459–464.
Bayley, J. M., King, J., and Gamborg, O. L. (1972). The effect of the source of inorganic nitrogen on growth and enzymes of nitrogen assimilation in soybean and wheat cells in suspension cultures. *Planta* **105**, 15–24.
Ben-Hayyim, G., and Neumann, H. (1983). Stimulatory effect of glycerol on growth and somatic embryogenesis in *Citrus* callus cultures. Z. *Pflanzenphysiol.* **110**, 331–337.
Berthelot, A. (1934). Nouvelles remarques d'ordre chimique sur le choix des milieux de culture naturels et sur la manière de formuler les milieux synthétiques. *Bull. Soc. Chim. Biol.* **16**, 1553–1557.
Boyes, C. J., Zapata, F. J., and Sink, K. C. (1980). Isolation, culture and regeneration to plants of callus protoplasts of *Salpiglossis sinuata* L. Z. *Pflanzenphysiol.* **99**, 471–474.
Brown, S., Wetherell, D. F., and Dougall, D. K. (1976). The potassium requirement for growth and embryogenesis in wild carrot suspension cultures. *Physiol. Plant.* **37**, 73–79.
Burkholder, P. R., and Nickell, L. G. (1949). Atypical growth of plants. I. Cultivation of virus tumors of *Rumex* on nutrient agar. *Bot. Gaz. (Chicago)* **110**, 426–437.
Caboche, M. (1980). Nutritional requirements of protoplast-derived haploid tobacco cells grown at low cell densities in liquid medium. *Planta* **149**, 7–18.
Chaleff, R. S., and Parsons, M. F. (1978). Isolation of a glycerol-utilizing mutant of *Nicotiana tabacum*. *Genetics* **89**, 723–728.
Chong, C., and Taper, C. D. (1974). *Malus* tissue cultures. II. Sorbitol metabolism and carbon nutrition. *Can. J. Bot.* **52**, 2361–2364.
Chu, C. (1978). The N_6 medium and its applications to anther culture of cereal crops. *In* "Proceedings of Symposium on Plant Tissue Culture," pp. 43–45. Science Press, Peking.

Clarkson, D. T., and Hanson, J. B. (1980). The mineral nutrition of higher plants. *Annu. Rev. Plant Physiol.* **31**, 239–298.

Cousson, A., and Tran Thanh Van, K. (1981). *In vitro* control of de novo flower differentiation from tobacco thin cell layers cultured on a liquid medium. *Physiol. Plant.* **51**, 77–84.

Debergh, P. C. (1983). Effects of agar brand and concentration on the tissue culture medium. *Physiol. Plant.* **59**, 270–276.

Dix, L., and van Staden, J. (1982). Auxin- and gibberellin-like substances in coconut milk and malt extract. *Plant Cell. Tissue Organ Cult.* **1**, 239–245.

Eriksson, T. (1965). Studies on the growth requirements and growth measurements of cell cultures of *Haplopappus gracilis*. *Physiol. Plant.* **18**, 976–993.

Eskew, D. L., Welch, R. M., and Cary, E. E. (1983). Nickel: An essential micronutrient for legumes and possibly all higher plants. *Science* **222**, 621–623.

Fridborg, G., Pedersen, M., Landstrom, L., and Eriksson, T. (1978). The effects of activated charcoal on tissue cultures: Adsorption of metabolites inhibiting morphogenesis. *Physiol. Plant.* **43**, 104–106.

Gamborg, O. L., and Shyluk, J. P. (1970). The culture of plant cells with ammonium salts as the sole nitrogen source. *Plant Physiol.* **45**, 598–600.

Gamborg, O. L., Miller, R. A., and Ojima, K. (1968). Nutrient requirements of suspension cultures of soybean root cells. *Exp. Cell Res.* **50**, 151–158.

Gautheret, R. J. (1937). Nouvelles récherches sur la culture du tissue cambial. *C. R. Hebd. Seances Acad. Sci.* **205**, 572–574.

Gautheret, R. J. (1938). Sur le repiquage des cultures de tissue cambial de *Salix caprea*. *C. R. Hebd. Seances Acad. Sci.* **206**, 125–127.

Gautheret, R. J. (1939). Sur la possibilité de réaliser a la culture indéfinie des tissus de tubercules de carotte. *C. R. Hebd. Seances Acad. Sci.* **208**, 118–120.

Gautheret, R. J. (1955). The nutrition of plant tissue cultures. *Annu. Rev. Plant Physiol.* **6**, 433–484.

Grout, B. W. W., Chan, K. W., and Simpkins, I. (1976). Aspects of growth and metabolism in a suspension culture of *Acer psuedoplatanus* (L.) grown on a glycerol carbon source. *J. Exp. Bot.* **27**, 77–86.

Haberlandt, G. (1902). Kulturversuche mit isolierten Pflanzenzellen. *Mat. Nat. Kais. Akad. Wiss.* **111**, 69–92.

Halperin, W., and Wetherell, D. F. (1965). Ammonium requirement for embryogenesis *in vitro*. *Nature (London)* **205**, 519–520.

Hangarter, R. P., and Good, N. E. (1981). Evidence that IAA conjugates are slow-release sources of free IAA in plant tissues. *Plant Physiol.* **68**, 1424–1427.

Hart, J. W., and Filner, P. (1969). Regulation of sulfate uptake by amino acids in cultured tobacco cells. *Plant Physiol.* **44**, 1253–1259.

Heller, R. (1953). Récherches sur la nutrition minerale des tissus végétaux cultives *in vitro*. *Ann. Sci. Nat. Bot. Biol. Veg.* **14**, 1–223.

Heller, R. (1965). Some aspects of the inorganic nutrition of plant tissue cultures. *In* "Plant Tissue Culture" (P. R. White and A. R. Grove, eds.), pp. 1–17. McCutchan Publ., Berkeley, California.

Hildebrandt, A. C. (1962). Tissue and single cell cultures of higher plants as a basic experimental method. *In* "Modern Methods of Plant Analysis" (K. Paech, M. V. Tracey, and H. F. Linskens, eds.), Vol. 5, pp. 383–421. Springer-Verlag, Berlin and New York.

Hildebrandt, A. C., and Riker, A. J. (1949). The influence of various carbon compounds on the growth of marigold, Paris-daisy, periwinkle, sunflower and tobacco tissue *in vitro*. *Am. J. Bot.* **36**, 74–85.

Hildebrandt, A. C., and Riker, A. J. (1953). Influence of concentrations of sugars and polysaccharides on callus tissue growth *in vitro*. *Am. J. Bot.* **40**, 66–76.

Hildebrandt, A. C., Riker, A. J., and Duggar, B. M. (1946). The influence of the composition of the medium on growth *in vitro* of excised tobacco and sunflower tissue cultures. *Am. J. Bot.* **33**, 591–597.

Ho, W. J., and Vasil, I. K. (1983). Somatic embryogenesis in sugarcane (*Saccharum officinarum* L.). II. Growth and plant regeneration from embryogenic cell suspension cultures. *Ann. Bot. (London)* **51**, 719–726.

Hoagland, D. R., and Snyder, W. C. (1933). Nutrition of strawberry plant under controlled conditions. *Proc. Am. Soc. Hortic. Sci.* **30**, 288–294.

Huxter, T. J., Thorpe, T. A., and Reid, D. M. (1981). Shoot initiation in light- and dark-grown tobacco callus: The role of ethylene. *Physiol. Plant.* **53**, 319–326.

Johansson, L. (1983). Effects of activated charcoal in anther cultures. *Physiol. Plant.* **59**, 397–403.

Kamada, H., and Harada, H. (1979). Studies on the organogenesis in carrot tissue cultures. II. Effects of amino acids and inorganic nitrogenous compounds on somatic embryogenesis. *Z. Pflanzenphysiol.* **91**, 453–463.

Kao, K. N. (1981). Plant formation from barley anther cultures with Ficoll media. *Z. Pflanzenphysiol.* **103**, 437–443.

Kao, K. N., and Michayluk, M. R. (1975). Nutritional requirements for growth of *Vicia hajastana* cells and protoplasts at a very low population density in liquid media. *Planta* **126**, 105–110.

Karlsson, S. B., and Vasil, I. K. (1985). Growth, cytology and flow cytometry of embryogenic cell suspension cultures of *Panicum maximum* Jacq. (Guinea grass) and *Pennisetum purpureum* Schum. (Napier grass). Submitted.

Katterman, F. R. H., Williams, M. D., and Clay, W. F. (1977). The influence of a strong reducing agent upon the initiation of callus from the germinating seedlings of *Gossypium barbadense*. *Physiol. Plant.* **40**, 98–100.

Kevers, C., Coumans, M., Coumans-Gilles, M-F., and Gaspar, T. (1984). Physiological and biochemical events leading to vitrification of plants cultured *in vitro*. *Physiol. Plant.* **61**, 69–74.

Knop, W. (1865). Quantitative Untersuchungen über den Ernährungsprozess der Pflanzen. *Landwirtsch. Vers. Stn.* **7**, 93–107.

Knop, W. (1884). Bereitung einer concentrierten Nährstofflösung für Pflanzen. *Landwirtsch. Vers. Stn.* **30**, 292–294.

Kochba, J., and Button, J. (1974). The stimulation of embryogenesis and embryoid development in habituated ovular callus from the Shamouti orange (*Citrus sinensis*) as affected by tissue age and sucrose concentration. *Z. Pflanzenphysiol.* **73**, 415–421.

Kochba, J., Spiegel-Roy, P., Neumann, H., and Saad, S. (1982). Effect of carbohydrates on somatic embryogenesis in subcultured nucellar callus of *Citrus* cultivars. *Z. Pflanzenphysiol.* **105**, 359–368.

Kohlenbach, H. W. (1978). Comparative somatic embryogenesis. *In* "Frontiers of Plant Tissue Culture 1982" (T. A. Thorpe, ed.), pp. 59–66. Univ. of Calgary Press, Calgary, Alberta, Canada.

Kohlenbach, H. W., and Wernicke, W. (1978). Investigations on the inhibitory effect of agar and the function of activated carbon in anther culture. *Z. Pflanzenphysiol.* **86**, 463–472.

Koop, H. U., Weber, G., and Schweiger, H-G. (1983). Individual culture of selected single cells and protoplasts of higher plants in microdroplets of defined medium. *Z. Pflanzenphysiol.* **112**, 21–34.

Kotte, W. (1922a). Wurzelmeristem in Gewebekulture. *Ber. Dtsch. Bot. Ges.* **40**, 269–272.

Kotte, W. (1922b). Kulturversuche mit isolierten Wurzelspitzen. *Beitr. Allg. Bot.* **2**, 413–434

Kuberski, C., Scheibner, H., Steup, C., Diettrich, B., and Luckner, M. (1984). Embryogenesis and cardenolide formation in tissue cultures of *Digitalis lanata*. *Phytochemistry* **23**, 1407–1412.

Linsmaier, E. M., and Skoog, F. (1965). Organic growth factor requirements of tobacco tissue cultures. *Physiol. Plant.* **18**, 100–127.

Lörz, H., Larkin, P. J., Thomson, J., and Scowcroft, W. R. (1983). Improved protoplast culture and agarose media. *Plant. Cell, Tissue Organ Cult.* **2**, 217–226.

Lu, C., Vasil, I. K., and Ozias-Akins, P. (1981). Somatic embryogenesis in *Zea mays* L. *Theor. Appl. Genet.* **62**, 109–112.

Minocha, S. C., and Halperin, W. (1974). Hormones and metabolites which control tracheid differentiation with or without concomitant effects on growth in cultured tuber tissue of *Helianthus tuberosus* L. *Planta* **116**, 319–331.

Murashige, T. (1974). Plant propagation through tissue cultures. *Annu. Rev. Plant. Physiol* **25**, 135–166.

Murashige, T., and Skoog, F. (1962). A revised medium for rapid growth and bioassays with tobacco tissue cultures. *Physiol. Plant.* **15**, 473–497.

Nash, D. T., and Boll, W. G. (1975). Carbohydrate nutrition of Paul's Scarlet rose cell suspensions. *Can. J. Bot.* **53**, 179–185.

Negrutiu, I., and Jacobs, M. (1978). Restoration of the morphogenetic capacity in long-term callus cultures of *Arabidopsis thaliana*. *Z. Pflanzenphysiol.* **90**, 431–441.

Nesius, K. K., Uchytil, L. E., and Fletcher, J. S. (1972). Minimal organic medium for suspension cultures of Paul's Scarlet rose. *Planta* **106**, 173–176.

Nickell, L. G., and Burkholder, P. R. (1950). Atypical growth of plants. II. Growth *in vitro* of virus tumors of *Rumex* in relation to temperature, pH and various sources of nitrogen, carbon and sulfur. *Am. J. Bot.* **37**, 538–547.

Nickell, L. G., and Maretzki, A. (1970). The utilization of sugars and starch as carbon sources by sugarcane cell suspension cultures. *Plant Cell Physiol.* **11**, 183–185.

Nitsch, C. (1974). La culture de pollen isolé sur milieu synthetique. *C. R. Hebd. Seances Acad. Sci., Ser. D.* **278**, 1031–1034.

Nitsch, J. P. (1951). Growth and development *in vitro* of excised ovaries. *Am. J. Bot.* **38**, 566–577.

Nitsch, J. P., and Nitsch, C. (1969). Haploid plants from pollen grains. *Science* **163**, 85–87.

Nobécourt, P. (1937). Cultures en série de tissus végétaux sur milieu artificiel *C. R. Hebd. Seances Acad. Sci.* **200**, 521–523.

Nobécourt, P. (1938). Sur le proliférations spontanées de fragments de tubercules de Carotte et leur culture sur milieu synthétique. *Bull. Soc. Bot. Fr.* **85**, 1–7.

Nobécourt, P. (1939). Sur la perennité et l'aumention de volume des cultures de tissus végétaux. *C. R. Seances Soc. Biol. Ses Fil.* **130**, 1270–1271.

Nobécourt, P. (1940). Synthèse de la vitamine B_1 dans les cultures de tissus végétaux. *C. R. Seances Soc. Biol. Ses Fil.* **133**, 530–532.

Nolan, R. A., and Nolan, W. G. (1972). Elemental analysis of vitamin-free casamino acids. *Appl. Microbiol.* **24**, 290–291.

Ohira, K., Ojima, K., and Fujiwara, A. (1973). Studies on the nutrition of rice cell culture. I. A simple defined medium for rapid growth in suspension culture. *Plant Cell Physiol.* **14**, 1113–1121.

Ohira, K., Ojima, K., Saigusa, M., and Fujiwara, A. (1975). Studies on the nutrition of rice cell culture. II. Microelement requirement and the effects of deficiency. *Plant Cell Physiol.* **16**, 73–81.

Okamura, M., Hayashi, T., and Miyazaki, S. (1984). Inhibiting effect of ammonium ion in protoplast culture of some Asteraceae plants. *Plant Cell Physiol.* **25**, 281–286.

Phillips, R. (1980). Cytodifferentiation. *Int. Rev. Cytol. Suppl.* **11A**, 55–70.

Pollard, J. K., Shantz, E. M., and Steward, F. C. (1961). Hexitols in coconut milk: Their role in nurture of dividing cells. *Plant Physiol.* **36**, 492–501.

Raghavan, V. (1977). Diets and culture media for plant embryos. *In* "Handbook Series in

Nutrition and Food, Section G, Diets, Culture Media and Food Supplements" (M. Rech-cigl, ed.), Vol. IV, pp. 361–413. CRC Press, Boca Raton, Florida.

Rao, P. S., and Ozias-Akins, P. (1985). Plant regeneration through somatic embryogenesis in protoplast cultures of sandalwood (*Santalum album* L.). *Protoplasma* **124**, 80–86.

Raquin, C. (1983). Utilization of different sugars as carbon source for *in vitro* anther culture of *Petunia. Z. Pflanzenphysiol.* **111**, 453–457.

Rashid, A., and Reinert, J. (1983). Factors affecting high-frequency embryo formation in *ab initio* pollen cultures of *Nicotiana. Protoplasma* **116**, 155–160.

Robbins, W. J. (1922a). Cultivation of excised root tips and stem tips under sterile conditions. *Bot. Gaz. (Chicago)* **73**, 376–390.

Robbins, W. J. (1922b). Effect of autoclaved yeast and peptone on growth of excised corn root tips in the dark. *Bot. Gaz. (Chicago)* **74**, 59–79.

Salisbury, F. B. (1959). Growth regulators and flowering. II. The cobaltous ion. *Plant Physiol.* **34**, 598–604.

Scala, J., and Semersky, F. E. (1971). An induced fructose-1,6-diphosphatase from cultured cells of *Acer pseudoplatanus* (English sycamore). *Phytochemistry* **10**, 567–570.

Schenck, R. U., and Hildebrandt, A. C. (1972). Medium and techniques for induction and growth of monocotyledonous and dicotyledonous plant cell cultures. *Can. J. Bot.* **50**, 199–204.

Shepard, J. F., and Totten, R. E. (1977). Mesophyll cell protoplasts of potato: Isolation, proliferation and plant regeneration. *Plant Physiol.* **60**, 313–316.

Sievert, R. C., and Hildebrandt, A. C. (1965). Variations within single cell clones of tobacco tissue cultures. *Am. J. Bot.* **52**, 742–750.

Singh, M., and Krikorian, A. D. (1980). Chelated iron in culture media. *Ann. Bot. (London)* **46**, 807–809.

Singh, M., and Krikorian, A. D. (1981). White's standard nutrient solution. *Ann. Bot. (London)* **47**, 133–139.

Skoog, F., and Miller, C. O. (1957). Chemical regulation of growth and organ formation in plant tissues cultured *in vitro. Symp. Soc. Exp. Biol.* **11**, 118–131.

Stehsel, M. L., and Caplin, S. M. (1969). Sugars: Effect of autoclaving vs. sterile filtration on the growth of carrot root tissue in culture. *Life Sci.* **8**, 1255–1259.

Steward, F. C., and Shantz, E. M. (1959). The chemical regulation of growth. Some substances and extracts which induce growth and morphogenesis. *Annu. Rev. Plant Physiol.* **10**, 379–404.

Steward, F. C., Mapes, M. O., Kent, A. E., and Holsten, R. D. (1964). Growth and development of cultured plant cells. *Science* **143**, 20–27.

Street, H. E. (1967). Excised root culture. In "Methods in Developmental Biology" (F. H. Wilt and N. K. Wessells, eds.), pp. 425–434. Crowell, New York.

Stuart, D. A., and Strickland, S. G. (1984). Somatic embryogenesis from cell cultures of *Medicago sativa* L. I. The role of amino acid additions to the regeneration medium. *Plant Sci. Lett.* **34**, 165–174.

Tazawa, M., and Reinert, J. (1969). Extracellular and intracellular chemical environments in relation to embryogenesis *in vitro. Protoplasma* **68**, 157–173.

Thomas, E., and Street, H. E. (1972). Factors influencing morphogenesis in excised roots and suspension cultures of *Atropa belladonna. Ann. Bot. (London)* **36**, 239–247.

Thorpe, T. A. (1980). Organogenesis in vitro: Structural, physiological and biochemical aspects. *Int. Rev. Cytol. Suppl.* **11A**, 71–111.

Trelease, S., and Trelease, H, M. (1933). Physiologically balanced culture solutions with stable hydrogen ion concentration. *Science* **78**, 438–439.

Tulecke, W., Weinstein, L. H., Rutner. A., and Laurencot, H. J., Jr., (1961). The biochemical composition of coconut water (coconut milk) as related to its use in plant tissue culture. *Contrib. Boyce Thompson Inst.* **21**, 115–128.

Uspenski, E. E., and Uspenskaia, W. J. (1925). Reinkultur und ungeschlechtliche Fort-pflanzung der *Volvox minor* und *Volvox globator* in einer synthetischen Nährlösung. *Z. Bot.* **17**, 273–308.

van Overbeek, J., Conklin, M. E., and Blakeslee, A. F. (1941). Factors in coconut milk essential for growth and development of very young *Datura* embryos. *Science* **94**, 350–351.

van Staden, J., and Drewes, S. E. (1975a). Identification of zeatin and zeatin riboside in coconut milk. *Physiol. Plant* **34**, 106–109.

van Staden, J., and Drewes, S. E. (1975b). Isolation and identification of zeatin from malt extract. *Plant Sci. Lett.* **34**, 391–394.

Vasil, I. K. (1973). Plants: Haploid tissue cultures. *In* "Tissue Culture: Methods and Applications" (P. F. Kruse, Jr., and M. K. Patterson, Jr., eds.), pp. 157–161. Academic Press, New York.

Vasil, I. K. (1977). Nutrient requirements of plant tissues in culture for growth and differentia-tion. *In* "Handbook Series in Nutrition and Food" (M. Rechcigl, Jr., ed.), Section D, Vol. 1, pp. 479–486. CRC Press, Boca Raton, Florida.

Vasil, I. K. (1985). Somatic embryogenesis and its consequences in the Gramineae. *In* "Tissue Culture in Forestry and Agriculture" (R. Henke, K. Hughes, and A. Hollaender, eds.), pp. 31–48. Plenum Press, New York.

Vasil, I. K., and Hildebrandt, A. C. (1966). Growth and chlorophyll production in plant callus tissues grown *in vitro. Planta* **68**, 69–82.

Verma, D. C., and Dougall, D. K. (1977). Influence of carbohydrates on quantitative aspects of growth and embryo formation in wild carrot suspension cultures. *Plant Physiol.* **59**, 81–85.

Walker, K. A., and Sato, S. J. (1981). Morphogenesis in callus tissue of *Medicago sativa:* The role of ammonium ion in somatic embryogenesis. *Plant Cell, Tissue Organ Cult.* **1**, 109–121.

Weatherhead, M. A., Burdon, J., and Henshaw, G. G. (1978). Some effects of activated charcoal as an additive to plant tissue culture media. *Z. Pflanzenphysiol.* **89**, 141–147.

Wetherell, D. F., and Dougall, D. K. (1976). Sources of nitrogen supporting growth and embryogenesis in cultured wild carrot tissue. *Physiol. Plant* **37**, 97–103.

White, P. R. (1932). Plant tissue culture: A preliminary report of results obtained in the culturing of certain plant meristems. *Arch. Exp. Zellforsch. Besonders Gewebezuecht.* **12**, 602–620.

White, P. R. (1933a). Plant tissue cultures: Results of preliminary experiments on the culturing of isolated stem-tips of *Stellaria media. Protoplasma* **19**, 97–116.

White, P. R. (1933b). Concentrations of inorganic ions as related to growth of excised root tips of wheat seedlings. *Plant Physiol.* **8**, 489–508.

White, P. R. (1934). Potentially unlimited growth of excised tomato root tips in a liquid medium. *Plant Physiol.* **9**, 585–600.

White, P. R. (1938). Accessory salts in the nutrition of excised tomato roots. *Plant Physiol.* **13**, 391–398.

White, P. R. (1943). "A Handbook of Plant Tissue Culture." Jacques Cattell Press, Tempe, Arizona.

White, P. R. (1951). Nutritional requirements of isolated plant tissues and organs. *Annu. Rev. Plant Physiol.* **2**, 231–244.

White, P. R. (1963). "The Cultivation of Animal and Plant Cells," 2nd ed. Ronald Press, New York.

Wolter, K. E., and Skoog, F. (1966). Nutritional requirements of *Fraxinus* callus cultures. *Am. J. Bot.* **53**, 263–269.

Ziebur, N. K., Brink, R. A., Graf, L. H., and Stahman, M. A. (1950). The effect of casein hydrolysate on the growth *in vitro* of immature *Hordeum* embryos. *Am. J. Bot.* **37**, 144–148.

CHAPTER 5

Cytodifferentiation

Hiroo Fukuda

Department of Biology
Faculty of Science
Osaka University
Toyonaka, Osaka, Japan

Atsushi Komamine

Department of Botany
Biological Institute
Tohoku University
Sendai, Japan

149

I. INTRODUCTION

In higher plants, a somatic cell can differentiate to a whole plant under appropriate experimental conditions. This potential for totipotency indicates that all genes responsible for differentiation to a whole plant are contained in a somatic cell and that many of the genes may remain inactive in differentiated tissue or organs, but they can be expressed if the somatic cell is cultured under adequate conditions. Perfect demonstration of totipotency has been performed only in higher plant cells (Vasil and Hildebrandt, 1965a,b), which can provide, therefore, an efficient tool for the study of differentiation in multicellular organisms. One of the important objectives in developmental biology is to understand the process of the development of an adult organism from a single cell, the zygote. Such a process can also be thought of as the integration of cell division and specialization of cells. The specialization of cells, which is the basic event of development in eukaryotes, is defined as cytodifferentiation.

What possible experimental systems in higher plants can we use for analyzing cytodifferentiation from the above two points of view, that is, cytoplasmic and genetic control? One of the most efficient systems is cytodifferentiation *in vitro* from parenchyma cells to cells that have localized secondary wall thickenings as seen in the xylem. Until now, a considerable amount of information about this type of cytodifferentiation has been accumulated through studies of vascular development, xylogenesis, and xylemlike cell differentiation itself (Roberts, 1969, 1976; Torrey *et al.*, 1971; Shininger, 1979a; Barnett, 1979; Phillips, 1980). These types of cells have been referred to as wound vessel members, vessel elements, tracheids, tracheary elements, etc. In accordance with Torrey *et al.* (1971) and Phillips (1980), the term "tracheary element" is used hereafter as a general term (Fig. 1).

Differentiation to tracheary elements has the following advantages for the study of cytodifferentiation in higher plants.

1. Tracheary elements have annular, spiral, reticulate, or pitted secondary wall thickening. Their morphological characteristics enable us to easily distinguish differentiated cells from undifferentiated cells under a light microscope.

2. The formation of tracheary elements can be induced in tissue and cell cultures of many species by control of culture conditions, including phytohormone concentrations. Synchronous cytodifferentiation has been reported in some culture systems such as *Zinnia* mesophyll cells (Fukuda and Komamine, 1980a) or Jerusalem artichoke tuber explants (Minocha and Halperin, 1974; Phillips and Dodds, 1977). Furthermore, cytodifferentiation to tracheary elements from single cells can occur *in vitro* without cell proliferation (Kohlenbach and Schmidt, 1975; Torrey, 1975; Fukuda and

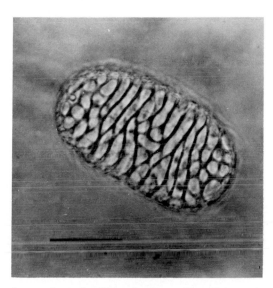

Fig. 1. A tracheary element formed at 72 hr of culture without intervening cell division from a single cell isolated from the mesophyll of *Zinnia elegans*. The bar represents 25 μm.

Komamine, 1980b; Kohlenbach and Schöpke, 1981; Kohlenbach *et al.*, 1982). This permits the analysis of cytodifferentiation without regard to direct cell interaction.

3. The cytodifferentiation of tracheary elements is a dramatic case of irreversible specialization of cells that is accompanied by the loss of nuclei and cell contents at maturity. Therefore, the possibility of dedifferentiation or redifferentiation to other types of cells is ruled out after termination of cytodifferentiation to tracheary elements or in the progression of cytodifferentiation.

4. It is important to know the marker proteins in order to analyze the cytodifferentiation sequence biochemically. Cytodifferentiation to tracheary elements, which is morphologically recognized as thickening of the walls in particular patterns, is closely associated with specific biochemical events, that is, active synthesis and deposition of wall polysaccharides and lignin. Therefore, marker proteins in cytodifferentiation to tracheary elements can easily be found.

Tracheary elements are the distinctive constituents of the xylem and are derived from cells of the procambium of the root and shoot in primary xylem or from cells produced by the vascular cambium in secondary xylem. Most studies on tracheary element differentiation in intact plants have been carried out in primary xylem because of the ease of induction and observation. Cells differentiating into tracheary elements lose their nuclei and cell contents and at maturity are aligned end to end, forming a sculptured, hollow, and continuous tubular system throughout the plant (Torrey *et al.*, 1971).

Tracheary element cytodifferentiation is composed of four ontogenetic stages (Torrey, 1953; Torrey *et al.*, 1971; Roberts, 1976), that is, acquirement of competence of the target cell to commence cytodifferentiation, cell enlargement, secondary wall formation and lignification, and autolysis of cell contents including the nucleus and selective dissolution of the wall. Tracheary element differentiation induced in cultured tissues or cells also proceeds through almost the same stages as *in situ*. In *in vitro* culture, commencement of cytodifferentiation is induced by phytohormones, especially auxin and cytokinin. There are factors other than phytohormones that are known to act as inducers of cytodifferentiation. For example, sucrose is an important factor in xylem formation in *Syringa* callus (Wetmore and Rier, 1963). Tracheary elements are formed *in situ* from cells that have arisen by division of a meristem or its derivatives. In cultured tissues, cell division is known to precede tracheary element differentiation (Roberts, 1976). Whether the cell cycle plays an important role in the induction of cytodifferentiation, therefore, has been studied with much interest. The period of cell enlargement is regarded as the period in which cells are accumulating various materials containing RNA, protein, and polysaccharides in preparation for consequent dramatic changes in morphology. The secondary wall deposits are in distinctive patterns of unique ridges (Roberts, 1976) composed of hemicellulose and cellulose that may be orientated by microtubules. A little later, lignin, which is a complex three-dimensional polymer of phenylpropanoid derivatives, is deposited on the ridges, conferring rigidity on the walls. Thereafter, differentiating cells lose their nuclei, contents, and parts of their primary walls, and become a hollow tube with a characteristic wall.

In this chapter, we will focus on tracheary element cytodifferentiation as a model system of cytodifferentiation in higher plants, especially tracheary element differentiation in cultured cells and induced primary xylem differentiation. The differentiation of the secondary xylem has been reviewed by Roberts (1976), Barnett (1979, 1981), and Shininger (1979a).

II. EXPERIMENTAL SYSTEMS

A. *In vitro* Culture

The development of *in vitro* culture systems brought about a marked improvement in the study of tracheary element differentiation. Plant tissue and cell culture systems offer many advantages in the analysis of cytodifferentiation:

1. Cytodifferentiation can be regulated by changing the composition of a medium, for example, changing known inducers of tracheary element differentiation such as auxin, cytokinin, gibberellin, and ethylene.

2. Homogeneous tissues or cells can be obtained as initial materials.

3. It should be possible to analyze the point of determination of cytodifferentiation. Cells that are not predetermined for the process of tracheary element differentiation can be induced to differentiate under experimental conditions. For example, *Zinnia* mesophyll cells, which have already differentiated as photosynthetic cells, can be redifferentiated to tracheary elements *in vitro* in the presence of auxin and cytokinin (Kohlenbach and Schmidt, 1975; Fukuda and Komamine, 1980a). Animal systems, for example, some stem cell systems, are not suitable for such analysis.

B. Tissue Culture

Internodes of intact *Coleus* plants have often been used for the study of tracheary element formation induced by wounding. Fosket and Roberts (1964) succeeded in the induction of wound-tracheary element differentiation in explants of *Coleus* stems *in vitro*. When 2-mm-thick slices of *Coleus* stem were cultured on semisolid media containing 2% sucrose, 1% agar, and 0.05 mg/liter IAA, wound-tracheary element differentiation occurred in sheetlike patterns or as branching strands or small groups of cells. In contrast to preexisting xylem tissues in *Coleus* stems, explants without any preexisting xylem tissues have also been used for the study of induced tracheary element differentiation. Clutter (1960) first reported that vascular tissue was induced in cultured explants of tobacco pith by phytohormones. Mizuno *et al.* (1971) succeeded in the induction of tracheary element differentiation at high frequency using explants of carrot phloem tissues containing no preexisting xylem tissue. This work also showed that different inducers produced different patterns of tracheary element differentiation, that is, light and kinetin, together with 2,4-dichlorophenoxyacetic acid (2,4-D), brought about tracheary element formation in the form of discrete nodules or as scattered groups, respectively. Tracheary element differentiation occurred in many other explant cultures in the form of small scattered groups rather than as discrete nodules (Phillips, 1980), although the absence of accompanying sieve elements with tracheary element formation was not ascertained.

By using explant systems, studies on the initiation point of cytodifferentiation can be performed, because explants are composed of fairly homogeneous parenchyma cells that have never been committed to cytodifferentiation of tracheary element. Tracheary element differentiation occurred partially synchronously after 48 hr of culture in explants of Jerusalem artichoke tubers (Phillips and Dodds, 1977). In this system, cells begin to divide synchronously, preceding tracheary element differentiation (Yeoman and Evans, 1967). Synchrony both in cell division and cytodifferentiation in explants of Jerusalem artichoke enabled Phillips and colleagues to analyze the relations between tracheary element differentiation and the cell cycle (Dodds and Phillips, 1977; Malawer and Phillips, 1979: Phillips, 1981a,b).

Although an increase in the number of tracheary elements occurred in the cultured artichoke explants in a short period (the first increase at 48 hr, the second at 56–58 hr of culture; Phillips and Dodds, 1977), the proportion of cells forming tracheary elements was only a small percentage of the total population. Therefore, until now, studies using this system have not focused on biochemical analysis of whole explants but rather on physiological studies of individual differentiating cells using techniques such as autoradiography (Phillips, 1980). Minocha and Halperin (1974) reported that as many as 35% of cells differentiated finally in explants of Jerusalem artichoke, and differentiation did not occur synchronously but gradually between 6 and 14 days of culture.

Lettuce pith explants and pea root explants have been used as other materials for the study of tracheary element differentiation. The former material has been used mainly for the analysis of hormonal effects on cytodifferentiation by Roberts and colleagues (auxin and cytokinin, Dalessandro and Roberts, 1971; ethylene, Roberts and Baba, 1978). Torrey and colleagues have studied the relationship between kinetin, cell division, and cytodifferentiation (Phillips and Torrey, 1973; Shininger, 1978) using explants of pea roots or pea root cortical tissues.

In these explant systems, the initial materials are under the influence of the physiological state of original explants or organs. For example, the maturity of the tubers affected the ability to differentiate tracheary elements in explants of Jerusalem artichoke tubers (Phillips, 1981a). It is desirable that all cells in tissues or organs from which explants are obtained should be in a uniform physiological state, but such materials are usually difficult to obtain.

C. Callus Culture

Calli are obtained by culturing various tissues of plants *in vitro*. Some calli, which show stable characteristics after subculture through many passages, offer the following advantages for the study of cytodifferentiation.

1. A uniform physiological state under known culture conditions.
2. Types of calli that are composed of fairly homogeneous cells.
3. Availability of large amounts of homogeneous calli as experimental materials.

In 1955, Wetmore and Sorokin (1955) reported the induction of vascular strands in *Syringa* callus derived from the cambial region of the stem by auxin or grafted apices of shoots. Since then, tracheary element differentiation has been induced in calli derived from many different tissues of various plants: *Fraxinus* (Doley and Leyton, 1970), *Parthenocissus* (Rier and Beslow, 1967), *Eucalyptus* (Sussex and Clutter, 1968), *Nicotiana tabacum* (Bornman and Ellis, 1971; Ronchi, 1981), *Allium sativum* (Havránek and Movák, 1973), *Vinca rosea* (Datta et al., 1979), *Phaseolus vulgaris* (Jeffs and Northcote, 1967; Haddon and Northcote, 1975), *Glycine max* (Fosket and Torrey, 1969), *Acer pseudoplatanus* (Wright and Northcote, 1973), etc. In many cases, the inducer was auxin alone or in combination with cytokinin. The ratio and concentrations of the two phytohormones required for the induction of cytodifferentiation are different between callus lines, as shown typically in sycamore callus (Wright and Northcote, 1972, 1973).

One disadvantage of callus systems for the study of cytodifferentiation is that tracheary elements very often preexist in calli subcultured in a maintenance medium. One must choose, therefore, a maintenance medium that does not support the formation of tracheary elements in the callus. Fosket and Torrey (1969) developed a maintenance medium for the proliferation of cells in the absence of tracheary element differentiation in soybean callus by choosing the kinds and concentrations of phytohormones. Kinetin and NAA were used at $5 \times 10^{-7} M$ and $10^{-7} M$ for maintenance and at $10^{-6} M$ and $10^{-5} M$ for induction of tracheary element differentiation, respectively. Using the same calli, the role of ethylene in tracheary element differentiation was also analyzed (Miller and Roberts, 1982).

Many cultures lose their potential for differentiation during continued subculture. Callus cultures isolated from various somatic tissues of *Phaseolus vulgaris* seedlings lost their potential for xylem differentiation after five to seven subcultures on a maintenance medium containing only auxin as the phytohormone (Haddon and Northcote, 1976c). On the other hand, the same callus could retain its potential for several years in the presence of cytokinin or coconut milk (Jeffs and Northcote, 1966; Haddon and Northcote, 1976c). The potential for differentiation was sometimes recovered spontaneously after 30 generations (Bevan and Northcote, 1979b). These results indicate that loss of the potential for cytodifferentiation may not be due to a genetic change in the cell but to epigenetic changes such as those in the hormonal requirements of the cell necessary for induction (Bevan and Northcote, 1979b) and that the hormonal composition of maintenance medium is important for retaining the potential for differentiation. Using bean callus subcultured on a medium containing coconut milk in which xylem differentiation occurred stably in response to the phytohormone, Northcote and colleagues (Jeffs and Northcote, 1966; Haddon and Northcote, 1976a; Bolwell and Northcote, 1981) showed that tracheary elements formed *in vitro* culture had the same composition of polysaccharide and lignin in the secondary wall as those formed *in situ* and that enzyme activities related to lignin and polysaccharide syntheses were regulated in close relation to xylem differentiation.

There are some disadvantages in the use of callus as an experimental system for the study of tracheary element differentiation. Tracheary elements are formed generally as discrete nests or nodules rather than as scattered groups or single elements, that is, cytodifferentiation is accompanied by tissue formation (Fosket and Torrey, 1969; Haddon and Northcote, 1975). It is difficult, therefore, to analyze the process of cytodifferentiation separately from tissue formation. Furthermore, tracheary element differentiation occurs asynchronously and at low frequency in many callus systems.

D. Suspension Culture

Tracheary element differentiation has been reported in suspension cultures of *Asparagus* (Albinger and Beiderbeck, 1983), bean (Bevan and Northcote, 1979a), tobacco (Kuboi and Yamada, 1978b), *Centaurea* (Torrey, 1975), *Pelargonium* (Reuther and Werckmeister, 1973), and peanut (Verma and van Huystee, 1970). In tobacco suspension culture, Kuboi and Yamada (1978a,b) analyzed the regulation of enzymes related to lignin synthesis during tracheary element differentiation, showing that enzymes in the shikimate and cinnamate pathways were coordinately enhanced during tracheary element differentiation. Association of phenylalanine ammonia-lyase and peroxidase with tracheary element differentiation was observed in bean (Bevan and Northcote, 1979a,b) and peanut (Verma and van Huystee, 1970) suspension cultures, respectively.

Use of suspension cultures may surmount some of the disadvantages of callus systems in the study of tracheary element differentiation. Tracheary element formation in callus tissues occurs in discrete limited areas, probably depending on the gradient of concentration of inducers such as auxin or sugars within the callus (Wetmore and Rier, 1963; Jeffs and Northcote, 1966). On the other hand, cells in a suspension culture forming small clusters in a defined liquid medium containing inducers can receive more homogeneous stimuli than cells in a callus culture. Therefore, the percentage of tracheary element formation and the synchrony may be expected to be higher in a suspension than in a callus culture system. In many suspension cultures, however, induced tracheary elements were observed internally in the clusters as nests or nodules (Sussex and Clutter, 1968: Verma and van Huystee, 1970: Wilbur and Riopel, 1971; Reuther and Werckmeister, 1973; Kuboi and Yamada, 1978b). Tracheary elements in *Centaurea* (Torrey, 1975) and *Nicotiana* suspension cultures (Ruether and Werckmeister, 1973) can differentiate as single elements or small colonies in the appropriate chemical and physical environment but at low frequency. Unfortunately, ideal experimental systems with high frequency and synchrony of tracheary element differentiation and unaccompanied by tissue formation have never been obtained using suspension cultures. Establishment of such ideal cell suspension systems will be a major requirement in analyzing cytodifferentiation by methods such as mutation analysis, cell cloning, and gene transfer.

E. Single-Cell Culture

At present, isolated single-cell systems are the nearest to an ideal system for the investigation of cytodifferentiation. Single cells have been obtained by mechanical isolation mainly from the mesophyll tissue. Ball and Joshi (1965) reported that single palisade cells isolated from the mesophyll of *Arachis hypogeae* leaves by scraping with the points of microscalpels can divide in a liquid medium. Mesophyll cells of *Macleaya cordata* could be isolated easily using a glass homogenizer and were shown to divide in White's medium containing coconut milk and 2,4-D (Kohlenbach, 1959). Using the same materials, Kohlenbach (1965) indicated that isolated mesophyll cells could differentiate to tracheary elements in a nestlike form in cell aggregates or as a fingerlike form projecting from the surface of the cell aggregates. Jones *et al.* (1960) also observed that a few single cells of tobacco differentiated to scalariform tracheids. Ronchi and Gregorini (1970), who studied adventitious bud formation in *Lactuca sativa* cotyledons cultured *in vitro*, observed that 4 days after culture, mesophyll cells were transformed to tracheary elements directly or after one division without any accompanying tissue formation, suggesting the possibility of establishment of a direct differentiation system using mesophyll cells.

In 1975, Kohlenbach and Schmidt (1975) first reported that mechanically isolated mesophyll cells of *Zinnia elegans* can differentiate into tracheary elements directly without a preceding cell division within 3 days of culture in a liquid medium containing 1 mg/liter 2,4-D and 1 mg/liter kinetin. This finding was confirmed and developed further by Fukuda and Komamine (1980a; Fig. 2). The important points of improvement were the use of the

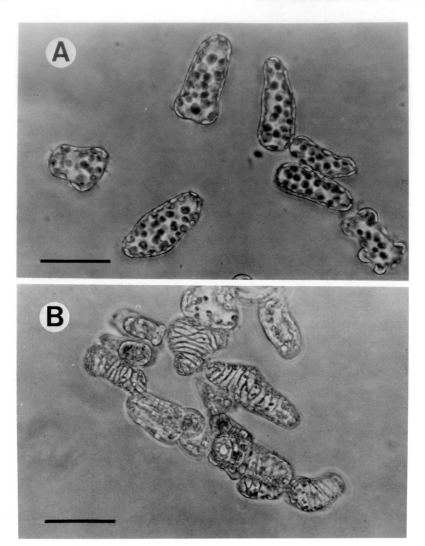

Fig. 2. (A) Phase-contrast photograph of single isolated *Zinnia* mesophyll cells. (B) Phase-contrast photograph of *Zinnia* cells cultured for 72 hr in the liquid medium containing NAA and BA, showing tracheary element formation at high frequency. The bars represent 50 μm in each figure.

first leaves of seedlings, a low concentration of ammonium salt, and the variety and amounts of hormones (NAA at 0.1 mg/liter and BA at 1 mg/liter) used. These improvements brought about a high frequency and synchrony of tracheary element formation, that is, 30–40% of mesophyll cells differentiated between 60 and 80 hr of culture (Fukuda and Komamine,

1980a, 1981a, 1982). This experimental system is considered to be useful for the study of tracheary element differentiation for the following reasons:

1. The initial cell population is composed of only single cells. Therefore, each cell can simultaneously receive equal stimulation through a given liquid medium to differentiate, and the effect of cell-to-cell interaction can be excluded in analyzing the process of cytodifferentiation. It is also possible to follow visually the sequence of cytodifferentiation and cell division in individual cells.

2. The starting materials are homogeneous cells composed mainly of palisade cells and partially of spongy cells without any tracheary elements. Furthermore, all cells are at the 2C level of DNA, that is, at the G_1 phase (Fukuda and Komamine, 1981a).

3. Isolated mesophyll cells differentiate to tracheary elements synchronously at high frequency.

4. Approximately 60% of the tracheary elements at 3 days of culture were formed without intervening mitosis, and the remaining tracheary elements were formed after undergoing one round of the cell cycle (Fukuda and Komamine, 1980b). It is possible, therefore, to analyze the process of cytodifferentiation without disturbance from tissue formation such as xylem or vascular differentiation.

Using the *Zinnia* mesophyll cell system, the relationship among cytodifferentiation, the cell cycle (Dodds, 1980; Fukuda and Komamine, 1980a,b, 1981a,b), and biochemical mechanisms of cytodifferentiation (Fukuda and Komamine, 1982, 1983) has been investigated. *Zinnia* mesophyll cells isolated by enzymatic maceration (Macerozyme R-10) can also differentiate to tracheary elements but only at low frequencies at the present time (H. Fukuda, unpublished data).

F. Protoplast Culture

Induction of tracheary elements using mesophyll protoplasts has been reported in *Zinnia elegans* (Kohlenbach and Schöpke, 1981). Protoplasts isolated from cultured haploid cells of *Brassica napus* have also been used successfully (Kohlenbach *et al.*, 1982). The protoplasts differentiated to tracheary elements without cell division in the presence of auxin and cytokinin. Development of these experimental systems may provide an answer to the question of whether the induced state of cytodifferentiation can persist through the process of removal and regeneration of cell walls (Phillips, 1980). However, these systems need to be improved because of the low rate of tracheary element formation and their asynchroneity. If an improvement can be achieved, these systems may offer the chance to examine the fundamental phenomenon of cytodifferentiation using cell fusion or transfer of foreign DNA (Kohlenbach and Schöpke, 1981).

III. CYTOLOGICAL AND CYTOCHEMICAL ASPECTS

A. Cytological and Cytochemical Changes Associated with Tracheary Element Differentiation

There have been many reports on cytological changes during tracheary element differentiation from procambial cells *in situ* (Torrey *et al.*, 1971; Barnett, 1979; Shininger, 1979a: Phillips, 1980; Hepler, 1981). Since the early 1960s ultrastructural studies on tracheary element differentiation have revealed changes in the distribution and density of organelles in the cytoplasm associated with secondary wall thickenings (Hepler and Newcomb, 1964; Wooding and Northcote, 1964; Cronshaw, 1965; Wardrop, 1965; O'Brien and Thimann, 1967; Pickett-Heaps, 1967, 1968; Hepler and Fosket, 1971; Maitra and De, 1971; Srivastava and Singh, 1972; Goosen-De Roo, 1973b; O'Brien, 1974, 1981; Hardham and Gunning, 1979). However, only a few ultrastructural observations (Cronshaw, 1967; Barnett, 1977b) have been made on the secondary wall thickenings using tissue or cell culture systems, and none have been made on the detailed process of tracheary element differentiation.

Ultrastructural changes during various stages of cytodifferentiation to tracheary elements from isolated mesophyll cells of *Zinnia elegans* have been studied to reveal early cytological changes (H. Fukuda and H. Shibaoka, unpublished data). Most parts of the cytoplasm of isolated *Zinnia* mesophyll cells are occupied by a line of lens-shaped chloroplasts, and in the center of the cells a large vacuole exists (Fig. 3A). Other organelles are hardly ever observed. When cells are cultured in a differentiation-inducing medium, the volume of the cytoplasm and the numbers of various organelles, except chloroplasts, increase (Figs. 3B and C). The arrangement of chloroplasts gradually falls into disorder, and the space between chloroplasts and plasma membrane becomes extended (Figs. 3B and C). These signs of alteration of intracellular organization indicate some of the earliest events in cytodifferentiation. Cell elongation, which has been considered an important process in tracheary element differentiation *in situ* (Torrey *et al.*, 1971), does not always occur in *Zinnia* mesophyll cells. Iso-

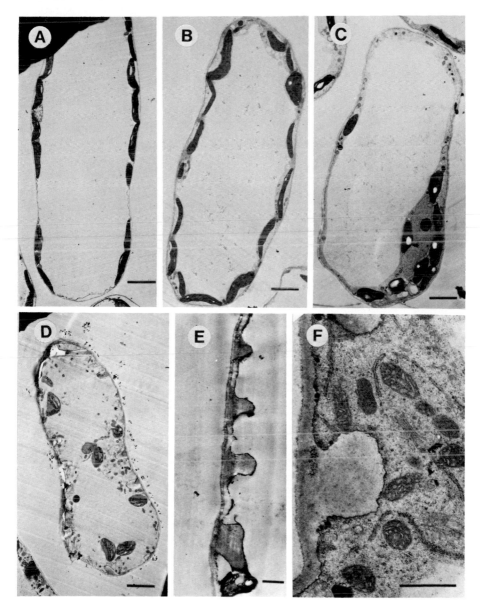

Fig. 3. Electron micrographs from longitudinal sections of *Zinnia* cells. (A) A freshly isolated *Zinnia* mesophyll cell. (B, C, and D) Cells cultured for 24 hr, 48 hr, 72 hr, respectively. Note secondary wall thickenings. (E) Portion of a mature tracheary element at 72 hr of culture. (F) Portion of a differentiating tracheary element at 58 hr of culture. The bars represent 5 μm in A–D and 1 μm in E and F.

lated palisade cells elongate only 15% on average and stop their elongation 12–24 hr before the secondary wall thickenings begin. This period is probably important as a period in which active synthesis occurs of various metabolites, which act as materials for the secondary wall or as regulators of the coming process of cytodifferentiation. Indeed, in this period, active synthesis of the RNA and protein essential for tracheary element differentiation occurs (Fukuda and Komamine, 1983). Srivastava and Singh (1972) investigated the early developmental process of the xylem in corn leaves, showing that procambial cells increased in cell volume and synthesized cytoplasmic protein that was observed under the electron microscope as masses composed of a fine fibrillar material and small granules. They also observed an increase in the rough endoplasmic reticulum and ribosomes at the same stage. An increase in the number of microtubules per unit of cell wall prior to the secondary wall thickenings also occurred during tracheary element differentiation in the roots of *Azolla* (Hardham and Gunning, 1979). In addition to the increase in endoplasmic reticulum, ribosomes, and mitochondria, dictyosomes and vesicles derived from them increased before the appearance of the banded secondary wall in differentiating cells of *Zinnia* mesophyll.

When cytological changes during cytodifferentiation are analyzed, two aspects of such changes, quantity and intracellular distribution, should be taken into consideration. The results described above show changes in quantity prior to secondary wall thickenings. Are there any intracellular changes in distribution preceding localized secondary wall deposition? About 40 years ago, Sinnott and Bloch (1944, 1945) reported that bands of densely granular cytoplasm appeared prior to the banded deposition of the secondary wall, as if they indicated the position of the subsequent localized secondary wall thickenings. Under a light microscope, Kirschner and Sachs (1978) observed that strands of cytoplasm were formed only in the tracheary element precursor cells in pea seedlings within a few hours after wounding. However, the strands of cytoplasm were different from the patterns of the subsequent secondary wall thickenings. Although much work using the electron microscope has been carried out in an attempt to discover cytoplasmic prepatterning, no such prepatterning has yet been found. Recently, microtubules were found to be distributed in the limited area where subsequent secondary wall deposition would occur in differentiating tracheary elements in *Azolla* root (Hardham and Gunning, 1979). Future investigations of prepatterning may benefit from this finding of alternation of microtubules, and this will be discussed in a latter part of this section.

The formation of the secondary wall is the most characteristic event in tracheary element differentiation and occurs in an annular, pitted, helical, or reticulate pattern, especially a helical or reticulate pattern in cultured tissues or cells (Figs. 1 and 3D). The secondary wall is composed of cellulose microfibrils arranged parallel to one another and to the banded secondary wall and of encrusting substances containing lignin, hemicellulose, pectin, and protein, adding strength and rigidity to the wall (Torrey et al., 1971). The ordered pattern of cellulose microfibrils enables detection of tracheary elements easily at early stages under a polarized light microscope. By means of an electron microscope, Wooding and Northcote (1964) traced the development of spirally thickened tracheary elements from a cambial initial of sycamore. They observed a proliferation of dictyosomes and Golgi vesicles during the period of secondary wall thickenings. An increase in dictyosomes and Golgi vesicles during the early stages of secondary wall thickenings has been reported in other species (Srivastava and Singh, 1972; Goosen-De Roo, 1973c; Esau and Charvat, 1978; H. Fukuda, unpublished data in Zinnia). Ultrastructural evidence of the fusion of vesicles with the secondary walls (Fig. 4; Cronshaw, 1965; Pickett-Heaps, 1968; Roberds and Kidwai, 1969; Maitra and De, 1971; Hardham and Gunning, 1980) suggests that materials for the developing thickenings stemmed from the Golgi vesicles. The Golgi vesicles changed in size and in electron density with the process of secondary wall thickening in primary xylem cells of cucumber hypocotyl (Goosen-De Roo, 1973b) and in differentiating tracheary elements from cultured Zinnia mesophyll cells (H. Fukuda, unpublished data). These results suggest that different

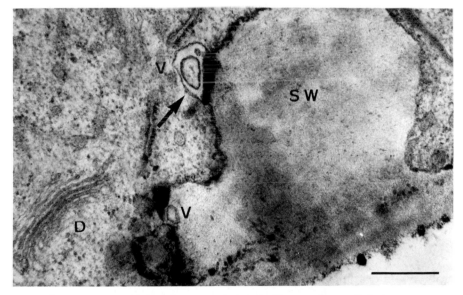

Fig. 4. Electron micrograph showing a vesicle fused with secondary wall thickening (arrow). SW, secondary wall; V, vesicle; D, dictyosome. The bar represents 0.4 μm.

types of vesicles derived from dictyosomes may transport different wall materials such as hemicellulose, pectin, and lignin precursors at different stages of secondary wall thickenings.

The endoplasmic reticulum has also been implicated in secondary wall thickenings (Cronshaw, 1965; Torrey et al., 1971; Roberts, 1976). It developed remarkably during the early stages of secondary wall formation in Zinnia (H. Fukuda, unpublished data) and in cucumber (Goosen-De Roo, 1973b), and its cisternae were found to include fine fibrils in wound-induced tracheary elements in Coleus (Wooding and Northcote, 1964). Morphological changes in endoplasmic reticulum during cytodifferentiation have also been reported (Goosen-De Roo, 1973b). During the period of active secondary wall deposition in corn tracheary elements, the endoplasmic reticulum showed a highly elaborate form and harbored intralamellar tubules (Srivastava and Singh, 1972). The endoplasmic reticulum often existed between the thickenings as if it blanketed those parts of the wall that remain unthickened (Pickett-Heaps, 1966; Srivastava and Singh, 1972; Goosen-De Roo, 1973c; Esau and Charvat, 1978). Although it seems certain that dictyosomes, Golgi vesicles, and the endoplasmic reticulum play important roles in secondary wall deposition, how each organelle functions or interacts in this process is still unknown.

Lignification is another characteristic event in cytodifferentiation to tracheary elements. In Zinnia, qualitative detection of lignin by the phloroglucinol test and quantitative measurement of lignin content according to the method of Morrison (1972) during tracheary element differentiation demonstrated a delay of approximately 5 hr in the initiation of lignification of tracheary elements as compared with the appearance of visible secondary wall thickenings, probably derived from the deposition of ordered cellulose microfibrils (Fukuda and Komamine, 1982). High activity of enzymes involved in lignification, such as peroxidase bound to the cell walls and phenylalanine ammonia-lyase, coincided with the active synthesis of lignin at the late stage of tracheary element differentiation (Fukuda and Komamine, 1982; Masuda et al., 1983). These results indicate that major deposition of lignin on the cell wall is preceded by localized deposition of cellulose, although the possibility cannot be rejected that minor lignin synthesis occurs at an early stage. Wardrop (1965) mentioned, on the basis of ultrastructural studies of woody plants, that the secondary wall thickenings of tracheary elements tended to precede lignification. Localized deposition of cellulose on the cell wall, as detected under a polarized light microscope, was also found to precede lignification, as detected by the phloroglucinol test, in developing protoxylem elements of Vicia faba (Gahan and Maple, 1966). Using $KMnO_4$ as a sensitive stain for lignification, Hepler et al. (1970) showed that lignification in wound-induced tracheary elements of Coleus had already begun in the earliest detectable stages of secondary wall formation, but lagged behind the deposition of cellulose. However, whether $KMnO_4$ is a reliable indicator for lignin has not yet been established (O'Brien, 1974). Since lignin is considered to be a major encrusting material that fills the spaces between cellulose microfibrils, it is probable that cellulose synthesis begins first and lignification occurs somewhat later and adds strength to the wall.

Aromatic alcohols, the direct precursors of lignin, which are produced in the cytoplasm, must be transported into the walls through the plasma membrane. In the cell walls, they polymerize autonomously, following the formation of their free radicals. The transportation and the synthesis of lignin precursors was investigated using electron microscopic autoradiography (Pickett-Heaps, 1968; Wooding, 1968). Pickett-Heaps (1968) found that tritiated phenylalanine and tyrosine, methyl-labeled methionine with ^3H, and especially tritiated cinnamic acid, were good precursors of lignin and that radioactivity was very markedly concentrated in the secondary wall thickenings. He also showed that labeled lignin precursors were incorporated in the endoplasmic reticulum, dictyosomes, and also Golgi vesicles, suggesting that these organelles were relevant to lignin deposition. The localization of peroxidase, which catalyzes the polymerization of aromatic alcohols resulting in lignin formation, has also been investigated in relation to tracheary element differentiation (Hepler et al., 1972; Minocha and Halperin, 1976; Fukuda and Komamine, 1982). Fukuda and Komamine (1982) offered evidence that a peroxidase bound to the cell wall was activated at the time of active lignin synthesis during tracheary element differentiation in a culture of Zinnia cells. By means of electron microscopic observation using diaminobenzidine staining, Hepler et al. (1972) revealed that peroxidase exists particularly in the secondary wall and in the primary wall adjacent to the secondary wall in wound-induced tracheary elements in Coleus. The stain was somewhat more intense in the outer or more recently formed portion of the secondary wall than in the inner core. In addition, the plasma membrane that overlaid the secondary wall thickenings was strongly stained. These results indicate that, in differentiating tracheary elements, peroxidase exists in those restricted regions where lignin synthesis would occur and functions at a precise time in the sequence of the secondary wall thickenings. It is worth noting that lignification seems to continue even during cell autolysis. In Zinnia, an increase in lignin deposition occurred even in mature tracheary elements whose nucleus had disappeared (H. Fukuda, unpublished data). Pickett-Heaps (1968) reported on the incorporation of labeled lignin precursors in the secondary wall of a mature tracheary element that had lost almost all of its cell content. The existence of peroxidase in the secondary wall of an empty tracheary element was also shown in Coleus (Hepler et al., 1972).

Cell autolysis follows secondary wall thickenings. Loss of nuclei of cultured *Zinnia* cells occurred within 10 hr after secondary wall thickenings began (H. Fukuda, unpublished data). Figure 3E shows part of a mature tracheary element that has lost its cellular contents. This process leading toward cell death seems to be programmed at the beginning of secondary wall thickening, because once the secondary wall thickenings begin, *Zinnia* mesophyll cells always go through cell autolysis to lose their cellular contents and nuclei.

It is not clear how the initiation of cell autolysis is controlled but some suggestions have been made. In the early stage of autolysis, tonoplast disappears, as shown by ultrastructural observation of corn xylem (Srivastava and Singh, 1972). Gahan and Maple (1966) suggested that the loss of the cell content was accompanied by a release of acid β-glycerophosphatase from the vacuole in maturing tracheary elements. Vacuoles may function as large lysosomes (Gahan, 1978), and their breakage may bring about the release of various types of hydrolases, resulting in disruption of the intracellular structure. On the other hand, Srivastava and Singh (1972) described how marked modification of the endoplasmic reticulum also occurred in the early stages of autolysis, at the time when the endoplasmic reticulum produced the hydrolyzing enzymes, suggesting the importance of the endoplasmic reticulum in autolysis. Following the modification of the endoplasmic reticulum, other organelles are also transformed. Although chloroplasts in differentiating tracheary elements derived from *Zinnia* mesophyll cells are not markedly modified until cell autolysis begins, the lamellar structure breaks down and the chlorophyll content decreases during this period (H. Fukuda, unpublished data). Finally, disappearance of all organelles and plasma membrane takes place. The removal of noncellulosic polysaccharides from primary walls unprotected by lignin is also a general phenomenon during the last stages of cytodifferentiation to tracheary elements (O'Brien and Thimann, 1967; O'Brien, 1970). Study of *Zinnia* cell differentiation using electron microscope by Burgess and Linstead (1984) revealed the changes of nucleus and nucleolus during tracheary element differentiation in addition to the results described in this section.

B. Possible Role of Microtubules
   in Secondary Wall Thickenings

Ledbetter and Porter (1963) first showed microtubules in plant cells using glutaraldehyde fixation, and then Hepler and Newcomb (1964) found that microtubules that appeared on the plasmalemma grouped specifically over the ridge of the secondary wall and oriented parallel to the cellulose microfibrils and to the bands of the secondary wall. Figures 5A

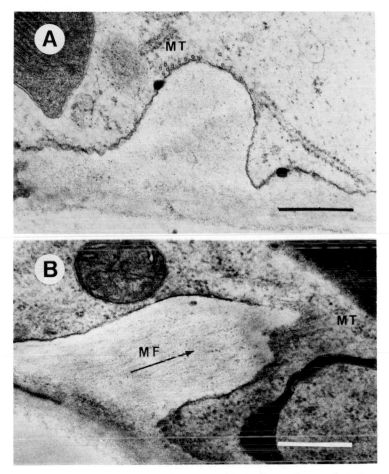

Fig. 5. (A) Electron micrograph from transverse section of secondary wall thickening of an immature tracheary element formed from an isolated *Zinnia* mesophyll cell. Microtubules (MT), circular in cross section, are clustered over the ridge of the secondary wall. The bar represents 0.5 μm. (B) Electron micrograph from an oblique section of secondary wall thickening of an immature tracheary element from an isolated *Zinnia* mesophyll cell. Microtubules (MT) are parallel to the microfibrils (MF) of the wall. The bar represents 0.5 μm.

and B, respectively, show the grouped microtubules over the secondary wall thickening and their parallelism with the microfibrils in a differentiating tracheary element from an isolated *Zinnia* mesophyll cell. During tracheary element differentiation, the distribution of microtubules changed markedly from an even to a grouped pattern along the longitudinal wall (Hardham and Gunning, 1979). These and other results suggest that microtubules play an important role in determining the pattern of secondary wall thickening; that is, microtubules are responsible for spatial control in differentiating cells (Wooding and Northcote, 1964; Cronshaw and Bouck, 1965; Esau et al., 1966; Pickett-Heaps, 1967; Hepler and Fosket, 1971; Maitra and De, 1971; Goosen-De Roo, 1973a; Hardham and Gunning, 1978).

Cytoplasmic microtubules in plants have a tubular structure about 24 nm in diameter and consist of 13 protofilaments that are formed from dimers of α- and β-tubulins (Gunning and Hardham, 1982). Hardham and Gunning (1978, 1979), in contrast to an earlier report in which very long microtubules surrounding the cells as "hoops" were predicted (Ledbetter and Porter, 1963), revealed by serial section analysis that microtubules in differentiating tracheary elements were 2–4 μm long and existed in extensively overlapping arrays in roots of Azolla pinnata.

Microtubules may function as a guide for the localized migration of vesicles filled with wall materials toward the thickenings (Roberds and Kidwai, 1969; Maitra and De, 1971; Hepler, 1981). Maitra and De (1971) observed that some Golgi vesicles lie in close proximity to microtubules oriented over the thickenings. Sometimes the vesicles are completely surrounded by microtubules in differentiating tracheary elements of Alfalfa. This suggests that, when the Golgi vesicles, which are moved by cytoplasmic streaming, meet with microtubules, the vesicles are trapped and fused to the site overlain by microtubules. Because participation of microtubules in intracellular movement of nuclei, chromosomes, and vesicles has been shown in plant cells (Gunning and Hardham, 1982), it is probable that microtubules act as a guide for the movement of Golgi vesicles. At present, however, there is no direct evidence for such guidance. Barnett (1979) stated in his review that microtubules may be involved in lignification itself, because an increase in the number of microtubules occurred simultaneously with lignification in Pinus radiata (Barnett, 1977a), and incomplete lignification was accompanied by abnormal distribution of microtubules in the "rubbery wood" of apple trees (Nelmes et al., 1973). However, the facts described below indicate that microtubules play an important role in the orientation of cellulose microfibrils rather than in lignification or cellulose synthesis.

Colchicine is an alkaloid that causes microtubule depolymerization. In the presence of a low concentration of colchicine, parenchyma cells differentiate to abnormal tracheary elements that possess irregular and highly aberrant secondary wall thickenings (Fig. 6; cf. Fig. 1). Observations using a light microscope in respect to disturbance of secondary wall thickenings by colchicine have been made in Coleus stem segments (Roberts and Baba, 1968a), pea root cortical tissues (Hammersley and McCully, 1980), and isolated Zinnia mesophyll cells (Fukuda and Komamine, 1980b). Ultrastructural analysis by Pickett-Heaps (1967) showed that colchicine treatment during primary xylem differentiation in Triticum vulgare brought about the disappearance of microtubules along the cell wall and massive irregular

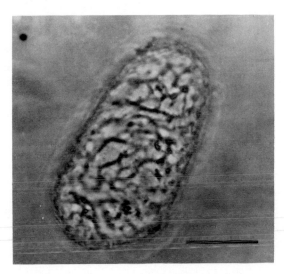

Fig. 6. Abnormal tracheary element formed at 96 hr from a single *Zinnia* mesophyll cell cultured in a liquid medium containing 10⁻⁴ M cholchicine. The bar represents 25 μm.

wall depositions, suggesting that colchicine may not prevent secondary wall synthesis itself but may disturb the organized deposition of the secondary wall (Pickett-Heaps, 1967: Hardham and Gunning, 1980). Cellulose microfibrillar orientation in colchicine-induced aberrant secondary wall thickenings did not occur parallel to the bands of the secondary wall thickenings but instead occurred in swirls (Hepler and Fosket, 1971). Similarly, in epidermal cell wall formations of *Vigna angularis*, colchicine treatment resulted in the changing of a crossed polylamellate structure into a wall structure in which microfibrils continued to run in the same direction (Takeda and Shibaoka, 1981). These results suggest that microtubules may not be involved in arranging individual microfibrils parallel to one another but may be involved in determining the direction of microfibrils as a mass.

There are a few reports (Goosen-De Roo, 1973b; Hardham and Gunning, 1979) of dynamic changes in microtubules during tracheary element differentiation. Hardham and Gunning (1979) showed changes in the distribution and increase of microtubules associated with secondary wall thickenings in *Azolla* roots. Microtubules were distributed relatively evenly along the longitudinal walls in tracheary element precursor cells in which secondary wall thickenings had not yet occurred. Microtubules became grouped in the cells a little before any formation of thickenings was evident, and deposition of secondary wall proceeded beneath the groups of microtubules. The average density of microtubules also increased from about 5 microtubules per micrometer of the cell wall in tracheary element precursor cells to about 7–8 microtubules/μm in immature tracheary elements. As autolysis began, the density of the microtubules decreased. Similarly, a preliminary electron microscopic analysis in the *Zinnia* system suggested that, in the process of tracheary element differentiation from isolated mesophyll cells, the number of microtubules per micometer of the cell wall increased 2–3 times compared to the initial cells (H. Fukuda, unpublished data).

These dynamic changes in microtubules during differentiation evoke the further question whether an increase in the number of microtubules per cell during tracheary element differentiation depends on new synthesis of tubulins that are constituents of microtubules. It has been reported that activated synthesis of tubulins during differentiation of various types of animal cells results from an increase in newly synthesized tubulin mRNA (Kemphues *et al.*, 1982; Raff *et al.*, 1982; Cleveland, 1983), and the change in tubulin synthesis could be one of the earliest events during differentiation (Spiegelman and Farmer, 1982). We have never discovered the early events responsible for cytodifferentiation to tracheary elements. Tubulin may be regarded as one of the marker proteins at the early stage of tracheary element differentiation, and the expression of tubulin genes can be a model system for the study of the regulatory mechanism of gene expression during cytodifferentiation. Therefore, there is an urgent need to investigate changes in tubulin synthesis during tracheary element differentiation. Since methods for isolation of tubulins are being established also in higher plants (Mizuno *et al.*, 1981; Morejohn and Fosket, 1982), it will become possible in the near future to analyze tubulin synthesis easily during tracheary element differentiation. Another question is how the organization of the cytoskeleton, especially of the microtubule, is controlled in a whole precursor cell of tracheary element in relation to the progression of cytodifferentiation. The alteration of microtubule arrays during tracheary element differentiation has been suggested by electron microscopic analysis (Goosen-De Roo, 1973b; Hardham and Gunning, 1979). Lloyd (1983) and Wick *et al.* (1981), using whole-cell immunofluorescence and antibodies to tubulin, showed microtubule arrays in higher plant cells. This technique is efficient for analyzing microtubule orientation in the whole cell. Application of this technique to differentiating cells would reveal how microtubule organization changes during tracheary element differentiation, especially before and after the initiation of secondary wall thickenings.

Microtubules exist on the inner side of the plasma membrane, whereas cellulose microfibrils are deposited on the opposite side over the membrane. How then can microtubules orient cellulose deposition over the plasma membrane? The mechanism of orientation is not clear, but some suggestions have been made. Brower and Hepler (1976) showed the existence of crossbridges between microtubules and the plasmalemma in a differentiating tracheary element in *Allium*. Freeze-fracture techniques showed an ordered rosettelike structure composed of six particles on the cell membrane in various plants (Hepler, 1981). It has been reported that the rosettes were distributed only on limited membrane areas over which active cellulose deposition occurred during early stages of expansion in *Closterium* (Hogetsu, 1983), and photographs showing the cellulose microfibrils may spin out from the rosettelike structure were offered by Mueller and Brown (1982a,b). From these results, it may be considered probable that microtubules regulate orientation of the rosettes, which are cellulose-synthesizing complexes, on the plasma membrane through the crossbridges between the microtubules and the plasma membrane, resulting in control of organized cellulose deposition in the cell walls. Although the interrelationship between the crossbridges and the rosettes is unknown, the crossbridges might enter into the membrane and function as a kind of rail over which the rosettes run.

The secondary wall pattern does not seem to be controlled only by microtubules, because even in the absence of microtubules the secondary wall thickenings of two adjacent tracheary elements occur directly opposite one another across the common primary wall in *Coleus* (Hepler and Fosket, 1971). However, it is certain that microtubules are one of the most important organelles for spatial control of secondary wall thickenings, and, therefore, the relationship between microtubules and secondary wall thickenings should be investigated further. Further information on microtubules can be obtained from several review articles (Newcomb, 1969; Hepler and Palevitz, 1974; Hepler, 1981, Gunning and Hardham, 1982).

IV. PHYSIOLOGICAL ASPECTS

A. Hormonal Control

The action of hormones on growth and differentiation in higher plants is an old but still thorny problem. Over the past 50 years, many hormone-regulating phenomena in plants have been found. However, the regulatory mechanisms are still unknown. Tracheary element differentiation is also induced by plant hormones, especially auxin and cytokinin, in many cultured cells and tissues (Jeffs and Northcote, 1966; Torrey, 1968; Fosket and Torrey, 1969; Dalessandro and Roberts, 1971; Mizuno et al., 1971; Phillips and Dodds, 1977; Shininger, 1978; Fukuda and Komamine, 1980a; Savidge and Wareing, 1981b). Other plant hormones such as gibberellic acid, abscisic acid, and ethylene also seem to play some role in cytodifferentiation (Dalessandro, 1973; Minocha and Halperin, 1974; Roberts and Miller, 1982).

1. Auxin and Cytokinin

Auxins are one of the most important factors in the initiation and the sequence of differentiation in higher plants. Somatic embryogenesis in carrot suspension cultures is induced by the removal of auxin from the culture medium (Fujimura and Komamine, 1979). In the induction of various types of organogenesis in cultured tissues, auxins are also a principal factor (Fosket, 1980). In regard to xylogenesis, Jacobs (1952, 1954, 1969) and Jacobs and Morrow (1957) suggested that auxin may be a chemical-limiting factor in wound-induced xylem formation in the stems of *Coleus blumei*. Tracheary element differentiation in *Syringa* callus was also found to be induced by exogenous auxin at physiological concentrations (Wetmore and Sorokin, 1955). Thereafter, the induction of tracheary element differentiation has been reported in many different cultured cells and tissues. In almost all of them, the supply of exogenous auxin is essential for cytodifferentiation, although the optimum concentration varies. The endogenous as well as the exogenous auxin levels seem to affect cytodifferentiation. For example, even in the absence of exogenous auxin, tracheary elements can be induced in habituated cells (Phillips, 1980) and crown galls (Basile *et al.*, 1973), which are known to have high endogenous levels of indoleacetic acid. Also, the study of xylem differentiation *in situ* by Sheldrake and Northcote (1968a,b) suggested that indoleacetic acid produced as a consequence of the autolysis of tracheary elements may act as an initiator of further cytodifferentiation.

In general, auxin induces elongation or division in plant cells. It can be considered that the primary action of auxin in tracheary element differentiation may be the induction of cell elongation or cell division, which may trigger cytodifferentiation. However, direct evidence against this possibility has been obtained from cultured *Zinnia* mesophyll cells (Fukuda and Komamine, 1980b). Many *Zinnia* cells differentiate to tracheary elements without cell division or marked elongation. Although auxin does not induce cytodifferentiation to tracheary elements through the induction of cell division or elongation, auxin is still a prerequisite for cytodifferentiation in this system, suggesting that auxin acts directly in the induction of cytodifferentiation and not through cell division or cell elongation, at least in *Zinnia* cells.

Another important plant hormone for cytodifferentiation is cytokinin. Bergmann (1964) found that exogenous cytokinin as well as auxin was essential for the induction of tracheary element differentiation in tobacco callus. In combination with auxin, exogenous cytokinin acts in the induction of cytodifferentiation in many *in vitro* cultures. However, the degree of requirement varied between cultures. Cytokinin is an absolute requisite for tracheary element differentiation in *Zinnia* mesophyll cells (Fukuda and Komamine, 1980a), in soybean callus (Fosket and Torrey, 1969), and in pea root cortical explants (Phillips and Torrey, 1973). On the other hand, in explants of Jerusalem artichoke tubers, exogenous cytokinin appears to act only in the stimulation of auxin-induced tracheary element differentiation (Dalessandro, 1973; Minocha and Halperin, 1974). The observation that exogenous cytokinin is a prerequisite for cytodifferentiation in Jerusalem artichoke explants by Phillips and Dodds (1977) brought about further complexity. A key to the elucidation of this discrepancy has been offered by Mizuno *et al.* (1971, 1974) and Mizuno and Komamine (1978a). They studied tracheary element differentiation in cultured phloem slices of carrot roots and found a marked difference in the cytokinin requirement for cytodifferentiation among various carrot cultivars when cultured on a medium containing auxin in the dark. Analysis of endogenous cytokinin in different cultivars revealed that cultivars such as Kuroda-gosun, in which tracheary element differentiation was induced only by auxin in the dark, possessed a high endogenous level of cytokinin in the form of zeatin ribonucleoside, while cultivars such as Hokkaido-gosun, in which cytodifferentiation was not induced solely by auxin in the dark but induced by an exogenous supply of both auxin and cytokinin, did not possess measurable endogenous cytokinin. Tracheary element differentiation in cultured slices of Hokkaido-gosun roots occurred in the presence of auxin if they were cultured in the light. Under this condition, the cultured slices were also found to produce cytokinin in the form of zeatin and zeatin ribonucleoside. These results indicate that cytokinin is essential for cytodifferentiation at least in the carrot system, and it is not important whether the source of cytokinin is exogenous or endogenous. From these results it appears that the difference in the cytokinin requirement for induction of tracheary element differentiation among many other cultures may be due to the difference in endogenous levels of cytokinin.

In the presence of auxin, cultured explants of carrot roots (Hokkaido-gosun) produced cytokinin by cAMP as well as by light (Mizuno and Komamine, 1978b). Cytokinin synthesis in carrot callus by supply of auxin was also shown by Linstedt and Reinert (1975). These results appear to show that auxin is a principal factor limiting tracheary element differentiation and that cytokinin acts in the succeeding process to the onset of cytodifferentiation caused by auxin. H. Fukuda and A. Komamine (unpublished data) analyzed this possibility by investigating the time sequence of hormonal effects on the induction of cytodifferentiation in *Zinnia* mesophyll cells that definitely required both exogenous auxin and cytokinin. Figure 7 shows the time course of tracheary element formation from

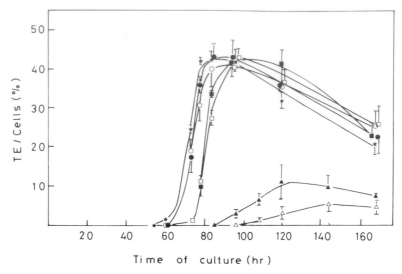

Fig. 7. Changes in ratios of tracheary element (TE) formation from single *Zinnia* mesophyll cells cultured in a liquid medium containing NAA and BA, which are added at various times of culture; both NAA and BA at the onset of culture (0 hr) (★), NAA at 0 hr and BA at 12 hr (○), BA at 0 hr and NAA at 12 hr (●), NAA at 0 hr and BA at 24 hr (□), BA at 0 hr and NAA at 24 hr (■), NAA at 0 hr and BA at 48 hr (△), and BA at 0 hr and NAA at 48 hr (▲).

Zinnia cells cultured in a liquid medium containing NAA, BA, or both, supplied at different times of culture. Cells which were cultured for 12 or 24 hr in a medium containing only NAA before the addition of BA differentiated in almost the same time sequence as those treated with BA and NAA in the reverse order, indicating that induction of cytodifferentiation occurs when both auxin and cytokinin are present and that the presence of both plant hormones may be essential for the initiation of cytodifferentiation. The effects of the two hormones on the sequence of cytodifferentiation seemed to be a little different in the *Zinnia* system. The presence of NAA up until the late stage of cytodifferentiation was necessary for the completion of tracheary element differentiation, while BA was a requisite during a brief period in the early stage. Minocha (1976) also showed that the BA requirement was restricted to the early stage of cytodifferentiation in cultured explants of Jerusalem artichoke tubers. Furthermore, he showed a rapid turnover in incorporated BA and suggested that this restriction in the BA requirement did not result from the carryover of incorporated BA but from the fact that it was unnecessary in the late stage of cytodifferentiation. These results indicate that the initiation of cytodifferentiation may be regulated by both auxin and cytokinin but that the succeeding progression of cytodifferentiation may be controlled mainly by auxin. However, this idea is still tentative because of ignorance surrounding the endogenous level of cytokinin.

Although the initial action of cytokinin and/or auxin to cytodifferentiation was often regarded as being the induction of cell division (Minocha, 1976; Phillips, 1980), this does not seem to be true, since direct differentiation without cell division was induced by both auxin and cytokinin as shown in *Zinnia* cells (Kohlenbach and Schmidt, 1975; Fukuda and Komamine, 1980b) and in *Brassica* protoplasts (Kohlenbach et al., 1982). What, then, is the primary action of these two plant hormones in cytodifferentiation? There is no direct evidence showing the primary role of auxin and cytokinin in cytodifferentiation. However, it has been reported that exogenous auxin can very rapidly initiate the expression of some specific genes (Zurfluh and Guilfoyle, 1982). It was shown that two differentiation-specific proteins were synthesized before morphological changes occurred in the presence of cytokinin together with auxin (Fukuda and Komamine, 1983), suggesting that the expression of some genes was induced by cytokinin and auxin during the early process of cytodifferentiation. Therefore, the primary hormonal action in cytodifferentiation may be considered to be the expression of some specific genes. The next important problem is whether the effect of auxin and cytokinin is additive or synergistic in gene expression related to the initiation of cytodifferentiation, or, in other words, whether the initial events of cytodifferentiation occur as a result of the cooperation of gene products expressed separately by auxin and cytokinin, or as a result of the products of one or a few differentiation-specific genes expressed only by simultaneous stimulation by the two hormones. The dual control of differentiation-specific gene expression by auxin and cytokinin should be investigated intensively.

In regard to the dual control of differentiation by auxin and cytokinin in plants, the following findings have been reported: Skoog and Miller (1957) indicated that differentiation of tobacco tissue in culture was determined by the cytokinin/auxin ratio in the medium, and high and low ratios led to shoot and root formation, respectively. In sycamore callus, a high cytokinin/auxin ratio was found to favor both xylem and phloem differentiation (Wright and Northcote, 1973). Different combinations of various kinds of auxin and cytokinin brought about different patterns of xylem formation in cultured lettuce pith explants (Dalessandro and Roberts, 1971).

2. Gibberellic Acid, Abscisic Acid, Ethylene, and cAMP

Cultured explants of Jerusalem artichoke tubers are materials that have often been used for the study of tracheary element differentiation. Application of 1 mg/liter gibberellic acid to these tissues in the presence of auxin and cytokinin was found to promote cytodifferentiation (Dalessandro, 1973). Phillips and Dodds (1977), using the same materials, confirmed this result, but the gibberellic acid concentration necessary for promotion was

20 times higher than that Dalessandro used. On the other hand, reports by Minocha and Halperin (1974) clearly showed marked inhibition of tracheary element differentiation by gibberellic acid in the range of 0.1–5 mg/liter in cultured artichoke explants. The discrepancy may reflect a difference in the endogenous levels of gibberellic acid, which is known to change during dormancy in tubers (Philips and Dodds, 1977). In bean callus, although gibberellic acid caused delays in auxin/cytokinin-induced tracheary element differentiation and in the increase of phenylalanine ammonia-lyase activity, it did not affect maximum tracheary element formation itself (Haddon and Northcote, 1976b). Gibberellic acid did not affect cytodifferentiation of *Zinnia* cells either (H. Fukuda and A. Komamine, unpublished data). Therefore, gibberellic acid does not seem to function directly in the induction of cytodifferentiation in cultured cells and tissues.

There are a few reports showing the effects of abscisic acid on cytodifferentiation. Abscisic acid prevented cytodifferentiation to tracheary element in artichoke explants (Minocha and Halperin, 1974) and in bean callus (Haddon and Northcote, 1976b). In bean callus, the process sensitive to abscisic acid was found to occur for 6 days before the appearance of tracheary elements (Haddon and Northcote, 1976b).

Wounding is closely involved in tracheary element differentiation as typically shown in wound vessel members. Cytodifferentiation in isolated cells or explants may also be affected by wounding. Ethylene production is one of the earliest reactions caused by wounding (Boller and Kende, 1980). Therefore, it would be worthwhile studying ethylene control in cytodifferentiation. Roberts and Baba (1978) and Miller and Roberts (1982, 1984) found that methionine, S-adenosylmethionine, and 1-aminocyclopropane-1-carboxylic acid, ethylene precursors, at low concentrations promoted tracheary element formation in lettuce pith explants and in soybean callus. Based on these results and the observation that inhibition of cytodifferentiation by ethylene inhibitors such as $AgNO_3$, $Co(NO_3)_2$ and aminoethoxyvinylglycine was reversed by the ethylene precursors and that there is parallelism between induction of cytodifferentiation and ethylene production, Roberts and Miller (1982) and Miller and Roberts (1984) suggested that ethylene has a positive role in xylogenesis. This idea was also supported by the fact that an ethylene-releasing agent, 2-chloroethylphosphonic acid, enhanced cytodifferentiation (Miller and Roberts, 1984). Interestingly, they also showed the partial substitution of ethylene precursors for cytokinin during auxin/cytokinin-induced tracheary element differentiation in lettuce (Miller *et al.*, 1984).

Cyclic AMP plays an important role as a secondary messenger following hormonal stimulus in animal cells. In higher plants, the existence of cAMP has been reported, but its function has not been clarified. The possibility that cAMP regulates cytodifferentiation to tracheary elements was first shown by Basile *et al.* (1973). They revealed that exogenous 8-bromo-cAMP as well as cytokinin could induce cytodifferentiation in the presence of auxin in lettuce pith explants and that application of theophillin, an inhibitor of phosphodiesterase, promoted cytodifferentiation. Results showing cAMP promoting cytodifferentiation were also obtained in carrot explants (Mizuno and Komamine, 1978b). Since in carrot explants exogenous cAMP raised endogenous cytokinin activity, cAMP may induce cytodifferentiation via cytokinin production and not act as a secondary messenger of cytokinin. Whether cAMP-controlling cytodifferentiation occurs in other materials needs to be investigated.

B. Other Factors Affecting Cytodifferentiation

There are many chemical and physical factors other than plant hormones that affect cytodifferentiation to tracheary elements, such as carbohydrates (Wetmore and Rier, 1963; Minocha and Halperin, 1974), inorganic nitrogen sources (Phillips and Dodds, 1977), amino acids (Roberts and Baba, 1968b), an unknown endogenous factor present in pine needles (Savidge and Wareing, 1981a), light (Mizuno *et al.*, 1971), temperature (Shininger, 1979b), and osmotic pressure (Fukuda and Komamine, 1980a). However, a detailed analysis of the inducing mechanism by such factors has never been made. Therefore, only a brief description of the effect of each such factor on cytodifferentiation, without intensive discussion, is provided here.

Sucrose, in addition to its role as an energy source, may act as a regulator of xylem differentiation in cultured tissues (Wetmore and Sorokin, 1955; Wetmore and Rier, 1963). The work of Jeffs and Northcote (1967) indicated that low and high concentrations of sucrose in combination with auxin favored xylem and phloem formation in bean callus, respectively. From the viewpoint of cytodifferentiation, however, the regulatory effect of sucrose is not always clear. The optimum concentration of sucrose for tracheary element formation is similar to that favoring cell proliferation, for example, 2% in artichoke explants (Phillips and Dodds, 1977) and 1–2% in isolated *Zinnia* cells (Fukuda and Komamine, 1980a), although there is one exception in that a high level of sucrose (8%) is effective for cytodifferentiation in *Parthenocissus* callus (Rier and Beslow, 1967). Substitutive carbohydrates for sucrose for the induction of cytodifferentiation have been found to be α-glucosyl disaccharides in bean callus (Jeffs and Northcote, 1967), glucose in carrot root explants (Mizuno *et al.*, 1971), glucose and trehalose in artichoke explants (Minocha and Halperin, 1974), and glucose, glycerol, and myo-inositol in lettuce explants (Roberts and Baba, 1982). Many of these substances are considered to be sources of energy in the medium, and differences in requirement may reflect differences in carbohydrate metabolism between species. In cytodifferentiation, therefore, carbohydrates, including sucrose, may act as energy sources rather than as regulating factors.

Phillips and Dodds (1977) reported that a reduction in the total nitrogen (NH_4^+ and NO_3^-) content of the medium gave an increase in the percentage of tracheary elements without affecting cell division in cultured explants of Jerusalem artichoke. In a culture of *Zinnia* mesophyll cells, NH_4^+ was found to be a regulating factor of cytodifferentiation (Fukuda and Komamine, 1980a). A low concentration of NH_4^+ (1 mM, about one-twentieth of that contained in the medium of Murashige and Skoog, 1962) stimulated cytodifferentiation, and the raised concentration gave a decrease in tracheary element formation but did not inhibit cell proliferation. The optimum concentration of NO_3^- for tracheary element formation, 20 mM, was almost the same as that for cell proliferation in *Zinnia* cells. These results seem to show that inorganic nitrogen, especially NH_4^+, may act directly in the induction of cytodifferentiation but not via cell division. Because a low level of NH_4^+ also favored the induction of embryogenesis in carrot suspension culture (Fujimura and Komamine, 1979), intracellular changes in nitrogen metabolism caused by the exogenous supply of a nitrogen source in cultured cells might play an important role in switching to a differentiation state from a quiescent state.

The effects of light on tracheary element differentiation vary between cultured tissues. Kleiber and Mohr (1967) suggested that tracheary element formation in *Sinapis* hypocotyls may be controlled by phytochrome. In carrot phloem slices, light was found to induce cytodifferentiation via the synthesis of cytokinin (Mizuno *et al.*, 1971; Mizuno and Komamine, 1978a).

Temperature is also important in tracheary element formation. Shininger (1978, 1979b) found in pea root segments that tracheary element differentiation was more sensitive to low temperatures than cell replication and that the early process of cytodifferentiation was more sensitive than the later one. A difference in the optimum temperature between that for cytodifferentiation and that for cell replication was also reported in cultured explants of artichoke tubers (Phillips and Dodds, 1977). These results suggested that temperature may be a tool separating a class of events associated with cytodifferentiation from a mixture including cell replication-associated events.

There are few reports showing the effects of pH in the medium on tracheary element differentiation. In *Zinnia* cells, the optimum pH in the medium for cytodifferentiation was found to be around 5.0 after autoclaving, and pH below or above 5.0 resulted in a decrease in the number of formed tracheary elements and also a delay in tracheary element formation. The pursuit of changes in pH in the medium during culture revealed that a rapid decrease in pH occurred 12–24 hr before tracheary elements appeared in every case, regardless of the initial pH (H. Fukuda, un-published data). The significance of this observation remains to be clari-fied.

C. The Relationship between Cytodifferentiation and the Cell Cycle

In multicellular organisms, cell replication and cytodifferentiation are like two wheels of a cart. Cell division yields a large number of cells, which differentiate to various mature forms with different tasks *in situ*. These two wheels should be regulated interdependently *in situ*. By using intact plants and animals, however, it is difficult to investigate the relationship between cytodifferentiation and cell replication. *In vitro* systems have offered a good tool for the study of this relationship. Cultured parenchyma cells differ-entiating to tracheary elements in plants as well as myoblasts (Bischoff and Holtzer, 1969) and melanocytes (Mayer, 1980) in animals are typical exam-ples (Torrey *et al.*, 1971; Roberts, 1976; Shininger, 1978; Dodds, 1979, 1981a; Phillips, 1980).

In this subsection, the regulation of tracheary element differentiation in relation to the cell cycle, with particular reference to the following impor-tant problems, is discussed.

1. Whether the cell cycle (DNA replication and/or mitosis) preceding cytodifferentiation is a prerequisite for tracheary element differentiation.

2. Whether the initiation of tracheary element differentiation occurs at a specific point in the cell cycle.

3. Whether the sequence of tracheary element differentiation is incom-patible or compatible with the progression of the cell cycle.

4. Whether minor DNA synthesis such as DNA amplification or satellite DNA synthesis plays an important role in the induction of tracheary ele-ment differentiation.

The first report suggesting a close relationship between tracheary ele-ment differentiation and the cell cycle was by Fosket (1968). He found that inhibitors of DNA synthesis such as mitomycin C, fluorodeoxyuridine, and an inhibitor of mitosis such as colchicine all prevented cytodifferentiation in cultured *Coleus* stem segments in which cell division preceded the ap-pearance of tracheary elements, suggesting that cell division is essential for the induction of tracheary element differentiation (Fosket, 1968, 1970). The analysis of cultured pea root explants with autoradiography using tritiated thymidine revealed that almost all tracheary elements were formed after DNA synthesis (Torrey and Fosket, 1970). In cultured explants of Jerusalem artichoke tubers, three rounds of cell cycle preceded cytodifferentiation and visible differentiation occurred from 7–10 hr after the last mitosis (Malawer and Phillips, 1979; Phillips, 1981b). The precedence of cell division over cytodifferentiation has been reported in many other cultured tissues, for example, lettuce cotyledons (Ronchi and Gregorini, 1970) and soybean callus (Fosket and Torrey, 1969). Furthermore, based on a kinetic analysis of cell proliferation and cytodifferentiation data during culture of pea root cortical explants (Phillips and Torrey, 1973), Meins (1974) stated that cell division is the rate-limiting step in tracheary element differentiation. It may be possible to consider the preceding cell cycle as a quantal cell cycle, which generates daughter cells that are different from the mother cell, in contrast to a proliferative cell cycle, which generates copies of the mother cell (Holtzer and Rubinstein, 1977).

On the other hand, there are many reports suggesting the occurrence of cytodifferentiation to tracheary elements without intervening cell division. The pioneering work of Foard and Haber (1961) using γ-irradiated wheat seedlings showed that cytodifferentiation occurred without DNA replication and mitosis. Similar results were obtained in cultured explants of γ-irradiated young artichoke tubers (Phillips, 1981a). Cytodifferentiation without a preceding cell division has also been indicated by an anatomical analysis of wound-induced tracheary element differentiation (Sinnott and Bloch, 1945; Robbertse and McCully, 1979) and by treatments with colchicine (Hammersley and McCully, 1980) or caffeine (Dalessandro and Roberts, 1971). However, we must take care to remember that experiments using metabolic inhibitors do not give direct evidence, because of their side effects. Plating experiments in a single-cell suspension of *Centaurea cyanus* revealed direct differentiation to single tracheary elements from single cells, although this occurred at low frequency. In this case, the possibility that single cells were committed to differentiation before plating was not ruled out. Kohlenbach and Schmidt (1975) reported that mature mesophyll cells of *Zinnia elegans* can redifferentiate to tracheary elements without cell division. Fukuda and Komamine (1980b) confirmed their finding by serial observation of the process of cytodifferentiation from single cells isolated from *Zinnia* leaves and provided the first direct evidence for cytodifferentiation without cell division (Fig. 8). This was further supported by an investigation with colchicine. Directly differentiating tracheary elements are the major type in the *Zinnia* system (60%).

Cytokinin, an essential factor in both cytodifferentiation and cell division, functions in the induction of both events with different dose-responses in pea root segments (Shininger, 1978) and *Zinnia* mesophyll cells (Fukuda and Komamine, 1982), that is, the minimum concentration of exogenous cytokinin necessary for the induction of cytodifferentiation is higher than that required for cell division. These results seem to indicate

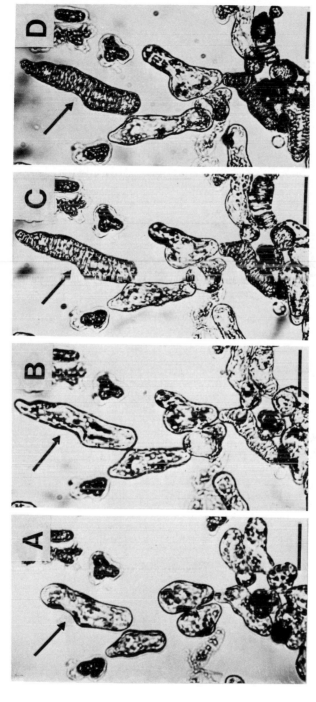

Fig. 8. Serial observation of the process of tracheary element differentiation from single cells isolated from the mesophyll of *Zinnia elegans*. Photographs were taken at 48 hr (A), 71 hr (B), 77 hr (C), and 96 hr (D) of culture. The bar represents 50 μm in each photograph. (From Fukuda and Komamine, 1980b.)

that tracheary element differentiation is not a simple and direct function of cell replication (Shininger, 1978). Furthermore, Shininger (1975) found that bromodeoxyuridine blocked cytodifferentiation without preventing cell replication. On the other hand, inhibition of cytodifferentiation by fluorodeoxyuridine has been reported in *Coleus* (Fosket, 1968), pea root segments (Shininger, 1975), and artichoke explants (Phillips, 1980). These results, according to Shininger (1978), showed that DNA replication preceding cytodifferentiation but not mitosis is essential and may function in the stabilization of transitory primary signals inducing cytodifferentiation at the molecular level.

In contrast, fluorodeoxyuridine treatments in cultured lettuce pith explants did not prevent tracheary element formation, suggesting that cytodifferentiation can occur without the preceding DNA synthesis (Turgeon, 1975). Direct evidence against the necessity of DNA replication was obtained in *Zinnia* single-cell system in which starting cells, which were exclusively in the G_1 phase, were shown (by using tritiated thymidine autoradiography and microspectrophotometry) to differentiate at high frequency to tracheary elements directly from the G_1 phase without intervening DNA synthesis (Fukuda and Komamine, 1981a). Did these contradictory findings about the relationship between the cell cycle and cytodifferentiation result from differences in species or tissues, or does any common rule lie behind these apparently different phenomena? Based on findings that direct differentiation without DNA replication occurred in γ-irradiated explants of immature and developing artichoke tubers, and that the capacity for direct differentiation declined with tuber maturity, Phillips (1981a) put forward an attractive hypothesis that the developmental plasticity of cells reaches a maximum immediately after cell division and declines with maturity into a parenchyma cell. Therefore, a cell in an immature state can respond directly to signals triggering cytodifferentiation, but one in a mature state loses the ability to respond directly and recovers the ability only after cell division. To elucidate whether direct differentiation in the *Zinnia* system can be explained according to this hypothesis, the differentiation ability of mesophyll cells derived from various ages of first leaves was investigated (H. Fukuda, I. Kawagishi, and A. Komamine, unpublished data). It was shown that the ability to differentiate to tracheary elements was highest in mesophyll cells isolated from mature leaves (14 days after sowing) and much lower in those from immature leaves (7 days) and senescent leaves (21 days), and the percentage of directly differentiated tracheary elements in total was almost the same in all cases. Therefore, the hypothesis offered by Phillips is not able to explain generally the mechanism of direct differentiation without DNA replication and/or mitosis.

For further understanding, we must ascertain the precise cell cycle time of the tracheary element precursor cells. Intensive work by Phillips and co-workers (Dodds and Phillips, 1977; Malawer and Phillips, 1979; Phillips, 1981b) has revealed that tracheary element precursor cells in artichoke explants start from the G_1 phase, spin along the cell cycle three times, and withdraw at the G_1 phase. In the *Zinnia* single-cell system, there are three types of tracheary elements: 60% are formed directly from the G_1 phase without an intervening S phase or mitosis (single tracheary elements), 30% are formed from the G_1 phase after one round of cell cycle (double tracheary elements), and less than 10% are formed, probably from the G_2 phase, after going through the S phase (Fukuda and Komamine, 1981a, cf. Fig. 10). In both the artichoke and the *Zinnia* systems, the cell cycle time of differentiating cells was almost the same as that of nondifferentiating cells.

Dodds (1981a,b) based on an idea of Vonderhaar and Topper (1974), stated that critical events for the induction of tracheary element differentiation occurred in the "early" G_1 phase. Parenchyma cells arrested in the early G_1 phase perceive stimuli inducing cytodifferentiation at that point and transform directly to tracheary elements as single tracheary elements in the *Zinnia* system, and cells arrested in the late G_1 phase do not perceive stimulus inducing cytodifferentiation until the early stage of the next G_1 phase after one round of cell cycle and then form tracheary elements as double tracheary elements in *Zinnia* and all tracheary elements in Jerusalem artichoke. For the latter cells, therefore, at least one round of cell cycle is a prerequisite for the induction of cytodifferentiation.

If Dodds's hypothesis is correct, the appearance of double tracheary elements that differentiate after one round of the cell cycle should be later than that of single tracheary elements that are initiated directly in the first G_1 phase. As shown in Fig. 9, however, these two types of tracheary elements appeared simultaneously after 58 hr of culture, and the pattern of increase in number of both types of tracheary elements was similar, showing that tracheary element precursor cells perceived the stimulus to differentiate simultaneously and differentiated at a constant time following the stimulus whether or not they went through the cell cycle (Fukuda and Komamine, 1981a). The results obtained in *Zinnia* cells are summarized in Fig. 10. The initiation of cytodifferentiation that may be controlled by auxin and cytokinin occurs in the first G_1 phase but not during the S phase or mitosis. The sequence of cytodifferentiation proceeds regardless of the cell cycle and can even proceed concurrently with the progression of the cell cycle, that is, cytodifferentiation is compatible with the cell cycle. Thereafter, there could be a second step at which differentiating cells withdraw from the cell cycle and are engaged in cytodifferentiation. This process may be irreversible. Also, making an assumption from the timing of mitosis, onset of the second step may occur, at the earliest, 20 hr before the ap-

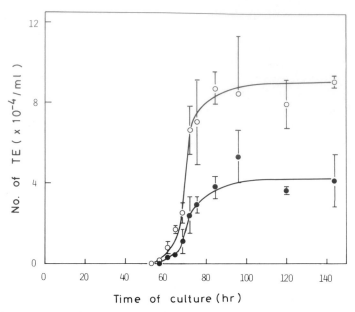

Fig. 9. Changes in number of single tracheary elements (○), which are formed without mitosis, and double tracheary elements (●), which are formed after undergoing mitosis and are composed of two cells, in a culture of single cells isolated from the *Zinnia* mesophyll. (From Fukuda and Komamine, 1981a.)

pearance of the secondary wall thickenings in *Zinnia* cells. Withdrawal from the cell cycle may be possible at any of the gap phases and, especially, the G_1 phase may be closely involved in withdrawing, because a higher frequency of withdrawal in the G_1 phase was observed in differentiating cells, which have been reported in cultured *Zinnia* mesophyll cells and

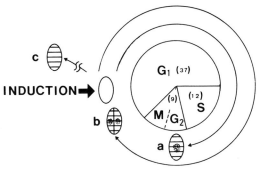

Fig. 10. Diagram of the relationship between cytodifferentiation into tracheary elements and the cell cycle in a culture of isolated single mesophyll cells of *Zinnia*. (From Fukuda and Komamine, 1981a.)

tuber cells of Jerusalem artichoke (Dodds and Phillips, 1977). In addition to this problem, the question of whether the initiation of cytodifferentiation is limited in the G_1 phase remains unanswered.

The hypothesis that the process of cytodifferentiation is independent of the progression of the cell cycle is not contradictory to the observation that many tracheary elements formed had already passed through cell division and/or DNA replication, as noted in Jerusalem artichoke (Dodds and Phillips, 1977; Malawer and Phillips, 1979; Phillips, 1981b), *Coleus* (Fosket, 1970), and pea systems (Shininger, 1978), if tracheary elements formed with intervening DNA synthesis and mitosis in these systems are regarded as "double tracheary elements" in the *Zinnia* system. If this is true, we may be able to consider in general that the induction of tracheary element differentiation may occur not after but before going along the cell cycle and the sequence of cytodifferentiation may proceed concurrently with the cell cycle. The most direct evidence can be provided by demonstrating an early marker of cytodifferentiation that continues to be produced as it goes along the cell cycle, such as melanin in melanocytes (Mayer, 1980). Cytochemical studies of cortical parenchyma cells of pea roots by Rana and Gahan (1982, 1983) have shown that esterase activity may be an early marker of xylem differentiation. Furthermore, they suggested that "switch-on" of programming of both esterase activity and secondary wall thickenings may occur prior to mitosis but that they are expressed after mitosis. This observation seems to support the idea mentioned above. Research into markers in the late stage incompatible with the cell cycle also would be important. Two newly synthesized proteins specific to cytodifferentiation found in cultured *Zinnia* mesophyll cells may become late marker proteins (Fukuda and Komamine, 1983).

DNA replication in the S phase does not seem to be necessary for cytodifferentiation. However, there have been some reports suggesting the involvement of minor DNA synthesis such as gene amplification and satellite DNA synthesis in differentiation (Nagl, 1976; Roberts, 1976). For example, differential replication of the DNA sequence has been observed during cartilage and neural retina differentiation (Strom *et al.*, 1978). In xylogenesis *in situ*, Avanzi *et al.* (1973) and Durante *et al.* (1977) reported amplification of ribosomal cistrons during the maturation of metaxylem in the root of *Allium cepa*, which indicates that only limited DNA families may be replicated in relation to cytodifferentiation.

In the *Zinnia* system, where DNA replication in the S phase was found to be unnecessary for cytodifferentiation, fluorodeoxyuridine, fluorouracil, mitomycin C, arabinosyl cytosine, and aphidicolin, which block DNA synthesis at different sites, all prevented cytodifferentiation at concentrations causing inhibition of the incorporation of tritiated thymidine into nucleic acid, suggesting the involvement of some DNA sequence in cytodifferen-

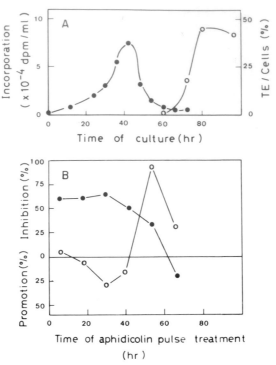

Fig. 11. (A) Time course of incorporation of tritiated thymidine into nucleic acid (●) and tracheary element (TE) formation (○) in a culture of isolated *Zinnia* mesophyll cells. (B) Inhibition or promotion of cell division (●) and tracheary element formation (○) by 12-hr pulse treatments of aphidicolin at various times of culture.

tiation (Fukuda and Komamine, 1981b). Figure 11 shows the effects of pulse-treated aphidicolin, which is a specific inhibitor of DNA polymerase-α (Ikegami *et al.*, 1978), on cell division and cytodifferentiation. Aphidicolin treatment between 48 and 60 hr of culture inhibited specifically the formation of both single and double tracheary elements, and the timing of the inhibition of cytodifferentiation was different from the S phase in double tracheary elements and from the timing of the inhibition of cell division. A preliminary analysis of the synthesized DNA by isopycnic centrifugation suggested possible synthesis of GC-rich satellite DNA at the time when aphidicolin prevented cytodifferentiation (Komamine and Fukuda, 1982). Satellite DNA synthesis during dedifferentiation has been reported in tobacco pith tissue (Parenti *et al.*, 1973) and in carrot root explants (Schäfer *et al.*, 1978; Hase *et al.*, 1979). These results may indicate that some satellite DNA may play a regulatory role in cytodifferentiation, although replication of the whole genome during the S phase is not a prerequisite for cytodifferentiation. From the timing of the synthesis of the satellite DNA involved in cytodifferentiation, it may be supposed that this minor DNA synthesis is not associated with the initiation of tracheary element differentiation but with a programming of the second step at which cytodifferentiation becomes irreversible. A small amount of satellite DNA synthesis occurring out of the S phase has also been observed in lily (Hotta and Stern, 1971).

5-Bromodeoxyuridine is an efficient tool for the study of the regulation of gene expression. This drug is incorporated into DNA and alters the structure of DNA (Luckner, 1982). It has been reported that bromodeoxyuridine inhibits the expression of the differentiated phenotype of several animal cell types (Bischoff and Holtzer, 1970; Levitt and Dorfman, 1972). Tracheary element differentiation from parenchyma cells in pea root (Shininger, 1975) and *Zinnia* mesophyll (H. Fukuda, unpublished data) is also prevented by this inhibitor, although cell division is not prevented. In developing chick cartilage, bromodeoxyuridine was preferentially incorporated into moderately repetitive DNA sequences, resulting in a blocking of cartilage differentiation without a blocking of cell division, total DNA synthesis, total RNA synthesis, and total protein synthesis (Levitt and Dorfman, 1972; Strom and Dorfman, 1976a). Furthermore, these specific sequences were found to be amplified during cartilage differentiation (Strom and Dorfman, 1976b; Strom et al., 1978). Therefore, it seems likely that the synthesis of specific DNA sequences but the replication of whole genomes is essential for cytodifferentiation. Recently, the association of extrachromosomal DNA with differentiation has been shown in pea root-tip cells (Krimer and Van't Hof, 1983). The inhibition of tracheary element formation by inhibitors of DNA synthesis observed in *Coleus* stem segments (Fosket, 1968) and pea root explants (Shininger, 1975) may not be a result of the blockage of progression of the S phase but may result from the prevention of the synthesis of some specific DNA sequences required for cytodifferentiation.

It may also be probable that the process of DNA repair is involved in cytodifferentiation, because 3-aminobenzamide, an inhibitor of ADP-ribosyltransferase, which is essential to DNA excision repair, prevented tracheary element differentiation without inhibitory effects on cell division in cultured artichoke explants (Hawkins and Phillips, 1983; Phillips and Hawkins, 1985) and *Zinnia* cells (M. Sugiyama and A. Komamine, unpublished data).

The role of endopolyploidy in cytodifferentiation has been a subject of debate for 20 years. It is unlikely that endopolyploidy functions positively in the induction of tracheary element differentiation, because cytodifferentiation occurs in haploid cells (DeMaggio, 1972; Kohlenbach et al., 1982), and the distribution of DNA content (2C, 4C, 8C, and 16C) in differentiat-

ing cells was found to be similar to that of nondifferentiating cells in pea
cultures (Phillips and Torrey, 1974), although the differentiation ability
may decrease with an increase in endopolyploidy in culture (Wright and
Northcote, 1973).

V. BIOCHEMICAL ASPECTS

Most of the studies on biochemical events occurring during tracheary
element differentiation have been made in relation to cell wall changes,
which are the most marked changes in the process of cytodifferentiation
(Northcote, 1979). The alteration of wall components during cytodifferentiation has been shown to be controlled by the induction or the repression
of the activities of various enzymes that occur at or after the onset of
secondary wall thickenings but not before it (Haddon and Northcote,
1976a; Kuboi and Yamada, 1978a; Bolwell and Northcote, 1981; Gahan,
1981b; Wardrop, 1981; Fukuda and Komamine, 1982). There are only a few
reports about biochemical changes preceding secondary wall formation,
although they are very important for the elucidation of the induction mechanism of cytodifferentiation or for research into biochemical markers of
cytodifferentiation at an early phase (Fosket and Miksche, 1966; Simpson
and Torrey, 1977; Fukuda and Komamine, 1983). In this section, the following biochemical events during cytodifferentiation are dsicussed: RNA
and protein synthesis preceding cytodifferentiation, lignin synthesis and
its related enzymes, polysaccharide synthesis and its related enzymes, and
other biochemical events.

A. RNA and Protein Synthesis Preceding
   Cytodifferentiation

Cytodifferentiation to tracheary elements was inhibited by actinomycin
D, an inhibitor of RNA synthesis, in cultured Jerusalem artichoke explants
(Phillips, 1980) and cultured *Zinnia* mesophyll cells (Fukuda and Komamine, 1983). Differing sensitivities to actinomycin D in different stages of
cytodifferentiation have also been observed in *Coleus* (Fosket and Miksche,
1966) and *Zinnia* cells (Fukuda and Komamine, 1983). In the latter, 12-hr
pulse treatments of actinomycin D between 24 and 60 hr of culture prevented tracheary element formation almost completely but treatments between 0 and 24 hr and between 60 and 72 hr resulted only in weak inhibi-

tion. Furthermore, a differentiation-specific increase in RNA synthesis between 24 and 60 hr was shown in the differentiation-induced culture in comparison with the control culture in which cell division occurred but cytodifferentiation did not (Fukuda and Komamine, 1983). Similarly, high activity of RNA synthesis during the early phase of xylem differentiation has been suggested in roots of *Allium cepa* (Jensen, 1957) and cultured pea root cortical explants (Shininger, 1980).

The next question is, to which species of RNA is the high activity of RNA synthesis during cytodifferentiation due? Shininger (1980) showed that major RNA synthesized preceding DNA synthesis, and cytodifferentiation in cultured pea root explants may be mRNA immediately after the onset of RNA synthesis and rRNA thereafter. The occurrence of a specific increase in rRNA synthesis during cytodifferentiation to tracheary elements is supported by the observation that cistrons coding for rRNA are amplified during the maturation of metaxylem in the roots of *Allim cepa* (Avanzi *et al.*, 1973; Durante *et al.*, 1977). However, the possibility that an increase in the synthesis of rRNA is associated with cell division and not with cytodifferentiation cannot be ruled out, because tracheary element differentiation in the tissues used was accompanied by cell division. It is also probable that mRNA coding for some differentiation-specific proteins, for example lignin- or polysaccharide-synthetic enzymes, are transcribed in the period of active RNA synthesis during cytodifferentiation.

A requirement for protein synthesis in cytodifferentiation was shown in cultured *Zinnia* cells using cycloheximide, an inhibitor of protein synthesis, and the period when protein necessary for cytodifferentiation was synthesized (24–60 hr of culture) coincided with that of active protein and RNA syntheses (Fukuda and Komamine, 1983). In this period, however, changes in the amounts of protein and the rate of [14C]leucine incorporation into protein in the differentiation-induced culture of *Zinnia* were almost the same as those in the control culture (Fukuda and Komamine, 1983). Similar results were obtained in cultured pea root explants (Simpson and Torrey, 1977).

An investigation of proteins extracted from [35S]methionine-labeled cells using two-dimensional electrophoresis revealed that most of proteins synthesized in differentiating cells preceding morphological changes were common to those in nondifferentiating cells and would function as "housekeeping" proteins responsible for general metabolism, cell division, and so on (Fukuda and Komamine, 1983). Proteins associated with tracheary element differentiation were limited to a few as shown in Fig. 12, that is, two differentiation-specific proteins were newly synthesized and the synthesis of two other proteins was shut off only in differentiation-induced culture between 48 and 60 hr, when protein synthesis is a prerequisite for cytodifferentiation.

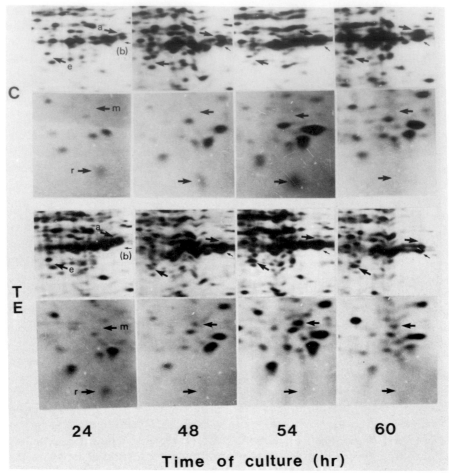

Fig. 12. Changes in polypeptides in differentiating (TE) or nondifferentiating cells (C) during culture period. *Zinnia* cells cultured for 23, 47, 53, or 59 hr in the tracheary element-inductive or the control medium were labeled with 1.5×10^{-6} Bq/ml [^{35}S]methionine for 2 hr. Exracted proteins were separated by two-dimensional electrophoresis. Note that polypeptides "a" and "r" were preferentially synthesized in nondifferentiating cells and "e" and "m" were in differentiating cells between 48 and 60 hr of culture. (From Fukuda and Komamine, 1983.)

In somatic embryogenesis of carrot, only two embryo-specific proteins have been detected by means of two-dimensional electrophoresis (Sung and Okimoto, 1981). The work of Fujimura and Komamine (1982) also showed that only a few different species of translatable mRNA were detectable at an early stage of embryogenesis in comparison with the control culture. Thus, there may be very few differentiation-specific genes expressed in an early stage, but their products, a few species of proteins, may play regulatory roles in differentiation. The functions of the differentiation-specific proteins, such as those shown in Fig. 12, are unknown at present, but it is certain that they can be regarded as biochemical markers preceding morphological changes for cytodifferentiation to tracheary elements.

B. Lignin Synthesis

Since cytodifferentiation to tracheary elements is associated with lignification, lignin is a characteristic biochemical marker of cytodifferentiation. The biosynthesis of lignin involves phenylalanine ammonia-lyase (PAL) at the initial step, four enzymes (hydroxylases, O-methyltransferases, CoA-ligases, and dehydrogenases) at the following steps, and peroxidase at the final step (see review by Hahlbrock and Grisebach, 1979).

PAL, which produces cinnamic acid from phenylalanine, is a key enzyme for the supply of lignin precursors and has been shown to be correlated with tracheary element differentiation (Rubery and Northcote, 1968; Rubery and Fosket, 1969; Haddon and Northcote, 1975; Durst, 1976; Kuboi and Yamada, 1978a; Fukuda and Komamine, 1982), although some reports suggested a negative correlation of changes in PAL activity with tracheary element differentiation (Minocha and Halperin, 1976). Figure 13 shows changes in PAL activity during cytodifferentiation in *Zinnia* cells. Although the first peak of PAL activity, which is probably a response to injury, is nonspecific for cytodifferentiation, the second peak is specific for differentiating cells and coincides with the time of active lignin synthesis, as shown in previous reports (Haddon and Northcote, 1975; Kuboi and Yamada, 1978a). In tissue localization, PAL activity was also correlated with xylem differentiation (Rubery and Northcote, 1968). Furthermore, the close association of PAL activity with the ability to form xylem tissue has been reported in bean callus and suspension cells subcultured continuously in a maintenance medium (Haddon and Northcote, 1976c; Bevan and Northcote, 1979b). From these results we can assume that PAL is a useful marker of tracheary element differentiation. It is probable that PAL activity during cytodifferentiation is regulated in the level of transcription, on the analogy of the control mechanism of PAL activity in light induced anthocyanin synthesis (Hahlbrock and Grisebach, 1979).

It is important for elucidation of the induction mechanism of cytodifferentiation to investigate whether cellulose deposition and lignification, which are two major biochemical events in secondary wall formation, are induced simultaneously by the same stimulus or separately by different stimuli. Separate induction is suggested by the fact that 75% of tobacco cells cultured on a medium containing 1 mg/liter zeatin and 0.1 mg/liter 2,4-D were

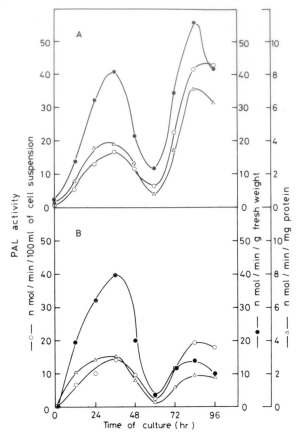

Fig. 13. Changes in phenylalanine ammonia-lyase (PAL) activity in isolated *Zinnia* mesophyll cells cultured in the tracheary element-inductive medium (A) and the control medium (B). (From Fukuda and Komamine, 1982.)

lignified, but cells differentiated to tracheary elements took up only a small percentage (K. Tateoka, personal communication). Also, mature tracheary elements with localized cellulose deposition but not lignin deposition occupied a given population in the *Zinnia* system, although the percentage of such tracheary elements in relation to the total population was small (Fukuda and Komamine, 1982). Clearer elucidation of the interrelationship between cellulose and lignin synthesis may be brought about by using a specific inhibitor of PAL, 2-aminooxy-3-phenylpropionic acid (Holländer *et al.*, 1979). Glyphosate, which interferes with the phenylpropanoid pathway (Holländer and Amrhein, 1980), or 2,6-dichlorobenzonitrile, an inhibitor of cellulose synthesis (Hogetsu *et al.*, 1974), may also be efficient tools for elucidating the interrelationship.

Kuboi and Yamada (1978a) indicated that the activities of shikimate dehydrogenase, cinnamate hydroxylase, caffeic acid-O-methyltransferase, and 5-hydroxyferulic acid-O-methyltransferase as well as PAL increased coordinately with cytodifferentiation in tobacco suspension culture. Caffeic acid-O-methyltransferase was also found to be involved in lignin synthesis during xylem differentiation in bean callus, in which changes in activities of this enzyme and PAL occurred coordinately under various growth conditions, suggesting that these two enzymes may be induced together to bring about the coordinated synthesis necessary for cytodifferentiation (Haddon and Northcote, 1976a). In the *Zinnia* system, however, an increase in O-methyltransferase activity was not correlated with lignin synthesis during cytodifferentiation (Fukuda and Komamine, 1982). This may indicate that lignin synthesized in *Zinnia* cells is methoxylated at a low level. We have little information about changes in CoA-ligases and dehydrogenases during tracheary element differentiation (Haddon and Northcote, 1976a). β-Glucosidases in the cell wall may also be involved in lignin synthesis during cytodifferentiation (Hösel *et al.*, 1982)

The last step of lignification, polymerization of aromatic alcohols, is catalyzed by peroxidase (Elstner and Heupel, 1976; Gross *et al.*, 1977; Mäder *et al.*, 1980). There have been conflicting reports about whether the peroxidase reaction is a rate-limiting step of lignification during tracheary element differentiation (Haddon and Northcote, 1976a; Kuboi and Yamada, 1978a). It should be noted, however, that many reports have focused only on peroxidase soluble in buffer. It has been shown biochemically (Mäder *et al.*, 1980) and cytochemically (Hepler *et al.*, 1972) that peroxidase bound to cell walls can function in lignification. Therefore, fractionation of peroxidase is necessary for determining whether or not peroxidase is really related to cytodifferentiation. Minocha and Halperin (1976) divided peroxidase activities into three fractions, one soluble in buffer, one extractable with $CaCl_2$ from cell walls, and one nonextractable with $CaCl_2$ but extractable from cell walls with wall-degrading enzymes. Their conclusion was that neither fraction was concerned with lignification during cytodifferentiation in cultured Jerusalem artichoke explants. However, the result obtained using wall-degrading enzymes is not relible, because the low activity of the enzyme can be considered to be a result of proteolysis of peroxidase through long treatment with wall-degrading enzymes (Masuda *et al*, 1983). On the other hand, in cultured *Zinnia* cells, the activities of wall-bound peroxidase extractable and nonextractable with $CaCl_2$ increased markedly in the early and middle stages, respectively, of lignification only in differentiating cells, whereas an increase in soluble peroxidase was not specific for differentiation (Fig. 14; Fukuda and Komamine, 1982; Masuda *et al.*, 1983).

Analysis of peroxidase isoenzyme patterns in the *Zinnia* system revealed

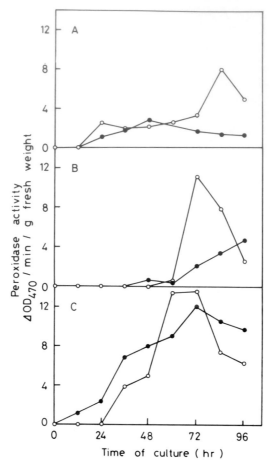

Fig. 14. Changes in peroxidase activity in isolated *Zinnia* mesophyll cells cultured in the tracheary element-inductive (○) and the control medium (●). (A) CaCl₂-nonextractable enzyme. (B) CaCl₂-extractable enzyme, (C) Soluble enzyme. (From Fukuda and Komamine, 1982.)

that in cathodic isoenzymes of CaCl₂-extractable peroxidase, a rapidly migrating isoenzyme (1, in Fig. 15) appeared at 84 hr of culture only in differentiating cells when lignin synthesis was most active, and other isoenzymes exhibited differentiation-associated increases in their activities in the various stages of cytodifferentiation (Masuda *et al.*, 1983). Mäder *et al.* (1980) reported for tobacco callus that two different peroxidase isoenzyme groups in cell walls play different roles in lignification, that is, the production of H_2O_2 and the polymerization of aromatic alcohols at the expense of H_2O_2. Therefore, it is probable that different isoenzymes of wall-bound peroxidase with different substrate-specificity may be induced at different stages of cytodifferentiation.

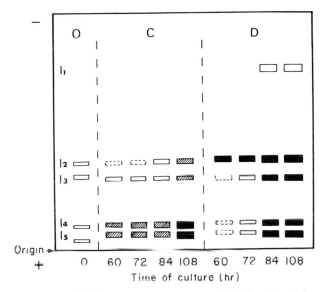

Fig. 15. Time course of cathodic isoenzyme patterns of CaCl$_2$-extractable peroxidase. D, Tracheary element-inductive culture; C, control culture; O, originally isolated mesophyll cells before culture. (From Masuda *et al.*, 1983.)

C. Polysaccharide Synthesis

During secondary wall thickening, there is an increase in the production of cellulose and hemicellulose and a cessation of pectin synthesis (North cote, 1963). Jeffs and Northcote (1966) reported that the ratios of the concentrations of xylose, which is a major component of hemicellulose in angiosperms, to arabinose, which is a major component of pectin, increased in xylem-differentiating bean callus, suggesting that enzymes involved in the synthesis of these wall materials may be regulated precisely during tracheary element differentiation.

Very little information is available about changes in cellulose synthetase activity during cytodifferentiation. As for hemicellulose, a principal precursor, UDP-xylose, is formed from UDP-glucose by UDP-glucose dehydrogenase and UDP-glucuronate decarboxylase. The activities of these enzymes have been shown to increase during xylem differentiation in sycamore and poplar cambium, and cultured Jerusalem artichoke explants (Dalessandro and Northcote, 1977a,c). This would bring about a greater flux of carbohydrates from UDP-glucose to give large amounts of UDP-xylose, which is required for the increase in hemicellulose during secondary wall formation (Northcote, 1979). Hydrogenase activity is also regulated by a feedback control according to the demand by the cell for UDP-xylose at any particular time (Northcote, 1979). In addition, hemicellulose synthesis during xylem differentiation is regulated by the induction or repression of synthetase activity that exists in membrane preparation (Dalessandro and Northcote, 1981a,b; Bolwell and Northcote, 1981). Measurement of changes in xylan synthetase activity during xylem differentiation in bean hypocotyl and callus showed that xylan synthetase activity was induced simultaneously with PAL activity during the period of secondary wall thickening (Bolwell and Northcote, 1981).

UDP-Galactose, UDP-galacturonic acid, and UDP-arabinose, the precursors of pectin, are formed by means of epimerases from UDP-glucose, UDP-glucuronic acid, and UDP-xylose, respectively (Northcote, 1979). Therefore, we would expect that any repression of pectin synthesis at the onset of secondary wall formation would be due to the decrease in epimerase activities. However, this is not so. Epimerase activities are maintained at a constant level during differentiation (Dalessandro and Northcote, 1977a,b,d). In contrast, arabinan synthetase activity in membrane preparation from bean hypocotyl and callus decreased remarkably at the onset of secondary wall thickening, although it was at a high level during cell division, and elongation preceding thickening, suggesting that the principal control of pectin synthesis during cytodifferentiation occurs at the synthetase reactions (Bolwell and Northcote, 1981).

Thus, polysaccharide synthesis during tracheary element differentiation is controlled through the induction or the regression of synthetase activity in membrane preparation as well as dehydrogenase and decarboxylase activity. This is also supported by the fact that division and tracheary element differentiation in cultured lettuce pith explants were accompanied by a remarkable increase in the incorporation of [^{14}C]glucose into polysaccharides in isolated membrane fractions composed of the endoplasmic reticulum and Golgi apparatus (Wright and Bowles, 1974).

D. Other Biochemical Events

The final step in tracheary element maturation, cell autolysis, is associated with an increase in the activities of hydrolytic enzymes, most of which exist in the vacuole or lysosomes (Gahan, 1978). Gahan and Maple (1966) showed the involvement of acid phosphatase in autolysis by histochemical analysis of developing protoxylem elements of *Vicia faba*. High activity of this enzyme has been reported in wound-induced tracheary elements in *Coleus* (Jones and Villiers, 1972) and differentiating water-conducting elements of various mosses (Hébant, 1973). In addition to soluble hydrolytic enzyme, plasma membrane-bound K^+-activated acyl phosphatase and endoplasmic reticulum-bound glucose-6-phosphatase may also be associated with cytodifferentiation (Gahan, 1978).

Cytochemical analysis of esterase activity using naphthol AS-D acetate as a substrate revealed that this enzyme was induced tissue specifically in pea roots, for example, at a high level in the root cap, rhizodermis, and stele, especially in vascular bundles, but at a low level in the meristem and cortex (Gahan, 1981a). In wound-induced xylem formation in the cortex of pea roots, induced high esterase activity indicated a region of future xylem formation (Rana and Gahan, 1983). This shows that esterase may be used as a marker of cytodifferentiation to tracheary elements at an early stage. Rana and Gahan (1983) suggested that in wound-induced xylem differentiation, determination of the induction of esterase activity and of the secondary wall thickenings occurred by 10 and 20 hr after wounding, respectively, and each determination preceded mitosis. This esterase was found to be a carboxylesterase (Gahan et al., 1983).

VI. CONCLUSIONS AND PERSPECTIVES

Figure 16 is a summary of various events occurring in the process of cytodifferentiation to tracheary elements from isolated mesophyll parenchyma cells of Zinnia (Fukuda and Komamine, 1980a,b, 1981a,b, 1982, 1983; Komamine and Fukuda, 1982; Masuda et al., 1983).

In the in vitro system, initiators of cytodifferentiation may be wounding and a cooperative action of auxin and cytokinin (Roberts, 1976; Fukuda and Komamine, 1980a). The processes following after receiving stimuli can be divided roughly into reversible and irreversible. A little before the onset of secondary wall thickening, the reversible process induced by wounding and phytohormones transits to the irreversible one, engaged in an intracellular reorganization that leads to death via the formation of precisely patterned secondary walls. Rana and Gahan (1983) suggested that wound-induced xylem differentiation of pea roots may proceed in two steps. The first process is from cortical parenchyma cells to stelelike cells characterized by their high esterase activity and subsequently the second process from stelelike cells to tracheary elements occurs. These two processes might correspond to the reversible and irreversible process, respectively.

Cytodifferentiation can proceed concurrently with the progression of the cell cycle during the reversible process of cytodifferentiation (Fukuda and Komamine, 1980b, 1981a). Mitosis and DNA replication during the S phase do not seem to be prerequisites for the induction of cytodifferentiation,

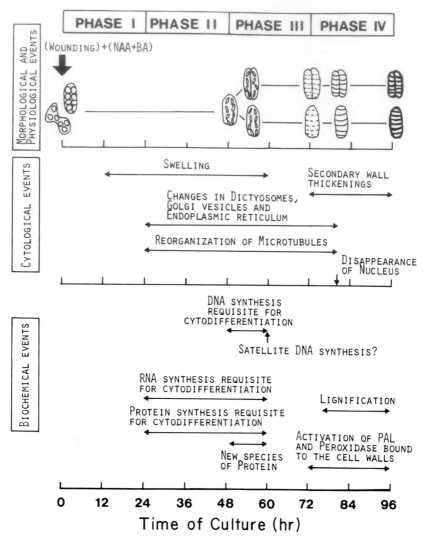

Fig. 16. Sequential events in the process of tracheary element cytodifferentiation from isolated single mesophyll cells of *Zinnia elegans*.

although it may be important for cytodifferentiation that cells perceive initial stimulation to differentiation at the G_1 phase (Dodds and Phillips, 1977; Fukuda and Komamine, 1981a).

The process of tracheary element cytodifferentiation is composed of four phases, phase I (0–24 hr of culture), II (24–48 hr), III (48–72 hr) and IV (72–96 hr), in the *Zinnia* system.

Phase I is a lag period. The first half of the phase proceeds even in the absence of phytohormones. In the latter half, for the progression of which auxin and cytokinin are required, swelling and an increase in RNA and protein synthesis begin (Fukuda and Komamine, 1983). Markers for cytodifferentiation in this phase have not yet been discovered.

Phase II is the period of the highest level of cell activity. Increases in the number of intracellular organelles and in protein and RNA synthesis are observed (Fukuda and Komamine, 1983). However, these increases seem to reflect mainly the activation of general "housekeeping" metabolism and there seems to be little activation of cytodifferentiation-specific events. In this phase, no clear marker for differentiation has been detected.

Phase III is a critical period for tracheary element differentiation. High metabolic activity is still kept up in the first half of this phase, when critical events seem to occur leading to the irreversible process of cytodifferentiation. Differentiation-specific proteins (18K and 36K) are synthesized during the first half of this phase, and these can be regarded as marker proteins for cytodifferentiation (Fukuda and Komamine, 1983). The first half of this phase is also very sensitive to any inhibitors of DNA synthesis, and satellite DNA that is synthesized specifically in this period may be associated with switching to the irreversible process (Komamine and Fukuda, 1982). In the latter half of phase III, RNA and protein synthesis decrease markedly, although this decrease does not always mean the repression of differentiation-specific metabolism. Throughout this phase (the first half and the latter half), dynamic changes in intracellular structure occur in order to give rise to localized secondary wall formation.

Phase IV starts from the onset of thickening of the sculptured secondary walls. Several hours after the appearance of detectable wall thickenings, the deposition of lignin into cell walls and cell autolysis involved in the loss of nuclei and cytoplasm occur (Fukuda and Komamine, 1982). Differentiation is completed around 96 hr of culture in the Zinnia system, resulting in the formation of conducting elements that have precisely patterned and lignified secondary walls. Increases in dictyosomes, Golgi vesicles, and the endoplasmic reticulum, and changes in their localization, are associated with secondary wall formation. Microtubules, whose array and density alter at the transition from phase III to IV, seem to play an important role in the spatial control of cytodifferentiation through the orientation of cellulose microfibrils, the movement or organelles, and so on. The process of autolysis has not been analyzed extensively. In addition to a characteristic morphological marker such as patterned secondary walls, in this final phase of cytodifferentiation, lignin deposition, phenylalanine ammonialysase and wall-bound peroxidase can be regarded as biochemical markers for cytodifferentiation (Fukuda and Komamine, 1982; Masuda et al., 1983). Other enzymes related to lignin synthesis, such as cinnamate hydroxylase (Kuboi and Yamada, 1978a), xylan synthetase (Bolwell and Northcote,

1981), and acid phosphatases (Gahan, 1978), may also be considered as marker proteins.

As described above, the information that we have obtained about cytodifferentiation until now is limited and is only fragmentary about events whose interrelationships are not clear. We are ignorant of the possible induction mechanism of cytodifferentiation and also of the key event that initiates the irreversible process of tracheary element differentiation. However, recent developments in some fields in developmental biology are throwing a new light on the investigation of cytodifferentiation.

As for the induction mechanism of cytodifferentiation, some suggestions may come from studies of the primary action of phytohormones. Two-dimensional gel electrophoretic analysis of *in vitro* translation products of isolated mRNA revealed that auxin-treatments gave rise to a selective increase in some mRNA sequences within a few minutes (Theologis and Ray, 1982; Zurfluh and Guilfoyle, 1982). A highly selective effect of auxin on the expresssion of a small number of mRNA sequences has also been indicated by RNA blot hybridization using auxin-responsive cDNA clones (Walker and Key, 1982; Hagen *et al.*, 1984). Although very little information is known about the selective effects of cytokinin on gene expression, it is probable that a primary action of cytokinin is involved in the activation of specific genes. In addition, we may also look at spatial control by phytohormones. Recently, Simmonds *et al.* (1983), using a *Vicia* suspension culture, showed a close association between auxin and/or gibberellin action and altered microtubule distribution, although a causal relationship has not been established. Possible regulation of microtubule organization by a phytohormone has also been suggested by Durnam and Jones (1982) and Mita and Shibaoka (1984). Through the control of microtubule organization, phytohormones might regulate morphological changes as well as cell wall synthesis, cell division, and elongation.

Recent studies on crown gall tumorigenesis are bringing forth essential information about hormonal control of plant morphogenesis (Leemans *et al.*, 1982; Akiyoshi *et al.*, 1983; Ream *et al.*, 1983). When the transferred (T) region of Ti plasmid harbored by *Agrobacterium tumefaciens* is integrated into the plant cell genome, plant cells become capable of growth in culture on a basal medium without added auxin and cytokinin and form crown gall tumors (Braun, 1958; Zaenen *et al.*, 1974). Recent studies showed that genetic loci in the T region affect tumor morphology such as the formation of shoots, roots, and abnormally large tumors, and these loci may play a role in cytokinin and auxin metabolism in host cells (Akiyoshi *et al.*, 1983; Ream *et al.*, 1983). Further genetic analysis suggested that two of the genes thought to be involved in phytohormone metabolism code an isopentenyl transferase that catalyzes the first step in cytokinin biosynthesis (Barry *et al.*, 1984) and an indoleacetamide hydrolase that converts indoleacetamide to indoleacetic acid, the natural auxin (Thomashow *et al.*, 1984; Inzé *et al.* (1984). Not only do these data suggest that plant morphogenesis may be controlled by endogenous levels of auxin and cytokinin, but they also offer a clue for the elucidation of the regulatory mechanism of endogenous levels of phytohormones during morphogenesis. We should pay attention also to the plasma membrane function in hormonal control of cytodifferentiation, since some kind of auxin receptors should be located on the plasma membrane (Batt and Venis, 1976; Vreugdenhil *et al.*, 1980).

Another important problem is to clarify the factor that initiates the transition from the reversible process to the irreversible process of tracheary element differentiation. It has been shown that in this transitory period minor DNA synthesis might relate the progression of cytodifferentiation to tracheary elements (Fukuda and Komamine, 1981b; Komamine and Fukuda, 1982). Levine *et al.* (1981) described that one of the possible regulatory processes of gene expression during development in eukaryotes is change in the gemonic DNA: (1) rearrangement of gene as shown in the mammalian immune system; (2) insertion or deletion of transposable DNA segments at various genomic sites of eukaryotes; (3) amplification of selective genes; and (4) modification of particular genes, typically by methylation of DNA. Since these changes in genomic DNA are often irreversible, it might be possible to assume that changes in organization of particular DNA sequences become a trigger of the irreversible process of cytodifferentiation.

Finally, one should remember that an application of new ideas obtained in other fields to the investigation of cytodifferentiation in higher plants will not be possible unless an adequate experimental system is used, and therefore further development of *in vitro* culture systems will be important in elucidating the mechanism of cytodifferentiation.

REFERENCES

Akiyoshi, D. E., Morris, R. O. Hinz, R., Mischke, B. S., Kosuge, T., Garfinkel, D. J., Gordon, M. P., and Nester, E. W. (1983). Cytokinin/auxin balance in crown gall tumors is regulated by specific loci in the T-DNA. *Proc. Natl. Acad. Sci. U.S.A.* **80**, 407–411.

Albinger, G., and Beiderbeck, T. (1983). Differentiation of tracheary elements and storage parenchyma cells in suspension cultures of *Asparagus plumosus* Baker. *Z. Pflanzenphysiol.* **112**, 443–448.

Avanzi, S., Maggini, F., and Innocenti, A. M. (1973). Amplification of ribosomal cistrons during the maturation of metaxylem in the root of *Allium cepa*. *Protoplasma* **76**, 197–210.

Ball, E., and Joshi, P. C. (1965). Division in isolated cells of palisade parenchyma of *Arachis hypogaea*. *Nature (London)* **207**, 213–214.

Barnett, J. R. (1977a). Tracheid differentiation in *Pinus radiata*. *Wood Sci. Technol.* **11**, 83–92.
Barnett, J. R. (1977b). Fine structure of parenchymatous and differentiated *Pinus radiata* callus. *Ann. Bot. (London)* **42**, 367–373.
Barnett, J. R. (1979). Current research into tracheary element formation. *Curr. Adv. Plant Sci.* **11**, 33.1–33.13.
Barnett, J. R. (1981). "Xylem Cell Development." Castle House Publ. Ltd., Tunbridge Wells, England.
Barry, G. F., Rogers, S. G., Fraley, R. T., and Brand, L. (1984). Identification of a cloned cytokinin biosynthesis gene. *Proc. Natl. Acad. Sci. U.S.A.* **81**, 4776–4780.
Basile, D. V., Wood, H. N., and Braun, A. C.(1973). Programming of cells for death under defined experimental conditions: Relevance to the tumor problem. *Proc. Natl. Acad. Sci. U.S.A.* **70**, 3055–3059.
Batt, S., and Venis, M. A. (1976). Separation and localization of two classes of auxin binding-sites in corn coleoptiles. *Planta* **130**, 15–21.
Bergmann, L. (1964). Der einfluss von kinetin auf die ligninbildung und Differenzierung im gewebekulturen von *Nicotiana tabacum*. *Planta* **62**, 211–254.
Bevan, M., and Northcote, D. H. (1979a). The interaction of auxin and cytokinin in induction of phenylalanine ammonia-lyase in suspension cultures of *Phaseolus vulgaris*. *Planta* **147**, 77–81.
Bevan, M., and Northcote, D. H. (1979b). The loss of morphogenic potential and induction of phenylalanine ammonia-lyase in suspension cultures of *Phaseolus vulgaris*. *J. Cell Sci.* **39**, 339–353.
Bischoff, R., and Holtzer, H. (1969). Mitosis and the processes of differentiation of myogenic cells *in vitro*. *J. Cell Biol.* **41**, 188–200.
Bischoff, R., and Holtzer, H. (1970). Inhibition of myoblast fusion after one round of DNA synthesis in 5-bromodeoxyuridine. *J. Cell Biol.* **44**, 134–150.
Boller, T., and Kende, l. (1980). Regulation of wound ethylene synthesis in plants. *Nature (London)* **286**, 259–260.
Bolwell, G. P., and Northcote, D. H. (1981). Control of hemicellulose and pectin synthesis during differentiation of vascular tissue in bean (*Phaseolus vulgaris*) callus and in bean hypocotyl. *Planta* **152**, 225–233.
Bornman, C. H., and Ellis, R. P. (1971). Differentiation of pseudothallial tracheary elements in *Nicotiana tabacum* tissue cultured *in vitro*. *J. S. Afr. Bot.* **37**, 281–289.
Braun, A. C. (1958). A physiological basis for the autonomous growth of the crown gall tumor cell. *Proc. Natl. Acad. Sci. U.S.A.* **44**, 344–349.
Brower, D. L., and Hepler, P. K. (1976). Microtubules and secondary wall deposition in xylem: The effects of isopropyl N-phenylcarbamate. *Protoplasma* **87**, 91–111.
Burgess, J., and Linstead, P. (1984). *In vitro* tracheary element formation: Structural studies and the effect of triiodobenzoic acid. *Planta* **160**, 481–489.
Cleveland, D. W. (1983). The tubulins: From DNA to RNA to protein and back again. *Cell* **34**, 330–332.
Clutter, M. E. (1960). Hormonal induction of vascular tissue in tobacco pith *in vitro*. *Science* **132**, 548–549.
Comer, A. E. (1978). Pattern of cell division and wound vessel member differentiation in *Coleus* pith explants. *Plant Physiol.* **62**, 354–359.
Cronshaw, J. (1965). Cytoplasmic fine structure and cell wall development in differentiating xylem elements. *In* "Cellular Ultrastructure of Woody Plants" (W. A. Côté, Jr., ed.), pp. 99–124. Syracuse Univ. Press, Syracuse, New York.
Cronshaw, J. (1967). Tracheid differentiation in tobacco pith cultures. *Planta* **72**, 78–90.
Cronshaw, J., and Bouck, G. B. (1965). The fine structure of differentiating xylem elements. *J. Cell Biol.* **24**, 415–431.

Dalessandro, G. (1973). Interaction of auxin, cytokinin and gibberellin on cell division and xylem differentiation in cultured explants of Jerusalem artichoke. *Plant Cell Physiol.* **14,** 1167–1176.

Dalessandro, G., and Northcote, D. H. (1977a). Changes in enzymic activities of nucleoside diphosphate sugar interconversions during differentiation of cambium to xylem in sycamore and poplar. *Biochem. J.* **162,** 267–279.

Dalessandro, G., and Northcote, D. H. (1977b). Changes in enzymic activities of nucleoside diphosphate sugar interconversions during differentiation of cambium to xylem in pine and fir. *Biochem. J.* **162,** 281–288.

Dalessandro, G., and Northcote, D. H. (1977c). Changes in enzymic activities of UDP-D-glucuronate decarboxylase and UDP-D-xylose 4-epimerase during cell division and xylem differentiation in cultured explants of Jerusalem artichoke. *Phytochemistry* **16,** 853–859.

Dalessandro, G., and Northcote, D. H. (1977d). Possible control of polysaccharide synthesis during cell growth and wall expansion of pea seedlings. *Planta* **134,** 39–44.

Dalessandro, G., and Northcote, D. H. (1981a). Xylan synthetase activity in differentiated xylem cells of sycamore trees (*Acer pseudoplatanus*). *Planta* **151,** 53–60.

Dalessandro, G., and Northcote, D. H. (1981b). Increase of xylan synthetase activity during xylem differentiation of the vascular cambium of sycamore and poplar trees. *Planta* **151,** 61–67.

Dalessandro, G., and Roberts, L. W. (1971). Induction of xylogenesis in pith parenchyma explants of *Lactuca. Am. J. Bot.* **58,** 378–385.

Datta, P. C., Mukherjee, S., Chakrabarti, S., and Saha, S. (1979). Tracheary cell types of cultured *Vinca rosea* L. tissue in different concentrations of sucrose and nitrate. *Indian J. Exp. Biol.* **17,** 46–49.

DeMaggio, A. E. (1972). Induced vascular tissue differentiation in fern gametophytes. *Bot. Gaz. (Chicago)* **133,** 311–317.

Dodds, J. H. (1979). Is cell cycle activity necessary for xylem cell differentiation? *What's New in Plant Physiol.* **10,** 13–16.

Dodds, J. H. (1980). The effect of 5-fluorodeoxyuridine and colchicine on tracheary element differentiation in isolated mesophyll cells of *Zinnia elegans* L. *Z. Pflanzenphysiol.* **99,** 283–285.

Dodds, J. H. (1981a). The role of the cell cycle and cell division in xylem differentiation. *In* "Xylem Cell Development" (J. R. Barnett, ed.), pp. 153–167. Castle House Publ. Ltd , Tunbridge Wells, England.

Dodds, J. H. (1981b). Relationship of the cell cycle to xylem cell differentiation: A new model. *Plant, Cell. Environ.* **4,** 145–146.

Dodds, J. H., and Phillips, R. (1977). DNA and histone content of immature tracheary elements from cultured artichoke explants. *Planta* **135,** 213–216.

Doley, D., and Leyton, L. (1970). Effects of growth-regulating substances and water potential on the development of wound callus in *Fraxinus. New Phytol.* **69,** 87–102.

Durante, L., Cremonini, R., Brunori, A., Avanzi, S., and Innocenti, A. M. (1977). Differentiation of metaxylem cell line in the root of *Allium cepa*. I. DNA heterogeneity and ribosomal cistrons of two different stages of differentiation. *Protoplasma* **93,** 289–303.

Durnam, D. J., and Jones, R. L. (1982). The effects of colchicine and gibberellic acid on growth and microtubules in excised lettuce hypocotyls. *Planta* **154,** 204–211.

Durst, F. (1976). The correlation of phenylalanine ammonia-lyase and cinnamic acid-hydroxylase activity changes in Jerusalem artichoke tuber tissue. *Planta* **132,** 221–227.

Elstner, E. F., and Heupel, A. (1976). Formation of hydrogen peroxide by isolated cell walls from horseradish (*Armoracia lapathifolia* Gilib.). *Planta* **130,** 175–180.

Esau, K., and Charvat, I. (1978). On vessel member differentiation in the bean (*Phaseolus vulgaris* L.). *Ann. Bot. (London)* **42,** 665–677.

Esau, K., Cheadle, V. I., and Gill, R. H. (1966). Cytology of differentiating tracheary elements. III. Structures associated with cell surfaces. *Am. J. Bot.* **53**, 765–771.

Foard, D. E., and Haber, A. H. (1961). Anatomic studies of gamma-irradiated wheat growing without cell division. *Am. J. Bot.* **48**, 438–446.

Fosket, D. E. (1968). Cell division and the differentiation of wound-vessel members in cultured stem segments of *Coleus*. *Proc. Natl. Acad. Sci. U.S.A.* **59**, 1089–1096.

Fosket, D. E. (1970). The time course of xylem differentiation and its relation to DNA synthesis in cultured *Coleus* stem segments. *Plant Physiol.* **46**, 64–68.

Fosket, D. E. (1980). Hormonal control of morphogenesis in cultured tissues. *In* "Plant Growth Substances 1979" (F. Skoog, ed.), pp. 362–369. Springer-Verlag, Berlin and New York.

Fosket, D. E., and Miksche, J. P. (1966). Protein synthesis as a requirement for wound xylem differentiation. *Physiol. Plant.* **19**, 982–991.

Fosket, D. E., and Roberts, L. W. (1964). Induction of wound vessel differentiation in isolated *Coleus* stem segments *in vitro*. *Am. J. Bot.* **51**, 19–25.

Fosket, D. E., and Torrey, J. G. (1969). Hormonal control of cell proliferation and xylem differentiation in cultured tissues of *Glycine max* var. Biloxi. *Plant Physiol.* **44**, 871–880.

Fujimura, T., and Komamine, A. (1979). Synchronization of somatic embryogenesis in a carrot cell suspension culture. *Plant Physiol.* **64**, 162–164.

Fujimura, T., and Komamine, A. (1982). Molecular mechanism of somatic embryogenesis. *In* "Plant Tissue Culture 1982" (A. Fujiwara, ed.), pp. 105–106. Maruzen, Tokyo.

Fukuda, H., and Komamine, A. (1980a). Establishment of an experimental system for the study of tracheary element differentiation from single cells isolated from the mesophyll of *Zinnia elegans*. *Plant Physiol.* **65**, 57–60.

Fukuda, H., and Komamine, A. (1980b). Direct evidence for cytodifferentiation to tracheary elements without intervening mitosis in a culture of single cells isolated from the mesophyll of *Zinnia elegans*. *Plant Physiol.* **65**, 61–64.

Fukuda, H., and Komamine, A. (1981a). Relationship between tracheary element differentiation and the cell cycle in single cells isolated from the mesophyll of *Zinnia elegans*. *Physiol. Plant.* **52**, 423–430.

Fukuda, H., and Komamine, A. (1981b). Relationship between tracheary element differentiation and DNA synthesis in single cells isolated from the mesophyll of *Zinnia elegans*. Analysis by inhibitors of DNA synthesis. *Plant Cell Physiol.* **22**, 41–49.

Fukuda, H., and Komamine, A. (1982). Lignin synthesis and its related enzymes as markers of tracheary element differentation in single cells isolated from the mesophyll of *Zinnia elegans*. *Planta* **155**, 423–430.

Fukuda, H., and Komamine, A. (1983). Changes in the synthesis of RNA and protein during tracheary element differentiation in single cells isolated from the mesophyll of *Zinnia elegans*. *Plant Cell Physiol.* **24**, 603–614.

Gahan, P. B. (1978). A reinterpretation of the cytochemical evidence for acid phosphatase activity during cell death in xylem differentiation. *Ann. Bot.* **42**, 755–758.

Gahan, P. B. (1981a). An early cytochemical marker of commitment to stelar differentiation in meristems from dicotyledonous plants. *Ann. Bot.* **48**, 769–775.

Gahan, P. B. (1981b). Biochemical changes during xylem element differentiation. *In* "Xylem Cell Development" (J. R. Barnett, ed.), pp. 168–191. Castle House Publ. Ltd., Tunbridge Wells, England.

Gahan, P. B., and Maple, A. J. (1966). The behaviour of lysosome-like particles during cell differentiation. *J. Exp. Bot.* **17**, 151–155.

Gahan, P. B., Rana, M. A., and Phillips, R. (1983). Activation of carboxylesterases in root cortical parenchymal cells of *Pisum sativum* during xylem induction *in vitro*. *Cell Biochem. Funct.* **1**, 109–111.

Goosen-De Roo, L. (1973a). The fine structure of the protoplast in primary tracheary elements of the cucumber after plasmolysis. *Acta Bot. Neerl.* **22,** 467–485.

Goosen-De Roo, L. (1973b). The relationship between cell organelles and cell wall thickenings in primary tracheary elements of the cucumber. I. Morphological aspects. *Acta Bot. Neerl.* **22,** 279–300.

Goosen-De Roo, L. (1973c). The relationship between cell organelles and cell wall thickenings in primary tracheary elements of the cucumber. II. Quantitative aspects. *Acta Bot. Neerl.* **22,** 301–320.

Gross, G. G., Janse, C., and Elstner, E. F. (1977). Involvement of-malate,monophenols and the superoxide radical in hydrogen peroxide formation by isolated cell walls from horseradish (*Armoracia lapathifolia* Gilib.). *Planta* **136,** 271–276.

Gunning, B. E. S., and Hardham, A. R. (1982). Microtubules. *Annu. Rev. Plant Physiol.* **33,** 651–698.

Haddon, L. E., and Northcote, D. H. (1975). Quantitative measurement of the course of bean callus differentiation. *J. Cell Sci.* **17,** 11–26.

Haddon, L. E., and Northcote, D. H. (1976a). Correlation of the induction of various enzymes concerned with phenylpropanoid and lignin synthesis during differentiation of bean callus (*Phaseolus vulgaris* L.). *Planta* **128,** 255–262.

Haddon, L. E., and Northcote, D. H. (1976b). The influence of gibberellic acid and abscisic acid on cell and tissue differentiation of bean callus. *J. Cell Sci.* **20,** 47–55.

Haddon, L. E., and Northcote, D. H. (1976c). The effect of growth conditions and origin of tissue on the ploidy and morphogenic potential of tissue cultures of bean (*Phaseolus vulgaris* L.). *J. Exp. Bot.* **27,** 189–209.

Hagen, G., Kleinschmidt, A., and Guilfoyle, T. (1984). Auxin regulated gene expression in intact soybean hypocotyl and excised hypocotyl sections. *Planta* **162,** 147–153.

Hahlbrock, K., and Grisebach, H. (1979). Enzymatic controls in the biosynthesis of lignin and flavonoids. *Annu. Rev. Plant Physiol.* **30,** 105–130.

Hammersley, D. R. H., and McCully, M. E. (1980). Differentiation of wound xylem in pea roots in the presence of colchicine. *Plant Sci. Lett.* **19,** 151–156.

Hardham, A. R., and Gunning, B. E. S. (1978). Structure of cortical microtubule arrays in plant cells. *J. Cell Biol.* **77,** 14–34.

Hardham, A. R., and Gunning, B. E. S. (1979). Interpolation of microtubules into cortical arrays during cell elongation and differentiation in roots of *Azolla pinnata. J. Cell Sci.* **37,** 411–442.

Hardham, A. R., and Gunning, B. E. S. (1980). Some effects of colchicine on microtubules and cell division in roots of *Azolla pinnata. Protoplasma* **102,** 31–51.

Hase, Y., Yakura, K., and Tanifuji, S. (1979). Differential replication of satellite and main band DNA during early stages of callus formation in carrot root tissue. *Plant Cell Physiol.* **20,** 1461–1469.

Havránek, P., and Movák, F. J. (1973). The bud formation in the callus cultures of *Allium sativum* L. *Z. Pflanzenphysiol.* **68,** 308–318.

Hawkins, S. W., and Phillips, R. (1983). 3-Aminobenzamide inhibits tracheary element differentiation but not cell division in cultured explants of *Helianthus tuberosus. Plant Sci. Lett.* **32,** 221–224.

Hébant, C. (1973). Acid phosphomonoesterase activities (β-glycerophosphatase and naphthol AS-MX phosphatase) in conducting tissues of bryophytes. *Protoplasm* **77,** 231–241.

Hepler, P. K. (1981). Morphogenesis of tracheary elements and guard cells. *In* "Cytomorphogenesis in Plants" (O. Kiermayer, ed.), pp. 327–347. Springer-Verlag, Berlin and New York.

Hepler, P. K., and Fosket, D. E. (1971). The role of microtubules in vessel member differentiation in *Coleus. Protoplasma* **72,** 213–236.

Hepler, P. K., and Newcomb, E. H. (1964). Microtubules and fibrils in the cytoplasm of *Coleus* cells undergoing secondary wall deposition. *J. Cell Biol.* **20**, 529–533.

Hepler, P. K., and Palevitz, B. A. (1974). Microtubules and microfilaments. *Annu. Rev. Plant Physiol.* **25**, 309–362.

Hepler, P. K., Fosket, D. E., and Newcomb, E. H. (1970). Lignification during secondary wall formation in *Coleus*. An electron microscopic study. *Am. J. Bot.* **57**, 85–96.

Hepler, P. K., Rice, R. M., and Terranova, W. A. (1972). Cytochemical localization of peroxidase activity in wound vessel members of *Coleus*. *Can. J. Bot.* **50**, 977–983.

Hogetsu, T. (1983). Distribution and local activity of particle complexes synthesizing cellulose microfibrils in the plasma membrane of *Closterium acerosum* (Schrank) Ehrenberg. *Plant Cell Physiol.* **24**, 777–781.

Hogetsu, T., Shibaoka, H., and Shimokoriyama, H. (1974). Involvement of cellulose synthesis in actions of gibberellin and kinetin on cell expansion. 2,6-dichlorobenzonitrile as a new cellulose-synthesis inhibitor. *Plant Cell Physiol.* **15**, 389–393.

Holländer, H., and Amrhein, N. (1980). The site of the inhibition of the shikimate pathway by glyphosate. I. Inhibition by glyphosate of phenylpropanoid synthesis in buckwheat *(Fagopyrum esculentum* Moench). *Plant Physiol.* **66**, 823–829.

Holländer, H., Kiltz, H.-H., and Amrhein, N. (1979). Interference of L-α-aminooxy-β-phenylpropionic acid with phenylalanine metabolism in buckwheat. *Z. Naturforsch.* **34C**, 1162–1173.

Holtzer, H., and Rubinstein, N. (1977). Binary decisions, quantal cell cycles and cell diversification. *In* "Cell Differentiation in Microorganisms, Plants and Animals" (L. Nover and K. Mothes, eds.), pp. 424–437. G. Fischer Verlag, Jena, and Elsevier/North Holland, Amsterdam.

Hösel, W., Preiss, A. F., and Borgmann, E. (1982). Relationship of coniferin β-glucosidase to lignification in various plant cell suspension cultures. *Plant Cell, Tissue Organ Cult.* **1**, 137–148.

Hotta, Y., and Stern, H. (1971). Analysis of DNA synthesis during meiotic prophase in *Lilium*. *J. Mol. Biol.* **55**, 337–355.

Ikegami, S., Taguchi, T., Ohashi, M., Oguchi, M., Nagano, H., and Mano, Y. (1978). Aphidicolin prevents mitotic cell division by interfering with the activity of DNA polymerase-α. *Nature (London)* **275**, 458–460.

Inzé, D., Follin, A., Van Lijsebettens, M., Simoens, C., Genetello, C., Van Montagu, M., and Schell, J. (1984). Genetic analysis of the individual T-DNA genes of *Agrobacterium tumefaciens:* Further evidence that two genes are involved in indole-3-acetic acid synthesis. *Mol. Gen. Genet.* **194**, 265–274.

Jacobs, W. P. (1952). The role of auxin in differentiation of xylem around a wound. *Am. J. Bot.* **39**, 301–309.

Jacobs, W. P. (1954). Acropetal auxin transport and xylem regeneration—a quantitative study. *Am. Nat.* **90**, 163–169.

Jacobs, W. P. (1969). Regeneration and differentiation of sieve tube elements. *Int. Rev. Cytol.* **28**, 239–273.

Jacobs, W. P., and Morrow, I. B. (1957). A quantitative study of xylem development in the vegetative shoot apex of *Coleus*. *Am. J. Bot.* **44**, 823–842.

Jeffs, R. A., and Northcote, D. H. (1966). Experimental induction of vascular tissue in an undifferentiated plant callus. *Biochem. J.* **101**, 146–152.

Jeffs, R. A., and Northcote, D. H. (1967). The influence of indol-3yl-acetic acid and sugar on the pattern of induced differentiation in plant tissue culture. *J. Cell Sci.* **2**, 77–88.

Jense, W. A. (1957). The incorporation of [¹⁴C]adenine and [¹⁴C]phenylalanine by developing root-tip cells. *Proc. Natl. Acad. Sci. U.S.A.* **43**, 1038–1046.

Jones, D. T., and Villiers, T. A. (1972). Changes in distribution of acid-phosphatase activity during wound regeneration in *Coleus*. *J. Exp. Bot.* **23**, 375–380.

Jones, L. E., Hildebrandt, A. C., Riker, A. J., and Wu, J. H. (1960). Growth of somatic tobacco cells in microculture. *Am. J. Bot.* **47**, 468–475.

Kemphues, K. J., Kaufman, T. C., Raff, R. A., and Raff, E. C. (1982). The testis-specific β-tubulin subunit in *Drosophila melanogaster* has multiple functions in spermatogenesis. *Cell* **31**, 655–670.

Kirschner, H., and Sachs, T. (1978). Cytoplasmic reorientation: An early stage of vascular differentiation. *Isr. J. Bot.* **27**, 131–137.

Kleiber, H., and Mohr, H. (1967). Vom einfluss des phytochroms auf die xylem-differenzierung im hypokotyl des senfkeimlings (*Sinapis alba* L.). *Planta* **76**, 85–92.

Kohlenbach, H. W. (1959). Streckungs und teilungswachstum isolierter mesophyllzellen von *Macleaya cordata* (Willd.). *Naturwissenschaften* **46**, 116–117.

Kohlenbach, H. W. (1965). Die entwicklungspotenzen explantierter und isolierter dauezellen. II. Das zelluläre differenzierungswachstum bei blattzellkulturen von *Macleaya cordata*. *Beitr. Biol. Pflanz.* **41**, 469–480.

Kohlenbach, H. W., and Schmidt, B. (1975). Cytodifferenzierung in form einer direkten umwandlung isolierter mesophyll-zellen zu tracheiden. *Z. Pflanzenphysiol.* **75**, 369–374.

Kohlenbach, H. W., and Schöpke, C. (1981). Cytodifferentiation to tracheary elements from isolated mesophyll protoplasts of *Zinnia elegans*. *Naturwissenschaften* **68**, 576–577.

Kohlenbach, H. W., Körber, M., and Li, L. (1982). Cytodifferentiation of protoplasts isolated from a stem embryo system of *Brassica napus* to tracheary elements. *Z. Pflanzenphysiol.* **107**, 367–371.

Komamine, A., and Fukuda, H. (1982). Biochemical mechanism of cytodifferentiation. *In* "Plant Tissue Culture 1982" (A. Fujiwara, ed.), pp. 91–92. Maruzen, Tokyo.

Krimer, D. B., and Van't Hof, J (1983). Extrachromosomal DNA of pea (*Pisum sativum*) root-tip cells replicates by strand displacement. *Proc. Natl. Acad. Sci. U.S.A.* **80**, 1933–1937.

Kuboi, T., and Yamada, Y. (1978a). Regulation of the enzyme activities related to lignin synthesis in cell aggregates of tobacco cell culture. *Biochim. Biophys. Acta* **542**, 181–190.

Kuboi, T., and Yamada, Y. (1978b). Changing cell aggregation and lignification in tobacco suspension cultures. *Plant Cell Physiol.* **19**, 437–443.

Ledbetter, M. C., and Porter, K. R. A. (1963). Microtubule in plant cell fine structure. *J. Cell Biol.* **19**, 239–250.

Leemans, J., Deblaere, R., Willmitzer, L., De Greve, H., Hernalsteens, J. P., Van Montagu, M., and Schell, J. (1982). Genetic identification of functions of TL-DNA transcripts in octopine crown galls. *EMBO J.* **1**, 147–152.

Levine, M., Garen, A., Lepesant, J. A., and Kejzlarova, J. L. (1981). Constancy of somatic DNA organization in developmentally regulated regions of the *Drosophila* gemone. *Proc. Natl. Acad. Sci. U.S.A.* **78**, 2417–2421.

Levitt, D., and Dorfman, A. (1972). The irreversible inhibition of differentiation of limb-bud mesenchyme by bromodeoxyuridine. *Proc. Natl. Acad. Sci. U.S.A.* **69**, 1253–1257.

Linstedt, D., and Reinert, J. (1975). Occurrence and properties of a cytokinin in tissue culture of *Daucus carota*. *Naturwissenschaften* **62**, 238–239.

Lloyd, C. W. (1983). Helical microtubular arrays in onion root hairs *Nature (London)* **305**, 311–313.

Luckner, M. (1982). Programmed gene expression during cell division cycles. *In* "Cell Differentiation" (L. Nover, M. Luckner, and B. Parthier, eds.), pp. 512–528. Springer-Verlag, Berlin and New York.

Mäder, M., Ungemach, J., and Schloss, P. (1980). The role of peroxidase isoenzyme groups of *Nicotiana tabacum* in hydrogen peroxide formation. *Planta* **147**, 467–470.

Maitra, S. C., and De, D. N. (1971). Role of microtubules in secondary thickenings of differentiating xylem element. *J. Ultrastruct. Res.* **34**, 15–22.

Malawer, C. L., and Phillips, R. (1979). The cell cycle in relation to induced xylem differentiation: Tritiated thymidine incorporation in cultured tuber explants of *Helianthus tuberosus* L. *Plant Sci. Lett.* **15,** 47–55.

Masuda, H., Fukuda, H., and Komamine, A. (1983). Changes in peroxidase isoenzyme patterns during tracheary element differentiation in a culture of single cells isolated from the mesophyll of *Zinnia elegans. Z. Pflanzenphysiol.* **112,** 417–426.

Mayer, T. C. (1980). The relationship between cell division and melanocyte differentiation in epidermal cultures from mouse embryos. *Dev. Biol.* **79,** 419–427.

Meins, F., Jr. (1974). Cell division and determination phase of cytodifferentiation in plants. *In* "Results and Problems in Cell Differentiation" (W. Berrmann, J. Reinert, and H. Ursprung, eds.), pp. 151–175. Springer-Verlag, Berlin and New York.

Miller, A. R., and Roberts, L. W. (1982). Regulation of tracheary element differentiation by exogenous L-methionine in callus of soya bean cultivars. *Ann. Bot. (London)* **50,** 111–116.

Miller, A. R., and Roberts, L. W. (1984). Ethylene biosynthesis and xylogenesis in *Lactuca* pith explants cultured *in vitro* in the presence of auxin and cytokinin: The effect of ethylene precursors and inhibitors. *J. Exp. Bot.* **35,** 691–698.

Miller, A. R., Pengelly, W. L., and Roberts, L. W. (1984). Introduction of xylem differentiation in *Lactuca* by ethylene. *Plant Physiol.* **75,** 1165–1166.

Minocha, S. C. (1976). Cytokinin metabolism in relation to differentiation of tracheids in cultured tuber tissue of Jerusalem artichoke *(Helianthus tuberosus). Proc. Int. Conf. Plant Growth Subst., 9th, 1976, Lausanne,* 258–260.

Minocha, S. C., and Halperin, W. (1974). Hormones and metabolites which control tracheid differentiation, with or without concomitant effects on growth, in cultured tuber tissue of *Helianthus tuberosus* L. *Planta* **116,** 319–331.

Minocha, S. C., and Halperin, W. (1976). Enzymatic changes and lignification in relation to tracheid differentiation in cultured tuber tissue of Jerusalem artichoke *(Helianthus tuberosus). Can. J. Bot.* **54,** 79–89.

Mita, T., and Shibaoka, H. (1984). Gibberellin stabilizes micortubules in onion leaf sheath cells. *Protoplasma,* **119,** 100–109.

Mizuno, K., and Komamine, A. (1978a). Isolation and identification of substances inducing formation of tracheary elements in cultured carrot-root slices. *Planta* **138,** 59–62.

Mizuno, K., and Komamine, A. (1978b). A possible role of cyclic AMP on tracheary element formation in cultured carrot-root slices. *Bot. Mag.* **91,** 213–219.

Mizuno, K., Komamine, A., and Shimokoriyama, M. (1971). Vessel element formation in cultured carrot-root phloem. *Plant Cell Physiol.* **12,** 823–830.

Mizuno, K., Komamine, A., and Shimokoriyama, M. (1974). Isolation of substances inducing vessel element formation in cultured carrot root slices. *Plant Growth Subst., Proc. Int. Conf. 8th, 1973, Tokyo,* 111–118.

Mizuno, K., Koyama, M., and Shibaoka, H. (1981). Isolation of plant tubulin from azuki bean epicotyls by ethyl-N-phenyl-carbamate-sepharose affinity chromatography. *J. Biochem. (Tokyo)* **89,** 329–332.

Morejohn, L. C., and Fosket, D. E. (1982). Higher plant tubulin identified by self-assembly into microtubules *in vitro. Nature (London)* **297,** 426–428.

Morrison, I. M. (1972). A semimicro method for the determination of lignin and its use in predicting the digestibility of forage crops. *J. Sci. Food Agric.* **23,** 455–463.

Mueller, S. C., and Brown, R. M., Jr. (1982a). The control of cellulose microfibril deposition in the cell wall of higher plants. I. Can directed membrane flow orient cellulose microfibrils? Indirect evidence from freeze-fractured plasma membranes of maize and pine seedlings. *Planta* **154,** 489–500.

Mueller, S. C., and Brown, R. M., Jr. (1982b). The control of cellulose microfibril deposition in the cell wall of higher plants. II. Freeze-fracture microfibril patterns in maize seedling tissues following experimental alteration with colchicine and ethylene. *Planta* **154,** 501–515.

Murashige, T., and Skoog, F. (1962). A revised medium for rapid growth and bioassays with tobacco tissue cultures. *Physiol. Plant.* **15,** 473–497.

Nagl, W. (1976). Nuclear organization. *Annu. Rev. Plant Physiol.* **27,** 39–69.

Nelmes, B. J., Preston, R. D., and Worth, D. (1973). A possible function of microtubules suggested by their abnormal distribution in rubbery wood. *J. Cell Sci.* **13,** 741–751.

Newcomb, E. H. (1969). Plant microtubules. *Annu. Rev. Plant Physiol.* **20,** 253–288.

Northcote, D. H. (1963). Changes in the cell walls during differentiation. *Symp. Soc. Exp. Biol.* **17,** 157–174.

Northcote, D. H. (1979). Biochemical mechanisms involved in plant morphogenesis. *In* "British Plant Growth Regulator Group" (E. C. George, ed.), Monograph 3, Control of Plant Development, pp. 11–20.

O'Brien, T. P. (1970). Further observations on hydrolysis of the cell wall in xylem. *Protoplasma* **69,** 1–14.

O'Brien, T. P. (1974). Primary vascular tissues. *In* "Dynamic Aspects of Plant Ultrastructure" (A. W. Robards, ed.), pp. 414–440. McGraw-Hill, New York.

O'Brien, T. P. (1981). The primary xylem. *In* "Xylem Cell Development" (J. R. Barnett, ed.), pp. 14–46. Castle House Publ. Ltd., Tunbridge Wells, England.

O'Brien, T. P., and Thimann, K. V. (1967). Observations on the fine structure of the oat coleoptile. III. Correlated light and electron microscopy of the vascular tissues. *Protoplasma* **63,** 443–478.

Parenti, R., Guillé, E., Grisvard, J., Durante, M., Giorgi, L., and Buiatti, M. (1973). Transient DNA satellite in differentiating pith tissue. *Nature (London), New Biol.* **246,** 237–239.

Phillips, R. (1980). Cytodifferentiation. *Int. Rev. Cytol. Suppl.* **11A,** 55–70.

Phillips, R. (1981a). Direct differentiation of tracheary elements in cultured explants of gamma-irradiated tubers of *Helianthus tuberosus*. *Planta* **153,** 262–266.

Phillips, R. (1981b). Characterization of mitotic cycles preceding xylogenesis in cultured explants of *Helianthus tuberosus* L. *Ann. Bot. (London)* **47,** 785–792.

Phillips, R., and Dodds, J. H. (1977). Rapid differentiation of tracheary elements in cultured explants of Jerusalem artichoke. *Planta* **135,** 207–212.

Phillips, R., and Hawkins, S. W. (1985). Characteristics of the inhibition of induced tracheary element differentiation by 3-aminobenzamide and related compounds. *J. Exp. Bot.* **36,** 119–128.

Phillips, R., and Torrey, J. G. (1973). DNA synthesis, cell division and specific cytodifferentiation in cultured pea root cortical explants. *Dev. Biol.* **31,** 336–347.

Phillips, R., and Torrey, J. G. (1974). DNA levels in differentiating tracheary elements. *Dev. Biol.* **39,** 322–325.

Pickett-Heaps, J. D. (1966). Incorporation of radioactivity into wheat xylem walls. *Planta* **71,** 1–14.

Pickett-Heaps, J. D. (1967). The effects of colchicine on the ultrastructure of dividing plant cells, xylem wall differentiation and distribution of cytoplasmic microtubules. *Dev. Biol.* **15,** 206–236.

Pickett-Heaps, J. D. (1968). Xylem wall deposition: Radioautographic investigations using lignin precursors. *Protoplasma* 65, 181–205.

Raff, E. C., Fuller, M. T., Kaufman, T. C., Kemphues, K. J., Rudolph, J. E., and Raff, R. A. (1982). Regulation of tubulin gene expression during embryogenesis in *Drosophila melanogaster*. *Cell* **28,** 33–40.

Rana, M. A., and Gahan, P. B. (1982). Determination of stelar elements in roots of *Pisum sativum* L. *Ann. Bot. (London)* **50,** 757–762.

Rana, M. A., and Gahan, P. B. (1983). A quantitative cytochemical study of determination for xylem-element formation in response to wounding in roots of *Pisum sativum* L. *Planta* **157,** 307–316.

Ream, L. W., Gordon, M. P., and Nester, E. W. (1983). Multiple mutations in the T region of the *Agrobacterium tumefaciens* tumor-inducing plasmid. *Proc. Natl. Acad. Sci. U.S.A.* **80**, 1660–1664.

Reuther, G., and Werckmeister, P. (1973). Bildung von sekundären wandstrukturen in frei suspendierten einzelzellen von pelargonien-kallus in flüssigkultur. *Z. Pflanzenphysiol.* **70**, 276–282.

Rier, J. P., and Beslow, D. T. (1967). Sucrose concentration and the differentiation of xylem in callus. *Bot. Gaz. (Chicago)* **128**, 73–77.

Robbertse, P. J., and McCully, M. E. (1979). Regeneration of vascular tissue in wounded pea roots. *Planta* **145**, 167–173.

Roberds, A. W., and Kidwai, P. (1969). Vesicular involvement in differentiating plant vascular cells. *New Phytol.* **68**, 343–349.

Roberts, L. W. (1969). The initiation of xylem differentiation. *Bot. Rev.* **35**, 201–250.

Roberts, L. W. (1976). "Cytodifferentiation in Plants: Xylogenesis as a Model System." Cambridge Univ. Press, London and New York.

Roberts, L. W., and Baba, S. (1968a). IAA-induced xylem differentiation in the presence of colchicine. *Plant Cell Physiol.* **9**, 315–321.

Roberts, L. W., and Baba, S. (1968b). Effect of proline on wound vessel member formation. *Plant Cell Physiol.* **9**, 353–360.

Roberts, L. W., and Baba, S. (1978). Exogenous methionine as a nutrient supplement for the induction of xylogenesis in lettuce pith explants. *Ann. Bot. (London)* **42**, 375–379.

Roberts, L. W., and Baba, S. (1982). Glycerol and *myo*inositol as carbon sources for the induction of xylogenesis in explants of *Lactuca. Can. J. Bot.* **60**, 1204–1206.

Roberts, L. W., and Miller, A. R. (1982). Ethylene and xylem differentiation. *What's New in Plant Physiol.* **13**, 13–16.

Ronchi, V. N. (1981). Histological study of organogenesis *in vitro* from callus culture of two *Nicotiana* species. *Can. J. Bot.* **59**, 1969–1977.

Ronchi, V. N., and Gregorini, G. (1970). Histological study of adventitious bud formation on *Lactuca sativa* cotyledons cultured *in vitro. G. Bot. Ital.* **104**, 443–455.

Ruberry, P. H., and Fosket, D. E. (1969). Changes in phenylalanine ammonia-lyase activity during xylem differentiation in *Coleus* and soybean. *Planta* **87**, 54–62.

Ruberry, P. H., and Northcote, D. H. (1968). Site of phenylalanine ammonia-lyase activity and synthesis of lignin during xylem differentiation. *Nature (London)* **219**, 1230–1234.

Savidge, R. A., and Wareing, P. F. (1981a). A tracheid-differentiation factor from pine needles. *Planta* **153**, 395–404.

Savidge, R. A., and Wareing, P. F. (1981b). Plant growth regulators and the differentiation of vascular elements. *In* "Xylem Cell Development" (J. R. Barnett, ed.), pp. 192–235. Castle House Publ. Ltd., Tunbridge Wells, England.

Schäfer, J. R., Blaschke, J. R., and Neumann, K. H. (1978). On DNA metabolism in carrot tissue cultures. *Planta* **139**, 97–101.

Sheldrake, A. R., and Northcote, D. H. (1968a). The production of auxins by tobacco internode tissue. *New Phytol.* **67**, 1–13.

Sheldrake, A. R., and Northcote, D. H. (1968b). Some continuents of xylem sap and their possible relationship to xylem differentiation. *J. Exp. Bot.* **19**, 681–689.

Shininger, T. L. (1975). Is DNA synthesis required for the induction of differentiation in quiescent root cortical parenchyma? *Dev. Biol.* **45**, 137–150.

Shininger, T. L. (1978). Hormone regulation of development in plant cells. *In Vitro* **14**, 31–50.

Shininger, T. L. (1979a). The control of vascular development. *Annu. Rev. Plant Physiol.* **30**, 313–337.

Shininger, T. L. (1979b). Quantitative analysis of temperature effects on xylem and nonxylem cell formation in cytokinin-stimulated root tissues. *Proc. Natl. Acad. Sci. U.S.A.* **76**, 1921–1923.

Shininger, T. L. (1980). Biochemical and cytological analysis of RNA synthesis in kinetin-treated pea root parenchyma. *Plant Physiol.* **65,** 838–843.

Simmonds, D., Setterfield, G., and Brown, D. L. (1983). Organization of microtubules in dividing and elongating cells of *Vicia hajastana* Grossh. in suspension culture. *Eur. J. Cell Biol.* **32,** 59–66.

Simpson, S., and Torrey, J. G. (1977). Hormonal control of deoxyribonucleic acid synthesis and protein synthesis in pea root cortical explants. *Plant Physiol.* **59,** 4–9.

Sinnott, E. W., and Bloch, R. (1944). Visible expression of cytoplasmic pattern in the differentiation of xylem strands. *Proc. Natl. Acad. Sci. U.S.A.* **30,** 388–392.

Sinnott, E. W., and Bloch, R. (1945). The cytoplasmic basis of intercellular patterns in vascular differentiation. *Am. J. Bot.* **32,** 151–156.

Skoog, F., and Miller, C. D. (1957). Chemical regulation of growth and organ formation in plant tissues cultured *in vitro. Symp. Soc. Exp. Biol.* **11,** 118–131.

Spiegelman, B. M., and Farmer, S. R. (1982). Decrease in tubulin and actin gene expression prior to morphological differentiation of 3T3 adipocytes. *Cell* **29,** 53–60.

Srivastava, L. M., and Singh, A. P. (1972). Certain aspects of xylem differentiation in corn. *Can. J. Bot.* **50,** 1795–1804.

Strom, C. M., and Dorfman, A. (1976a). Distribution of 5-bromodeoxyuridine and thymidine in the DNA of developing chick cartilage. *Proc. Natl. Acad. Sci. U.S.A.* **73,** 1019–1023.

Strom, C. M., and Dorfman, A. (1976b). Amplification of moderately repetitive DNA sequences during chick cartilage differentiation. *Proc. Natl. Acad. Sci. U.S.A.* **73,** 3428–3432.

Strom, C. M., Moscona, M., and Dorfman, A. (1978). Amplification of DNA sequences during chicken cartilage and neural retina differentiation. *Proc. Natl. Acad. Sci. U.S.A.* **75,** 4451–4454.

Sung, Z. R., and Okimoto, R. (1981). Embryonic proteins in somatic embryos of carrot. *Proc. Natl. Acad. Sci. U.S.A.* **78,** 3683–3687.

Sussex, I. M., and Clutter, M. E. (1968). Differentiation in tissues, free cells, reaggregated plant cells. *In Vitro* **3,** 3–12.

Takeda, K., and Shibaoka, H. (1981). Effects of gibberellin and colchicine on microfibril arrangement in epidermal cell walls of *Vigna angularis* Ohwi et. Ohashi epicotyls. *Planta* **151,** 393–398.

Theologis, A., and Ray, P. M. (1982). Early auxin-regulated polyadenylylated mRNA sequences in pea stem tissue. *Proc. Natl. Acad. Sci. U.S.A.* **79,** 418–421.

Thomashow, L. S., Reeves, S., and Thomashow, M. F. (1984). Crown gall oncogenesis: Evidence that a T-DNA gene from the *Agrobacterium* Ti plasmid pTiA6 encodes an enzyme that catalyzes synthesis of indoleacetic acid. *Proc. Natl. Acad. Sci. U.S.A.* **81,** 5071–5075.

Torrey, J. G. (1953). The effect of certain metabolic inhibitors on vascular tissue differentiation in isolated pea roots. *Am. J. Bot.* **40,** 525–533.

Torrey, J. G. (1968). Hormonal control of cytodifferentiation in agar and cell suspension cultures. *In* "Biochemistry and Physiology of Plant Growth Substances" (F. Wightman and G. Setterfield, eds.), pp. 843–855. Runge Press, Ottawa, Ontario, Canada.

Torrey, J. G. (1975). Tracheary element formation from single isolated cells in culture. *Physiol. Plant.* **35,** 158–165.

Torrey, J. G., and Fosket, D. E. (1970). Cell division in relation to cytodifferentiation in cultured pea root segments. *Am. J. Bot.* **57,** 1072–1080.

Torrey, J. G., Fosket, D. E., and Hepler, P. K. (1971). Xylem formation: A paradigm of cytodifferentiation in higher plants. *Am. Sci.* **59,** 338–352.

Turgeon, R. (1975). Differentiation of wound vessel members without DNA synthesis, mitosis or cell division. *Nature (London)* **257,** 800–808.

Vasil, V., and Hildebrandt, A. C. (1965a). Growth and tissue formation from single isolated tobacco cells in microculture. *Science* **147,** 1454–1455.

Vasil, V., and Hildebrandt, A. C. (1965b). Differentiation of tobacco plants from single isolated cells in microcultures. *Science* **150,** 889–890.

Verma, D. P. S., and van Huystee, R. B. (1970). Cellular differentiation and peroxidase isozyme in cell culture of peanut cotyledons. *Can. J. Bot.* **48,** 429–431.

Vonderhaar, B. K., and Topper, Y. J. (1974). Role of the cell cycle in hormone-dependent differentiation. *J. Cell Biol.* **63,** 707–712.

Vreugdenhil, D., Harkes, P. A. A., and Libbenga, K. R. (1980). Auxin-binding by particulate fractions from tobacco leaf protoplasts. *Planta* **150,** 9–12.

Walker, J. C., and Key, J. L. (1982). Isolation of cloned cDNAs to auxin-responsive poly(A)$^+$RNAs of elongating soybean hypocotyl. *Proc. Natl. Acad. Sci. U.S.A.* **79,** 7185–7189.

Wardrop, A. B. (1965). Cellular differentiation in xylem. *In* "Cellular Ultrastructure of Woody Plants" (W. A. Côté, Jr., ed.), pp. 61–97. Syracuse Univ. Press, Syracuse, New York.

Wardrop, A. B. (1981). Biochemical changes during xylem element differentiation. *In* "Xylem Cell Development" (J. R. Barnett, ed.), pp. 168–191. Castle House Publ. Ltd., Tunbridge Wells, England.

Wetmore, R. H., and Rier, J. P. (1963). Experimental induction of vascular tissues in the callus of angiosperms. *Am. J. Bot.* **50,** 418–430.

Wetmore, R. H., and Sorokin, S. (1955). On the differentiation of xylem. *J. Arnold Arbor., Harv. Univ.* **36,** 305–317.

Wick, S. M., Seagull, R. W., Osborn, M., Weber, K., and Gunning, B. E. S. (1981). Immunofluorescence microscopy of organized microtubule arrays in structurally stabilized meristematic plant cells. *J. Cell Biol.* **89,** 685–690.

Wilbur, F. H., and Riopel, J. L. (1971). The role of cell interaction in the growth and differentiation of *Pelargonium hortorum* cells *in vitro*. II. Cell interaction and differentiation. *Bot. Gaz. (Chicago)* **132,** 193–202.

Wooding, F. B. P. (1968). Radioautographic and chemical studies of incorporation into sycamore vascular tissue walls. *J. Cell Sci.* **3,** 71–80.

Wooding, F. B. P., and Northcote, D. H. (1964). The development of the secondary wall of the xylem of *Acer pseudoplatanus*. *J. Cell Biol.* **23,** 327–337.

Wright, K., and Bowles, D. J. (1974). Effects of hormones on the polysaccharide-synthesizing membrane systems of lettuce pith. *J. Cell Sci.* **16,** 433–443.

Wright, K., and Northcote, D. H. (1972). Induced root differentiation in sycamore callus. *J. Cell Sci.* **11,** 319–337.

Wright, K., and Northcote, D. H. (1973). Differences in ploidy and degree of intercellular contact in differentiating and non-differentiating sycamore calluses. *J. Cell Sci.* **12,** 37–53.

Yeoman, M. M., and Evans, P. K. (1967). Growth and differentiation of plant tissue culture. II. Synchronous cell divisions in developing callus cultures. *Ann. Bot. (London)* **31,** 323–332.

Zaenen, I., van Larebeke, N., Teuchy, H., van Montagu, M., and Schell, J. (1974). Supercoiled circular DNA in crown gall inducing *Agrobacterium* strains. *J. Mol. Biol.* **86,** 109–127.

Zurfluh, L. L., and Guilfoyle, T. J. (1982). Auxin-induced changes in the population of translatable messenger RNA in elongating maize coleoptile sections. *Planta* **156,** 525–527.

CHAPTER 6

Photoautotrophic Growth of Cells in Culture

Wolfgang Hüsemann

Department of Plant Biochemistry
University of Münster
Münster, Federal Republic of Germany

I. INTRODUCTION

Photosynthesis is a characteristic feature of many plant cells. Thus, the development of fully functional chloroplasts in plant cell cultures will greatly increase their applicability for diverse studies.

Gautheret (1959) reported the formation of chlorophyll in callus cultures. Later, Hildebrandt *et al.* (1963) and Venketeswaran (1965), as well as Vasil and Hildebrandt (1966), studied the cultural and nutritional conditions necessary to induce chlorophyll formation in cell cultures of a variety of different plant species. The authors were successful in establishing chlorophyllous callus cultures from carrot, tomato, endive, lettuce, potato, tobacco, and some rose varieties. Later, callus as well as suspension cultures were used to follow the development of chloroplasts and photosynthetic activities under the influence of light (Laetsch and Stetler, 1965; Bergmann and Berger, 1966; McLaren and Thomas, 1967; Sunderland and Wells, 1968). At the time, it was considered doubtful that the chlorophyllous cell cultures were capable of sustained photosynthetic growth. As Yeoman and Aitchison (1973) pointed out, this was presumably due to a lack of organization, rather than a lack of expression of full synthetic potential. Since the process of greening was always accompanied by a rapid decrease in the rate of cell division (Stetler and Laetsch, 1965; Sunderland and Wells, 1968), it was thought that intercellular differentiation (chloroplast development) and rapid cell proliferation were in general inversely correlated and more or less incompatible.

Bergmann (1967) and Corduan (1970) succeeded in growing cell cultures of *Nicotiana tabacum* and *Ruta graveolens* in light in the absence of an exogenous sugar but in CO_2-enriched (1% CO_2; v/v) atmospheres photosynthetically as well as photoautotrophically. These examples clearly demonstrated that the expression of photosynthetic competence in cells for photosynthetic self-sufficiency and hence photoautotrophic growth (physiological totipotency) may be realized by physicochemical factors imposed by the culture medium and by environmental conditions. By such methods it should be possible to realize the prediction made by Haberlandt (1902) of successfully cultivating higher plant cells as photoautotrophic elementary organisms.

There are now numerous reports of sustained photoautotrophic growth of plant cell cultures in the absence of an exogenous sugar but in the presence of enriched CO_2 levels in the atmosphere of the culture vessels. The culture technique has advanced to a stage that it is possible to propagate the cells under photoautotrophic conditions in small vials (Yamada and Sato, 1978), in two-tier culture flasks (Hüsemann and Barz, 1977), and in various fermenters and continuous-culture systems (Dalton, 1980a; Bender *et al.*, 1981; Yamada *et al.*, 1981; Peel, 1982; Hüsemann, 1982, 1983). Nevertheless, there are still various problems in establishing highly chlorophyllous cells from pigment-free heterotrophic cell cultures. Our knowledge regarding the "greening" of cultured plant cells is still unsatisfactory in spite of considerable research in this field (Yamada *et al.*, 1978; Dalton, 1980b; Barz and Hüsemann, 1982; Dalton and Peel, 1983; Horn and Widholm, 1984).

The experimental data concerning the chlorophyll formation, development of fully functional chloroplasts, induction of photoautotrophism in cultured plant cells, and their growth and metabolism are reviewed in this chapter.

II. INDUCTION OF CULTURED PLANT CELLS TO PHOTOAUTOTROPHIC GROWTH

A. Culture Conditions Favoring Chlorophyll Formation, Chloroplast Development, and Photoautotrophic Growth

Major prerequisites for establishing photoautotrophic cell cultures that are capable of growing in a carbohydrate-free nutrient medium with CO_2 as the sole carbon source are a high chlorophyll content and photosynthetic competence of the cells.

Callus cells may turn green when they are exposed to light. In general, *in vitro*—propagated cells contain much less chlorophyll than the mesophyll cells from the same plant species, the ratios being often in the region of $1:10$.

1. Intensity and Quality of the Light

Chlorophyll formation is favored by light intensities up to 10,000 lx (Yamada et al., 1978). For example, increasing the light intensity from 600 to 6000 lx at low sugar concentrations raised the chlorophyll content in photoheterotrophic tobacco callus from the initial 30 µg to 60 µg chlorophyll per gram fresh weight (Yamada and Sato, 1978). In regreening photoheterotrophic callus cultures from *Arachis hypogaea* and *Kalanchoe fedtschenkoi* (Seeni and Gnanam, 1981) chlorophyll formation was saturated at 5000 lx, but optimal development of photosynthetic functions in these cultured cells required more irradiance of 14,000 lx.

Light quality is another important factor for chloroplast development in cultured plant cells. In the intact plant, the pigment system responsible for chloroplast differentiation responds to both blue and red light. In cell cultures from higher plants, chlorophyll formation and chloroplast development only proceed in blue light. This is the case for callus cultures from tobacco (Bergmann and Berger, 1966; Beauchesne and Poulain, 1966) and for potato pith tissue (Berger and Bergmann, 1967), as well as for callus cultures from *Crepis capillaris* (Hüsemann, 1970). Richter and co-workers have clearly shown a blue light–dependent synthesis of specific types of mRNAs and proteins (light-harvesting chlorophyll a,b–protein, ribulose-bisphosphate carboxylase) in tobacco cell suspensions (Richter *et al.*, 1980, 1982, 1984; Hundrieser and Richter, 1982). Additionally, the formation of 5-aminolevulinic acid was specifically enhanced by blue light in tobacco callus cells, while red light was ineffective (Kamiya *et al.*, 1983).

2. Carbohydrates

The reports of inhibitory effects of exogenous sugars on chlorophyll formation and photosynthesis in plant cell cultures are numerous (Edelman and Hanson, 1971; Kaul and Sabharwal, 1971; Neumann and Rafaat, 1973; Pamplin and Chapman, 1975; Gross and Richter, 1982). In many batch-propagated cell suspensions, for example, *Atropa belladonna* (Davey *et al.*, 1971), *Spinacia oleracea* (Dalton and Street, 1977), and *Nicotiana tabacum* (Nato *et al.*, 1977), chlorophyll formation was enhanced only after culture growth had ceased markedly. In other words, chlorophyll formation was inversely proportional to culture growth.

The stimulation of chlorophyll formation at reduced growth rates of the cell cultures coincided with sugar depletion from the medium. This led to the assumption that chlorophyll formation is inhibited by sugar excess but enhanced during sugar depletion (Dalton, 1980b). The effect of sugar supply on chlorophyll formation was examined in *Spinacia oleracea* cells grown in a continuous culture at constant dilution rates at reduced oxygen partial pressure but in excess of fructose. Halving the fructose-feed concentration resulted in a new steady state without change in the specific growth rate. Under these conditions of sugar famine both the chlorophyll content and the photosynthetic capacities of chlorophyllous cell cultures of *Spinacia oleracea* and *Ocimum basilicum* were significantly enhanced (Dalton, 1980a, 1983, 1984). These findings support the hypothesis that chlorophyll formation in cultured cells is inhibited in conditions of excess sugar but promoted during sugar famine. Similar observations were made with *Nicotiana tabacum* callus cultures (Yamada and Sato, 1978) and cell suspensions of *Solanum tuberosum* (LaRosa *et al.*, 1984).

The nature of the carbohydrate provided in the culture medium can affect chlorophyll formation. This was clearly shown by Dalton and Street (1977), Dalton (1980a), and Dalton and Peel (1983). For example, cultured cells from *Asparagus officinalis* had 15 times higher chlorophyll content when supplied with lactose than with sucrose.

Based on the published work, we know that chlorophyll formation in cultured cells is permitted as sugar concentration in the batch-culture medium is depleted and the specific growth rate declines. The mechanism of this effect is still poorly understood.

3. Phytohormones

Phytohormones are essential for cell growth and differentiation. In many cases, obviously, major emphasis has been laid on establishing culture conditions that favor vigorous growth over cytodifferentiation. This may have led to a selection against chloroplast development. For example, in cell cultures from *Oxalis dispar* (Sunderland and Wells, 1968) auxin concentrations, optimal for growth, suppressed chlorophyll formation. Similar auxin effects on chlorophyll formation were reported for cell cultures from *Atropa belladonna* (Davey *et al.*, 1971). Some other workers found the synthetic auxin 2,4-dichlorophenoxyacetic acid (2,4-D) mainly inhibitory for chlorophyll formation (Bergmann, 1967; Hüsemann and Barz, 1977; Yamada *et al.*, 1978). The substitution of 2,4-D by α-naphthaleneacetic acid (NAA) often proved beneficial for chlorophyll formation and enhanced activity of the ribulosebisphosphate carboxylase in cell cultures (Bergmann, 1967; Yamada *et al.*, 1978). But there are exceptions. Chlorophyll formation in cultured cells from *Marchantia polymorpha* (Katoh, 1983) and *Peganum harmala* (Herzbeck, 1979) was not restricted by 2,4-D. The use of indoleacetic acid (IAA) should be avoided, because this phytohormone is readily decomposed under illumination (Yamada and Sato, 1978).

Cytokinins have been reported to be essential for chlorophyll synthesis in tobacco cell cultures (Stetler and Laetsch, 1965; Seyer *et al.*, 1975). Kinetin exerts its beneficial effect on chloroplast differentiation mainly in cases in which it is not required for culture growth (Neumann and Rafaat, 1973; Seyer *et al.*, 1975) because it can stimulate chloroplast replication and maturation without inducing cell division.

Ethylene has been reported to inhibit chlorophyll formation in spinach cell suspensions (Dalton and Street, 1976). The inhibitory effect could partially be abolished by elevated CO_2 concentrations in the culture atmosphere.

4. Composition of the Gas Phase
in the Culture Vessels

By elevating the CO_2 partial pressure to 1% (v/v), Bergmann (1967) succeeded in establishing highly chlorophyllous tobacco cell cultures capable of photoautotroyhic growth.

Dalton and Street (1976) found for photoheterotrophic cell cultures of *Spinacia oleracea* that the composition of the gas phase developed in the culture vessels severely affected cell growth and chlorophyll formation. They observed that ethylene accumulation and high dissolved oxygen concentrations (about air saturation) inhibited greening. Reducing the oxygen content to about 20% of air saturation resulted in considerable chlorophyll accumulation. In order to obtain green cell suspensions from S. oleracea, the cultures were continuously flushed with a gas mixture with reduced O_2 but increased CO_2 (1%, v/v) concentrations. CO_2 was found to abolish the inhibitory effect of ethylene.

This is in agreement with observations on cell cultures from Ocimum basilicum (Dalton, 1980b), Asparagus officinalis (Peel, 1982), and Nicotiana tabacum (Yamada et al., 1981) that lowered O_2 partial pressures (20–50% of air saturation) and enriched CO_2 levels (1–2%, v/v) in the culture atmosphere were necessary prerequisites for chlorophyll formation, for increased photosynthetic competence of the cells, and hence for photoautotrophic growth.

The reason that cultured plant cell suspensions are only capable of sustained photoautotrophic growth at highly elevated CO_2 partial pressures is still unknown. Meanwhile, there are two reports that photoautotrophic cell cultures of Arachis hypogaea (Bender et al., 1981) and Gossypium hirsutum (L. C. Blair and J. M. Widholm, unpublished observation) will continue to grow photoautotrophically at atmospheric CO_2 concentrations only if they are cultured on agar plates. But since it cannot be ruled out completely that agar may serve as a carbon source, the plate work should also be done with inert material, for example, sponge material on liquid media. The requirements of higher CO_2 concentrations for photosynthesis and growth of cell cultures, as compared to the leaves, may reflect higher diffusive resistance toward CO_2 exchange as well as a low CO_2 content in the nutrient solution at pH values varying between 4.5 and 5.8 during culture growth and restrictive mass transfer of CO_2 from the medium to the cells, because of the absence of stomata and the size of the cells (Berlyn and Zelitch, 1975; Dalton and Peel, 1983). The failure of cultured cells to grow photoautotrophically in normal air may be related to higher CO_2 compensation points, as compared to intact leaves. But it could not be decided whether the observed high CO_2 compensation point was mainly due to extremely low carbonic anhydrase activity or to high dark respiration in cultured tobacco cells (Tsuzuki et al., 1981).

5. Nutritional Requirements

Vasil and Hildebrandt (1966) found that the Murashige and Skoog medium (1962) proved most satisfactory for growth and chlorophyll formation in a variety of callus cultures from different plant species.

Decreased chlorophyll formation during nitrogen starvation is a well-known phenomenon. Ammonia as well as nitrate as the sole nitrogen sources inhibited growth and greening in cell suspensions from Marchantia polymorpha (Katoh et al., 1980) and Chenopodium rubrum (W. Hüsemann, unpublished observation). A $2:1$ ratio (NO_3^-/NH_4^+) was optimum for growth and chlorophyll formation in both cell cultures.

The beneficial effects of increased phosphate levels in the culture medium on growth and photosynthetic activity of the cells have been found for a variety of cell cultures (Katoh *et al.*, 1979; Sato *et al.*, 1981; Miginiac-Maslow *et al.*, 1981; Dalton, 1983). Ferric phosphate may precipitate in the Murashige and Skoog medium (Dalton *et al.*, 1983). Therefore, care must be taken that iron does not become limiting for cell growth in case of elevated phosphate concentrations.

B. Induction of Photoautotrophic Growth

The establishment of photoautotrophic cell cultures represents a process of serial selection with special consideration of those factors that will affect chlorophyll formation and chloroplast development. The selection procedure used replaces sugar as the normal carbon and energy source from the culture medium with elevated levels of carbon dioxide and light. Only those cells that can meet their carbon and energy demand by photosynthesis will survive this selection pressure. Thus, careful visual monitoring during subculturing will result in dark-green, photoautotrophically growing cell cultures. In general, the establishment of photoautotrophic cell cultures is a long-lasting process. For example, it took about 6 months to obtain stable, fast-growing photosynthetic and photoautotrophic cell suspensions from *Chenopodium rubrum* (Hüsemann and Barz, 1977) and *Peganum harmala* (Herzbeck, 1979), as well as *Spinacia oleracea* (Dalton, 1980a) and *Glycine max* (Horn *et al.*, 1983).

Usually, the change from the photoheterotrophic to the phosynthetic mode of nutrition (sugar-free culture medium supplemented with organic growth factors) at elevated CO_2 partial pressures is accompanied by a drastic reduction in the growth rate and chlorophyll content of the cultures. Therefore, rapidly growing and highly chlorophyllous cell cultures (50–80 μg chlorophyll per gram fresh weight, equivalent to 5–8 μg chlorophyll per 10^6 cells) should be chosen as the starting material for subsequent transfer to photosynthetic or photoautotrophic (mineral salt medium without any organic constituents) growth conditions with CO_2 as the sole carbon source.

However, high chlorophyll content may not always be a reliable index for photosynthetic competence of the cells as was shown for chlorophyllous callus cultures from *Philodendron amurense* (Sato *et al.*, 1979), *Arachis hypogaea*, and *Kalanchoe fedtschenkoi* (Seeni and Gnanam, 1981) because of reduced photosystem II activities. Therefore, the photosynthetic oxygen evolution or the ratio of photosynthetic oxygen evolution and respiratory oxygen uptake as proposed by Dalton and Peel (1983) will give a more reliable impression on the photosynthetic competence of the cells.

Elevated CO_2 partial pressures (1–2% CO_2; v/v) are absolute prerequisites for selecting and maintaining photosynthethic/photoautotrophic cell cultures. Two different methods have proven satisfactory for establishing high CO_2 partial pressures in the gaseous atmosphere above the cells.

The cells are flushed with a gas mixture of known composition enriched with CO_2. By this method, the oxygen concentration can be kept low at the desired level and an accumulation of volatile metabolites is prevented (Yamada and Sato, 1978; Katoh et al., 1979; Dalton, 1980a).

High CO_2 concentrations from 0.5 to 2% CO_2 (v/v) may be established by a $KHCO_3/K_2CO_3$ buffer solution in the closed system of a two-tier culture vessel (Hüsemann and Barz, 1977; Hüsemann, 1984). Under these conditions the oxygen concentrations will be adjusted by photosynthesis of the cells and volatile metabolites may accumulate.

Several different methods for selecting cells capable of sustained photosynthetic growth in sugar-free culture medium with elevated CO_2 concentrations are described following.

1. Direct Change from Photoheterotrophic to Photosynthetic Mode of Nutrition

The chlorophyllous cells were transferred into sugar-free nutrient medium that still contained organic compounds (vitamins, phytohormones) and grown in the light at elevated CO_2 concentrations (1–2%, v/v). The organic growth factors may be omitted from the culture medium after the cells are capable of sustained photosynthetic growth in sugar-free nutrient medium.

2. Sequential Change from Photoheterotrophic to Photosynthetic Mode of Nutrition

The cells were exposed to the appropriate gaseous atmosphere (enriched CO_2 and reduced O_2 partial pressures) while the sugar concentration in the culture medium was sequentially reduced to zero. This triggered the chlorophyllous cells to photosynthetic growth.

The weaning method for a step-wise reduction of carbohydrates in the culture medium under high CO_2 and low O_2 levels in the culture atmosphere thus provides conditions for carefully adapting the cells to reduced growth rates under sugar famine. This procedure favors cytodifferentiation (chloroplast development) over cell growth.

The use of continuous cultures for selecting cells with increased photosynthetic competence by establishing different steady states of growth under sugar famine has been reported for *Spinacea oleracea* and *Ocimum basilicum* (Dalton, 1980a, 1983, 1984).

3. Callus Induction under Photosynthetic Conditions

Callus induction under photosynthetic conditions was reported by Yasuda *et al.* (1980) and is based on callus induction from explanted leaves in light at elevated CO_2 concentrations on a sugar-free culture medium still containing organic growth factors. The authors were successful in directly isolating photosyntheticaly growing callus from *Hyoscyamus niger* and *Datura stramonium*.

iii. PHOTOSYNTHETIC AND PHOTOAUTOTROPHIC GROWTH OF CULTURED PLANT CELLS

The empirical experience of the past many years allows the *in vitro* culture of plant cells in sugar-free culture medium at elevated CO_2 concentrations. Recent results on chlorophyllous cell cultures growing in a simple mineral salts medium with CO_2 as the sole carbon source document complete photoautotrophic nutrition and the "physiological totipotency" of the plant cell.

To avoid any confusion in terminology it will be advantageous to distinguish between cells that exhibit complete photoautotrophic growth and others that are photosynthetically self-sufficient in regard to their carbon and energy supply but still require some organic growth factors.

I will use the following definitions, which are similar to those of Dalton and Peel (1983).

1. Photoautotrophic cell cultures. Chlorophyllous cells that are grown in light on a simple mineral salts medium without any organic constituents and with CO_2 as the sole carbon source.

2. Photosynthetic cell cultures. Chlorophyllous cells that are growing in light on a carbohydrate-free nutrient medium containing mineral salts and organic components such as vitamins, amino acids, and phytohormones. Photosynthetic growth is regarded as a subset of photoautotrophic growth.

A list of species in which photosynthetic and photoautotrophic cell cultures have been reported is given in Table I.

TABLE I

Photosynthetic and Photoautotrophic Plant Cell Cultures

Species	Culture medium	Mode of growth	Light intensity	Gas phase	References and remarks
Amaranthus retroflexus	MS–mineral salts Nicotinamide Pyridoxine D-Biotin Choline Ca–pantothenate Thiamine Folic acid p-Aminobenzoic acid Riboflavin Cyanocobalamin NAA Kinetin	Cell suspension	$230 \mu E\ m^{-2}\ sec^{-1}$	5% CO_2 in air	C. Blair and J. Widholm, personal communication
Arachis hypogaea	MS–mineral salts Myo-inositol IAA Kinetin	Batch fermenter	7000 lx	Air	Bender *et al.* (1981). Photoautotrophic growth of the cell suspensions was limited for a 6-week culture period
Asparagus officinalis	MS–mineral salts Thiamine Nicotinic acid Pyridoxine Glycine Myo-inositol Glutamine NAA Kinetin	Continuous culture	$120\ \mu E\ m^{-2}\ sec^{-1}$	2% CO_2 + 10% O_2 + 88% N_2	Peel (1982). The cells are grown as turbidostat cultures

Species	Medium	Culture type	Light	Gas	Reference
Chenopodium rubrum	MS–mineral salts Thiamine Pyridoxine Myo-inositol Glycin Nicotinic acid	Cell suspension	$19W/m^2$	2% CO_2 in air	Hüsemann and **Barz** (1977). The cells were grown in two-tier culture vessels. CO_2 was supplied by a $KHCO_3/K_2CO_3$ buffer solution
Chenopodium rubrum	MS–mineral salts	Cell suspension	$19W/m^2$	2% CO_2 in air	Hüsemann (1981)
Chenopodium rubrum	MS–mineral salts	Batch fermenter	$19W/m^2$	2% CO_2 in air	Hüsemann (1982)
Chenopodium rubrum	MS–mineral salts	Continuous culture	$19W/m^2$	2% CO_2 in air	Hüsemann (1983)
Cytisus scoparius	MS–mineral salts 4-fold phosphate Thiamine Myo-inositol NAA BA	Callus	8000 lx	1% CO_2 + 99% N_2	Sato et al. (1981)
Datura stramonium	MS–mineral salts Thiamine Myo-inositol NAA BA	Callus	3000–5000 lx	1% CO_2 in air	Yasuda et al. (1980)
Daucus carota	MS–mineral salts Myo-inositol IAA Kinetin	Batch fermenter	7000 lx	Air	Bender et al. (1981). Photosynthetic growth was limited for 5 weeks after sugar famine
Digitalis purpurea	MS–mineral salts Thiamine IAA	Cell suspension	4×10^5 erg cm^{-2} sec^{-1}	1% CO_2 in air	Hagimori et al. (1984)
Glycine max	MS–mineral salts Nicotinamde Pyridoxine	Cell suspension	200–300 µE m^{-2} sec^{-1}	5% CO_2 in air	Horn et al. (1983)

(continued)

TABLE I (*Continued*)

Species	Culture medium	Mode of growth	Light intensity	Gas phase	References and remarks
	Biotin, choline p-Aminobenzoic acid Ca–pantothenate Thiamine Cyanocobalamin Riboflavin NAA Kinetin				
Gossypium hirsutum	MS–mineral salts Nicotinamide Pyridoxine Biotin, folic acid Choline p-Aminobenzoic acid Ca–pantothenate Thiamine Cyanocobalamin Riboflavine NAA Kinetin	Cell suspension	230 μE m^{-2} sec^{-1}	5% CO_2 in air	C. Blair and J. Widholm, unpublished personal communication
Hyoscyamus niger	MS–mineral salts Thiamine Myo-inositol NAA BA	Callus	3000–5000 lx	1% CO_2 in air	Yasuda *et al.* (1980)
Marchantia polymorpha	MS–mineral salts Increased phosphate 2,4-D	Suspension	23.2 W/m^2	1% CO_2 in air	Katoh *et al.* (1979)

224

Species	Medium components	Culture type	Light	CO_2	Reference
Morinda lucida	MS–mineral salts Thiamine Pyridoxine Myo-inositol NZ-amine, type A 2,4-D NAA IAA Kinetin	Cell suspension	15 W/m²	2% CO_2 in air	Igbavboa et al. (1985). The cells were grown in two-tier culture vessels. CO_2 was supplied by a $KHCO_3/K_2CO_3$ buffer solution
Nicotiana tabacum	MS–mineral salts Nicotinic acid Thiamine Myo-inositol Pyridoxine NAA	Cell suspension	5000 lx	1% CO_2 in air	Bergmann (1967)
Nicotiana tabacum	MS–mineral salts Nicotinic acid Thiamine Pyridoxine Myo-inositol Glycine Casein hydrolysate 2,4-D Kinetin	Cell suspension	2500 lx	2% CO_2 in air	Chandler et al. (1972)
Nicotiana tabacum	MS–mineral salts Thiamine Myo-inositol NAA	Callus	200 μE m^{-2} sec^{-1}	1% CO_2 in air	Berlyn and Zelitch (1975)
Nicotiana tabacum	LS–mineral salts Thiamine Myo-inositol NAA BA	Callus	600–14000 lx	1% CO_2 in air	Yamada and Sato (1978)

(continued)

TABLE I (Continued)

Species	Culture medium	Mode of growth	Light intensity	Gas phase	References and remarks
Nicotiana tabacum	MS–mineral salts Thiamine Myo-inositol NAA Kinetin	Batch fermenter	8000 lx	1% CO_2 + 14% O_2 + 85% N_2	Yamada et al. (1981)
Peganum harmala	MS–mineral salts Thiamine Pyridoxine Myo-inositol Glycine Nicotinic acid NAA Kinetin	Cell suspension	19 W/m^2	2% CO_2 in air	Barz et al. (1980). The cells were grown in two-tier culture vessels. CO_2 was supplied by a $KHCO_3/K_2CO_3$ buffer solution
Ruta graveolens	H–mineral salts	Callus	2000 lx	1% CO_2 + 99% N_2	Corduan (1970)
Solanum tuberosum	MS–mineral salts Thiamine Glycine Myoinositol Folic acid Biotin 2,4-D Kinetin	Cell suspension	90–110 μmol m^{-2} sec^{-1}	2% CO_2 in air	LaRosa et al. (1984)
Spinacia oleracea	MS–mineral salts	Continuous culture	41–50 mE/hr per fermenter	1% CO_2 in air at reduced oxygen content	Dalton (1980a)

A. Cell Suspension Cultures in Small Batch-Culture Volumes

1. Kinetic Studies on the Dependency of Photosynthetic Growth on CO_2 Partial Pressure and Photon Flux Density

The patterns of growth in photosynthetic cell suspensions from *Marchantia polymorpha* (Katoh, 1983) were characterized by an extended exponential and linear growth phase. Highly chlorophyllous cells (23 µg chlorophyll per 10^6 cells) were grown in a sugar-free mineral salts medium with 2,4-D as the sole organic component. Light intensity for standard culture conditions was adjusted to 90 µE m^{-2} sec^{-1}. Flat, oblong 700-ml culture flasks were used for propagating the cells by flushing with air/CO_2 mixtures (1% CO_2; v/v). Maximum growth rate during exponential growth phase corresponded to a mean doubling time of the cells of 1.08 days. During exponential growth the specific growth rates $\mu_{cell} = 0.64$, $\mu_{DW} = 0.42$, and $\mu_{chl} = 0.61$ documented unbalanced growth.

The specific growth rate (μ) of the cells for dry weight (DW) increased from 0.121/day at normal air level CO_2 concentration to maximum values of $\mu_{DW} = 0.42$ for 1% CO_2 (v/v). Higher CO_2 concentrations significantly reduced the specific growth rate (Fig. 1). The saturation constant K_s for CO_2 (at which μ_{DW} attained half its maximum value) was found to be 0.132% CO_2. The inhibition constant K_i was calculated and found to be 4.5% CO_2.

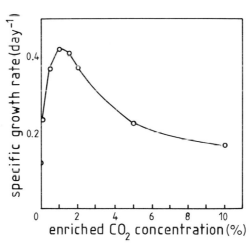

Fig. 1. Specific growth rate (μ_{DW}) at different CO_2 enrichments in the atmosphere during exponential growth phase of *Marchantia polymorpha* cell suspensions. Temperature, 25°C; photon flux density, 90 µmol m^{-2} sec^{-1}; aeration rate, 50 ml/min. From Katoh, 1983.)

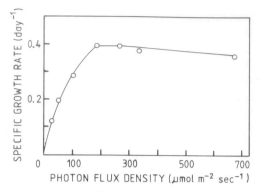

Fig. 2. Specific growth rate (μ_{DW}) at different photon flux densities during the exponential growth phase of *Marchantia polymorpha* cell suspensions. Temperature, 25°C; 1% CO_2 in air (50 ml/min). (From Katoh, 1983.)

The effect of photon flux density on the specific growth rate of photosynthetic culture growth was determined (Figs. 2 and 3). The specific growth rate μ_{DW} at low culture densities during exponential growth increased in a linear fashion with photon flux densities to maximum values of 0.39/day at 165 $\mu E\ m^{-2}\ sec^{-1}$. On the opposite, the linear growth rate (daily increase in dry weight per milliliter suspension) at high culture densities (more than 7.0×10^5 cells per milliliter) was a logarithmic function of photon flux density, representing a saturation curve.

These results correspond to the theoretical consideration of the functional relationship between photon flux density and growth found for *Chlorella ellipsoida* (Tamiya *et al.*, 1953).

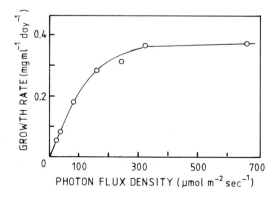

Fig. 3. Growth rate (dQ_{DW}/dt) at different photon flux densities during linear growth phase of *Marchantia polymorpha* cell suspensions. Temperature, 25°C; 1% CO_2 in air (50 ml/min). (From Katoh, 1983.)

Katoh (1983) further calculated from his experimental data that at high culture densities (approximately 7.9×10^5 cells per milliliter suspension) photon flux density became growth limiting because only the peripheral cell layers received full incident light. Growth limitation by photon flux density was also reported for continuous photosynthetic turbidostat cultures of *Asparagus officinalis* (Peel, 1982).

2. Growth Patterns in Photoautotrophic Cell Suspension Cultures

The patterns of cell number multiplication, protein, and chlorophyll accumulation were measured for photoautotrophic cell suspensions of *Chenopodium rubrum* (Hüsemann, 1981). These cells were grown under continuous light (19 W/m^2) in a mineral salts medium (Murashige and Skoog, 1962) without any organic constituents but with 2% CO_2 (v/v) in the gaseous atmosphere using the two-tier culture method (Hüsemann and Barz, 1977; Hüsemann, 1984). The cell suspensions mainly consisted of small cell groups (10–20 cells) and many single cells containing 25–30 μg chlorophyll per 10^6 cells. Photoautotrophic growth accounted for 500–600% increase in fresh weight and cell number during a 14-day growth cycle. The photosynthetic rate was 80–90 μmol CO_2 fixed per milligram chlorophyll per hour. The typical patterns of cell growth are given in Figs. 4–6.

Culture growth followed a typical sigmoid curve. The mean doubling time of the cells varied from 55 hr during exponential growth to more than 400 hr during the stationary growth phase. Inoculation of stationary growth phase cells into fresh culture medium resulted in enhanced protein synthesis, followed by an exponential phase of cell division, whereas the onset of rapid chlorophyll formation was delayed by about 2–4 days. Protein and chlorophyll accumulation finally was linear until stationary growth was reached (Fig. 4).

The changes in protein and chlorophyll content of the cells during growth cycle resulted from different rates of cell multiplication and protein as well as chlorophyll synthesis (Fig. 5). Lowest cellular contents of protein and chlorophyll were always found during exponential culture growth, later recovering to initial values (Fig. 6).

Changes in the cell composition during growth under photoautotrophic conditions were also measured for starch and sugar accumulation. Drastic reduction in starch and sugar content of the cells occurred, whereas stationary growth phase cells underwent rapid cell division after transfer to fresh culture medium. The reduced starch and sugar contents of cells during exponential growth could be attributed to lower synthesis of starch and rapid turnover of photosynthetically formed sugars (Section IV,A).

These fluctuations in the accumulation of certain cell components are

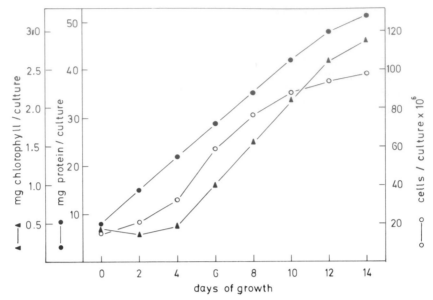

Fig. 4. Growth pattern in photoautotrophic cell suspension cultures from *Chenopodium rubrum*. Cells from stationary growth (day 14) were serially transferred into fresh culture medium and propagated under 19 W/m^2 of illumination in the presence of 2% (v/v) CO$_2$ at 26°C. The values are the means of five replicates. (From Hüsemann, 1981.)

indicative of unbalanced growth of batch-propagated photoautotrophic cells.

B. Batch-Fermenter and Continuous-Culture Growth

For most physiological and biochemical studies the use of uniform cell material at a defined stage of growth and development is desired. In view of this experimental goal, several culture units were developed that fulfill these requirements. Plant cells can be grown as mass cultures in batch fermenters or as continuous cultures under steady-state conditions in a chemostat or turbidostat (Martin, 1980; see also Chapter 3, this volume).

Plant cells are very sensitive to shearing forces generated by mechanical stirring of the suspended cells. Air-bubble turbulence, sometimes supported by low stirring, obviously is most suitable for agitating and aerating the cells in suspension if large culture volumes of 1–3 liters are desired (Wilson, 1980).

Meanwhile, batch-fermenter growth of photosynthetic and photoautotrophic cell suspensions has also been reported. Bender *et al.* (1981) culti-

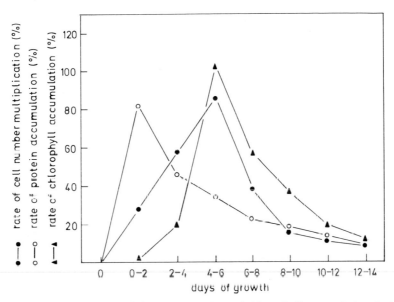

Fig. 5. Changes in the rate of cell division, protein, and chlorophyll accumulation during the growth cycle of photoautotrophic cell suspension cultures of *Chenopodium rubrum*. The total increase in cell number as well as in protein and chlorophyll content of the cell suspension cultures is determined for a 2-day interval and expressed as the percentage increase. The values are the means of five replicates. (From Hüsemann, 1981.)

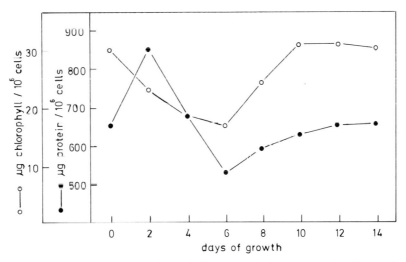

Fig. 6. Changes in the protein and chlorophyll content of photoautotrophically growing cell suspension cultures of *Chenopodium rubrum* during the growth cycle. The values are the means of five replicates. (From Hüsemann, 1981.)

vated cell suspensions of *Daucus carota* and *Arachis hypogaea* under photosynthetic conditions in a 5-liter laboratory fermenter (Biostat V, B. Braun, Melsungen, Federal Republic of Germany). Yamada *et al.* (1981) succeeded in growing tobacco cells photosynthetically in sugar-free nutrient solution in a jar fermenter by continuously flushing with a gas mixture (1% CO_2, 14% O_2, 85% N_2) to keep the CO_2 partial pressure high and oxygen concentration far below normal air level. Finally, photoautotrophic cell suspensions of *Chenopodium rubrum* were propagated in a 2-liter airlift fermenter. The growth patterns measured were identical to photoautotrophic cell suspensions cultured in small volumes in two-tier flasks (Hüsemann, 1982).

Continuous cultures of photoautotrophic spinach cells have been reported by Dalton (1980a). The author succeeded in establishing highly chlorophyllous cell suspensions capable of photosynthetic and finally photoautotrophic growth by the weaning method for gradual simplification of the culture medium. The concentration of a certain sugar, fructose, that is most beneficial for chlorophyll formation was successively reduced to establish definite steady-state conditions at reduced growth rates in continuous culture with sugar as the limiting substrate. The cells were continuously flushed with a gas mixture enriched with CO_2 (1%, v/v) but with reduced oxygen concentrations (below 5%, v/v). Thus, the cells were triggered to photosynthetic growth under conditions of sugar famine. Omitting the organic components finally resulted in true photoautotrophic growth for spinach cells under steady-state conditions. Growth under photoautotrophic conditions was found to be limited by the quantum flux density (Table II).

Peel (1982) reported the continuous photosynthetic growth of cells of *Asparagus officinalis* in a turbidostat. Steady states were maintained for more than 500 hr. A specific growth rate of 0.36/day was achieved, equivalent to a mean doubling time of 1.9 days for biomass. This is the highest growth rate obtained so far for higher plant cells propagated under photosynthetic conditions in continuous culture. It was found that quantum flux density was severely limiting growth in high-density cultures.

Photoautotrophic cell suspensions from *Chenopodium* have successfully been grown in continuous culture (Hüsemann, 1983) using an airlift culture system. The cells were grown at 2% CO_2 (v/v) generated by a $KHCO_3/K_2CO_3$ buffer. Oxygen concentration resulting from photosynthesis of the cells was maintained at a constant level of 24% (v/v). Using a dilution rate of 0.16/day, the mean generation time of the cells during steady-state growth accounted for about 100 hr. Photosynthetic CO_2 assimilation of steady-state cells was about 100 μmol CO_2 fixed per milligram chlorophyll per hour. Dark CO_2 fixation was 2% of the light values.

During continuous-culture growth, steady states were achieved with regard to nutrient uptake (phosphate, ammonia, and nitrate) and cell

TABLE II

Chloroplast Pigments, Potential Rates of Photosynthesis, and Growth of Spinach Cells in Continuous Culture for Steady States 1–5 during Transition from Photoheterotrophic to Photoautotrophic Nutrition[a]

Parameter (units)	Steady state number[b]				
	$\bar{X}_1$	$\bar{X}_2$	$\bar{X}_3$	$\bar{X}_4$	$\bar{X}_5$
Chlorophyll $a + b$(μg/g dry wt)	181	1123	1390	1160	1050
Photosynthesis (μmol O_2 per milligram chlorophyll per hour)	145	425	406	546	425
Photosynthesis/respiration	0.09	2.1	3.2	2.4	2.3
Packed cell volume (μl/ml)	112	228	213	100	108
Dilution rate (10^{-3} hr^{-1})	4.21	4.17	2.67	4.67	4.96

[a] Modified from Dalton (1980a).

[b] The mean values ($\bar{X}$) for each steady state (subscripts 1–5) are presented. The only difference between steady states 1 and 2 was the sugar (fructose) concentration. Fructose content was reduced to one-half in steady state 2 compared to steady state 1. In steady states 3–5 the medium was free of any organic components, so that photoautotrophic growth of the cells was achieved.

culture density, as well as cell protein and chlorophyll content, documenting balanced growth as shown in Fig. 7.

Equilibrium *in vitro* activities of enzymes related to different metabolic pathways were measured during continous-culture growth. Steadiness was documented by a percentage deviation of the standard error ranging below 10% as given in Table III.

The data for steady-state growth of higher plant cells in continuous culture under photoautotrophic conditions document that differentiation and multiplication of chloroplasts may occur in actively dividing cells. Therefore, the hypothesis of Sunderland and Wells (1968) and Laetsch (1971) of an inverse relationship between cell growth and chloroplast differentiation obviously cannot be applied to those cells that exclusively meet their energy and carbon requirement by photosynthesis. This is supported by electron microscopic examinations of the cells.

C. Fine Structure of Photoautotrophically Cultured Cells

Recently, the development of fully functional chloroplasts in photoheterotrophically grown cell suspensions of *Nicotiana tabacum* has been extensively studied by Nato *et al.* (1977) and Brangeon and Nato (1981). It has been shown that the cells as well as the chloroplasts underwent step-

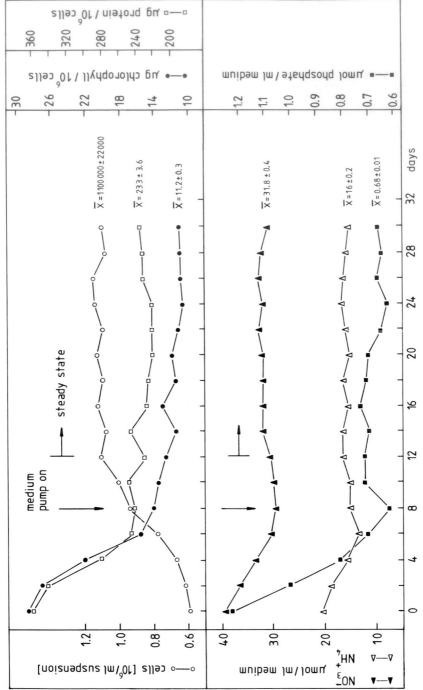

Fig. 7. Development of a steady state of several growth parameters in continuous photoautotrophic cell cultures of *Chenopodium rubrum*. The cells were propagated at 2% (v/v) CO_2 under 10,000 lx continuous illumination at 25°C. Continuous operation of the culture system was started at day 8. The values are the means of three replicates. Steadiness is documented by a standard error (SE $\bar{x}$) below 10%. (From Hüsemann, 1983.)

TABLE III

Steady-State Activities of Enzymes from Different Metabolic Pathways in Photoautotrophic Cells of Chenopodium rubrum Maintained in Continuous Culture[a]

Type of enzyme	nkat/10^6 cells
Ribulosebisphosphate carboxylase	0.287 ± 0.02
Phosphoenolpyruvate carboxylase	0.333 ± 0.03
Nitrate reductase	0.062 ± 0.01
Nitrite reductase	0.088 ± 0.01
Isocitrate dehydrogenase (NADPH-dependent)	0.650 ± 0.03
Malate dehydrogenase (NADH-dependent)	28.800 ± 2.30
Pyruvate kinase	0.180 ± 0.02

[a] From Hüsemann (1983).

[b] The data presented are the means plus or minus standard error of at least five independent determinations.

wise morphological as well as biochemical changes during the progress of culture growth (Lescure, 1978). It was assumed that during growth cycle of photosynthesizing cell cultures cytoplasmic activities, including the formation of cytoplasmic proteins, will precede chloroplast multiplication and maturation. These studies were performed with chlorophyllous cells growing photoheterotrophically in the presence of an exogenous sugar. Therefore, they do not necessarily reflect the situation in cells growing photoautotrophically.

Electron microscopic examination of the fine structure of photoautotrophically cultured cells of Chenopodium rubrum (Hüsemann et al., 1984; Figs. 8a and b) has been carried out. The chloroplasts possessed stroma and grana lamellae with starch grains. In exponential growth phase cells, the chloroplasts were surrounded by numerous mitochondria. This picture might indicate potential metabolic interactions between these organelles. The fine structure of the cells changed considerably after the culture entered stationary growth. Markedly less mitochondria were seen adjacent to the chloroplasts, which obviously contained more starch.

Measurements of metabolic activities in photoautotrophic cell suspensions of C. rubrum are in agreement with the fine structure of the cells at different growth stages. For example, higher mitochondrial activities were found during exponential growth compared to the stationary growth phase. On the other hand, substantially higher amounts of starch were accumulated in cells from stationary compared to exponential growth phase (Hüsemann et al., 1984).

An electron microscopic examination of cells from continuous photoautotrophic cultures of C. rubrum shows that a characteristic feature of

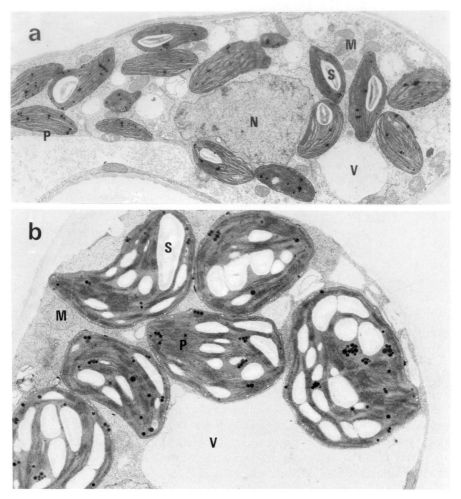

Fig. 8. Ultrastructure of cells from photoautotrophic cell suspension cultures of *Chenopodium rubrum*. (a) Exponential growth phase cell with numerous mitochondria (M) surrounding the chloroplasts (P). Magnification ×5600. (b) Stationary growth phase cell with numerous starch grains (S) accumulated in the chloroplasts (P). Magnification ×7700. (From Hüsemann *et al.*, 1984.)

the cells is the occurrence in the surroundings of the chloroplasts of numerous extremely elongated mitochondria (Fig. 9). The chloroplasts contain well-developed internal membrane systems but without any visible starch. The close grouping of large mitochondria around the chloroplasts obviously indicates morphological preconditions for strong metabolic interactions and substrate fluxes between these organelles.

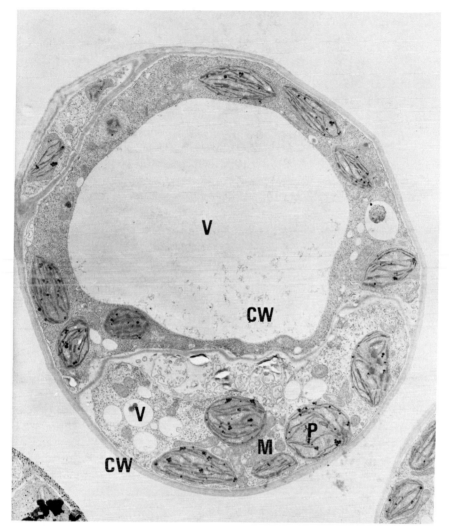

Fig. 9. Ultrastructure of *Chenopodium rubrum* cell growing photoautotrophically during steady state in continuous culture. M, mitochondria; P, chloroplast, CW, cell wall; V, vacuole. Magnification ×5000. (Photo courtesy of H. Robenek, Münster.)

D. Studies on Phytohormones

Besides *Spinacia oleracea* (Dalton, 1980a) and *Chenopodium rubrum* (Hüsemann, 1981), *Marchantia polymorpha* (Katoh, 1983) can also be grown as photoautotrophic cell cultures, documenting cytokinin as well as auxin autotrophy. For *C. rubrum* cell suspensions, reversion of photoautotrophic (green cells) to heterotrophic (colorless cells) growth conditions revealed that auxin as well as cytokinin autotrophy of the cells was not associated with fully functional chloroplasts. Furthermore, auxin-autotrophic culture growth was not reversed by exogenous auxin (W. Hüsemann, unpublished observation).

IV. METABOLIC ACTIVITIES

A. Photosynthetic CO_2 Assimilation

Photosynthesis and growth capacities of cell cultures propagated under photosynthetic and photoautotrophic conditions are documented in Table IV.

The photosynthetic CO_2 assimilation rates calculated on the chlorophyll basis were identical to the values known for mesophyll cells. In general, dark fixation rates for CO_2 varied between 2 and 5% of light-driven CO_2 assimilation.

The great similarity between photosynthetically cultured plant cells and leaves is further shown by values of pigment composition as obtained with *Peganum harmala* (Barz and Hüsemann, 1982). The data from cell cultures or leaves in micrograms per gram dry weight are as follows: Chlorophyll *a*, 3537/5100; chlorophyll *b*, 1130/1812; carotin, 357/376; lutein, 461/600; violaxanthine, 200/276; neoxanthine, 143/116; and lutein-5,6-epoxide and antheraxanthine, 60/44.

As shown by short-term $^{14}CO_2$ fixation studies in photoautotrophic cell cultures derived from C_3 plants, CO_2 is predominantly assimilated by $C_1 \rightarrow C_5$ carboxylation through ribulosebisphosphate carboxylase via the Calvin cycle. But there is a marked increase in ^{14}C label of C_4 carboxylic acids in exponentially dividing cells (10%) as compared to cells from stationary growth (5%), as shown for photosynthetic and photoautotrophic cell suspensions of *Chenopodium rubrum* (Hüsemann et al., 1979, Hüsemann, 1981).

Similar results were obtained with photosynthetic callus cultures of C_3 plants *Cytisus scoparius* and *Nicotiana tabacum* (Nishida et al., 1980). The incorporation of $^{14}CO_2$ into C_4 acids, especially malate, accounted for up to 30% of total fixed ^{14}C radioactivity. In this case, the high rates of carbon incorporation into malate via the phosphoenolpyruvate (PEP-) carboxylase were not obtained under natural culture conditions but in phosphate buffer pH 7.7. High pH values are known to shift CO_2 fixation toward increased incorporation into C_4 acids as shown for spinach cells (Böcher and

Kluge, 1977) and photoautotrophic cell cultures of *Chenopodium rubrum* (Herzbeck and Hüsemann, 1985).

Obviously, it seems to be a rule that chlorophyllous cell cultures of C_3 plants growing photoheterotrophically in the presence of an exogenous sugar will incorporate unusually high amounts of CO_2 into C_4 acids, probably via PEP-carboxylase (Nato *et al.*, 1977; Hüsemann *et al.*, 1979; Nishida *et al.*, 1980; Sato *et al.*, 1980; Seeni and Gnanam, 1982).

Thus, the mode of nutrition (photoheterotrophic growth in the presence of an exogenous sugar) and the physiological state of photoautotrophically cultured cells (exponential compared to stationary growth phase) will regulate the pathway of photosynthetic CO_2 assimilation.

In vitro measurements of carboxylation reactions partially corroborate the results from $^{14}CO_2$ incorporation studies. As so far measured, photosynthesizing cell cultures possessed substantially higher *in vitro* activities of the PEP-carboxylase compared to the plants from which they were derived (Nato and Mathieu, 1978). As found for photoautotrophically cultured cells of *Chenopodium rubrum*, during growth cycle the ratio of RuDP carboxylase to PEP-carboxylase changed from nearly 1.0 at the phase of maximum cell division to about 4.0 during stationary growth.

In agreement with some other workers (Yamada *et al.*, 1982; Nato *et al.*, 1981), I assume that PEP-carboxylase opens an anapleurotic pathway for providing additional organic acids to the tricarboxylic acid cycle as carbon skeletons for synthetic reactions, such as amino acid and chlorophyll synthesis and acetyl-CoA formation for fatty acid synthesis.

In photoautotrophic cell suspensions of *C. rubrum*, $^{14}CO_2$ fixation rates were markedly lowered during transition from exponential (85 μmol CO_2 per milligram chlorophyll per hour) to late stationary growth phase (60 μmol CO_2 per milligram chlorophyll per hour). This might be due to some kind of end-product inhibition, since the reduction in CO_2 assimilation was accompanied by an increase in starch and sugar accumulation, reaching values of 1–2 mg carbohydrates per gram fresh weight. High cellular sugar concentrations are known to inhibit photosynthesis in cell cultures (Neumann and Rafaat, 1973; LaRosa *et al.*, 1984) and in mesophyll cells (Herold *et al.*, 1980; Azcon-Bieto, 1983).

Moreover, at late stationary growth phase during growth cycle of photoautotrophic cell suspensions of *C. rubrum*, the culture medium is nearly depleted of phosphate and nitrogen (Hüsemann, 1981); thus, growth and metabolism may be limited by nutrient depletion. Recovery of the cells to full photosynthetic activity was realized only after 2-day incubation of stationary phase cells in fresh culture medium, corresponding to the lag phase for culture growth.

Recent studies on photoautotrophic cell suspensions of *C. rubrum* have provided definite examples for marked changes in photosynthetic product

TABLE IV

Growth, Chlorophyll Content, and Photosynthetic Activities in Photoautotrophic Cell Cultures

Species	Mode of Culture	Chlorophyll content	Growth rate[a]	Photosynthetic activity	References
Asparagus officinalis[b]	Continuous culture	2300 μg/g DW	2 days (doubling time)	230 μmol O_2 per milligram chlorophyll per hour	Peel (1982)
Chenopodium rubrum	Cell suspension	360 μg/g FW	6.0 (2 weeks)	90 μmol CO_2 per milligram chlorophyll per hour	Hüsemann (1981)
Chenopodium rubrum	Batch fermenter	30 μg/10^6 cells	2.7 (2 weeks)	90 μmol CO_2 per milligram chlorophyll per hour	Hüseman (9182)
Chenopodium rubrum	Continuous culture	11 μg/10^6 cells	4.2 days (doubling time)	90 μmol CO_2 per milligram chlorophyll per hour	Hüsemann (1983)
Cytisus scoparius[b]	Callus	70 μg/g FW	6.4 (6 weeks)	64 μmol CO_2 per milligram chlorophyll per hour	Yamada and Sato (1978)
Digitalis purpurea[b]	Cell suspension	220 μg/g FW	3.0 (3 weeks)	19 μmol O_2 per gram fresh weight per hour	Hagimori et al. (1984)

Species	Culture type	Yield	Growth rate[a]	Photosynthetic rate	Reference
Glycine max[b]	Cell suspension	400–600 µg/g FG	10–14 (2 weeks)	83 µmol CO_2 per milligram chlorophyll per hour	Horn et al. (1983)
Marchantia polymorpha[b]	Cell suspension	23 µg/10⁶ cells	37.0 (25 days)	66 µmol CO_2 per milligram chlorophyll per hour	Katoh (1983)
Nicotiana tabacum[b]	Callus	30 µg/g FW	3.3 (3 weeks)	200 µmol CO_2 per milligram chlorophyll per hour	Berlyn and Zelitch (1975)
Nicotiana tabacum[b]	Cell suspension	17 µg/g FW	1.7 (2 weeks)	500 µmol O_2 per milligram chlorophyll per hour	Chandler et al. (1972)
Peganum harmala[b]	Cell suspension	300 µg/g FW	3.0 (2 weeks)	150 µmol CO_2 per milligram chlorophyll per hour	Herzbeck (1979)
Solanum tuberosum[b]	Cell suspension	190 µg/g FW	4.0 (2 weeks)	80 µmol CO_2 per milligram chlorophyll per hour	LaRosa et al. (1984)
Spinacia oleracea	Continuous culture	1160 µg/g DW	6 days (doubling time)	425 µmol O_2 per milligram chlorophyll per hour	Dalton (1980a)

[a] Growth rate = ratio : harvest/inoculum. FW = Fresh weight; DW = Dry weight

[b] Cell culture growth was defined to be photosynthetically following the definition in the text (Section III). Cells were grown at enriched CO_2 concentrations in carbohydrate-free culture medium supplemented with organic growth factors.

formation and metabolic carbon flow during different growth stages (Hüsemann *et al.*, 1984). In each growth phase, most $^{14}CO_2$ was incorporated into soluble sugars and starch. During transition from exponential to stationary growth, the decrease in carbon incorporation into sugars was accompanied by an increased flow of carbon into starch. As compared to nondividing cells from stationary growth, rapidly dividing cells incorporated higher amounts of photosynthetically assimilated CO_2 into amino acids, proteins, organic acids, lipids, and structural components of the cells (Fig. 10).

Metabolic flow among the photosynthetic intermediates was measured by pulse–chase experiments and was found to be directly correlated to the different growth stages of cells. This was best exemplified by the turnover of photosynthetically formed sugars and organic acids and by the amount of carbon channeled into protein and structural components of the cells, both being highest during active culture growth.

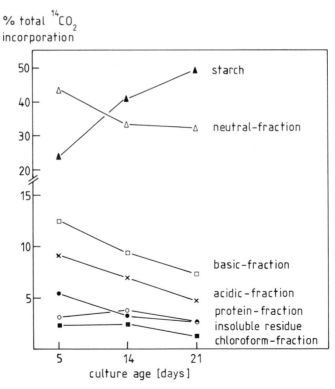

Fig. 10. Carbon-14 incorporation into photosynthetic products after 1 hr $^{14}CO_2$ photosynthesis in photoautotrophic cell suspensions of *Chenopodium rubrum* at different growth stages. (Modified from Hüsemann *et al.*, 1984.)

TABLE V

Development of Enzyme Activities in Cells from Photoautotrophic Cell Suspensions of Chenopodium rubrum at Different Stages of the Growth Cycle[a]

	nkat/10^6 cells[b]		
	Day 5	Day 14	Day 21
RuDP-carboxylase	0.80 ± 0.05	1.67 ± 0.09	1.88 ± 0.20
PEP-carboxylase	0.65 ± 0.04	0.52 ± 0.02	0.42 ± 0.05
NADP-glyceraldehyde-3-phosphate dehydrogenase	0.76 ± 0.08	1.07 ± 0.20	1.30 ± 0.25
NAD-glyceraldehyde-3-phosphate dehydrogenase	3.17 ± 0.40	2.20 ± 0.35	1.97 ± 0.60
Pyruvate kinase	0.88 ± 0.04	0.87 ± 0.04	0.32 ± 0.05
NADP-malate dehydrogenase	0.45 ± 0.05	0.90 ± 0.18	0.75 ± 0.23
NAD-malate dehydrogenase	24.00 ± 5.00	23.30 ± 4.30	14.10 ± 3.30
NADP-isocitrate dehydrogenase	0.78 ± 0.03	0.72 ± 0.08	0.37 ± 0.03
NAD-isocitrate dehydrogenase	0.10 ± 0.01	0.11 ± 0.01	0.05 ± 0.01
Cytochrome c oxidase	0.33 ± 0.10	0.09 ± 0.01	0.10 ± 0.01

[a] From Hüsemann et al. (1984).
[b] Variations of the average are expressed as the standard error of the mean.

Differences in carbon metabolism in photoautotrophic cell cultures of C. rubrum during different growth stages were equally expressed by the in vitro activities of enzymes related to different metabolic pathways as shown in Table V. Cytoplasmic and mitochondrial activities dominated during exponential culture growth, while chloroplast-related activities increased after transition to stationary growth.

These findings confirm the proposal of sequential development of cytoplasmic and chloroplastic activities during the growth cycle of cultured plant cells (Nato et al., 1981).

B. Photorespiratory Activities

Since photorespiration is closely related to photosynthesis, it has been investigated whether photorespiratory activities are expressed in photoautotrophic cell cultures. Berlyn et al. (1978) studied photorespiratory properties of photosynthetic callus cultures of Nicotiana tabacum. They found [14C]glycolate formation after $^{14}CO_2$ photosynthesis to be three times higher in photosynthetic than in photoheterotrophic cells. External [14C]glycolate was rapidly metabolized to $^{14}CO_2$, [14C]serine, and [14C]-glycine. Photorespiration in this system was also observed by oxygen

TABLE VI

In Vitro Activities of Enzymes from Photorespiration[a]

	nmol substrate per milligram protein per minute[b]		
	Glycolate-oxidase	Hydroxypyruvate reductase	Serine-glyoxylate aminotransferase
Chenopodium rubrum			
Intact plant	95 ± 10	144 ± 12	116 ± 6
Cell suspension culture[c]			
Photoheterotrophic	8 ± 2	47 ± 5	N.D.[d]
Photoautotrophic			
Auxin-requiring	18 ± 4	74 ± 5	29 ± 4
Auxin-autotrophic	21 ± 4	80 ± 8	33 ± 4
Flushed with air	32 ± 4	130 ± 15	43 ± 7

[a] Modified from Storck (1979).
[b] The data are the means plus or minus standard errors of five independent determinations.
[c] The photoheterotrophic cell cultures were grown in Murashige and Skoog medium with 2% sucrose. The photoautotrophic cell suspensions were propagated in sugar-free Murashige and Skoog mineral salts medium (supplemented with 2,4-D) at 1% (v/v) CO_2. Light intensity was adjusted to 19 W/m².
[d] N.D. = not detected.

inhibition of CO_2 assimilation. The activities of some photorespiratory enzymes were measured in *C. rubrum* cell suspensions (Table VI). Enzyme activities increased progressively while the cells attained higher degrees of photoautotrophism. Lowering the CO_2 partial pressure to normal air level resulted in a further enhancement of enzyme activities.

Although photoautotrophic cell cultures obviously possess an active photorespiratory system, their full expression under natural culture conditions, i.e., 1–2% CO_2 (v/v) and approximately 25% O_2 (v/v) in the atmosphere above the cells, might be restricted. It is known that high CO_2 concentrations will competitively inhibit the oxygenase function of the ribulosebisphosphate carboxylase.

C. Studies on Membrane Transport

Photoautotrophically cultured cells of *C. rubrum* resemble a mesophyll cell in their photosynthetic capacities (Hüsemann *et al.*, 1979) and their membrane-bound lipids (Hüsemann *et al.*, 1980). Membrane transport studies have documented that this relationship also exists for the water transport properties (Büchner *et al.*, 1980), for the protonmotive force-dependent transport of amino acids (Steinmüller and Bentrup, 1981) and hexoses (Gogarten and Bentrup, 1984), and for the basic electrophysiological membrane properties (Ohkawa *et al.*, 1981). A high membrane potential and resistance, both indicating a high degree of membrane selectivity and integrity, were found for photoautotrophically cultured *C. rubrum* cells.

D. Secondary Metabolism

Photoautotrophic cell cultures can be a valuable tool for studies on secondary metabolism as a consequence of the biosynthetic capacity of fully functional chloroplasts (Barz and Hüsemann, 1982; Dalton and Peel, 1983; Horn and Widholm, 1984). Comparative measurements of secondary metabolites in heterotrophic, photoheterotrophic, and photosynthetic/photoautotrophic cell cultures can reveal how the mode of nutrition together with the inherent specific pattern of the biosynthetic pathway will determine the potential of the cultured cells for the production of secondary metabolites.

1. Lipids

Lipids and constituent fatty acids have been extensively studied in photosynthetic cell cultures of *Chenopodium rubrum* (Hüsemann *et al.*, 1980), *Peganum harmala* (Barz *et al.*, 1980, and *Glycine max* (Martin *et al.*, 1984). The biochemical transformations associated with the change of cell cultures of *C. rubrum* from the heterotrophic to the photosynthetic mode of nutrition are reflected in the composition of the lipid classes and the patterns of their constituent fatty acids (Radwan *et al.*, 1979a,b). Due to the differentiation of chloroplasts, photosynthetic cell cultures were characterized by much higher amounts of monogalactosyldiacylglycerols and digalactosyldiacylglycerols as well as fair amounts of sulfoquinovosyldiacylglycerols and diacylglycerophosphoglycerols compared to heterotrophic cells from which they have been derived. Photosynthetic cell cultures are therefore similar to green leaves in regard to their lipid composition. The lipids of photosynthetic cells always contained higher proportions of linolenic acid but lower concentrations of linoleic acid than those of heterotrophic cultures (Hüsemann *et al.*, 1980).

These data show that photosynthetic cell cultures are suitable systems for research in the role of lipids in photosynthesis and in activating the biosynthesis of specific metabolites.

2. Other Metabolites

It has been shown that the presence of photosynthetically active chloroplasts may be coupled with striking changes in the quinone and alkaloid formation in cell cultures. Cell cultures from Rubiaceae, when grown heterotrophically in the presence of exogenous sugar and naphthalenic acid (NAA) as the auxin, produced large amounts of anthraquinones (Zenk, 1978). Chlorophyllous, photosynthetically cultured cells of *Morinda lucida* produced no anthraquinones but substantial amounts of lipophilic quinones, mainly plastoquinone and ubiquinone. The content in phylloquinone (vitamin K_1) and tocopherols accounted for 30–40% of the leaf values. The synthesis of vitamin K_1 is known to be correlated to the chloroplasts. After transition of the cells from the photosynthetic to the heterotrophic mode of nutrition, anthraquinones were produced, while at the same time the lipophilic quinones of the chloroplasts disappeared. Thus, in cultured cells photoautotrophy correlates with lipoquinone synthesis, and heterotrophy correlates with anthraquinone synthesis. This reflects the situation in the intact plants in which lipoquinones are chloroplast associated, whereas anthraquinones are synthesized in the roots (Igbavboa *et al.*, 1985). The underlying reasons for the observed change in the regulation of quinone biosynthesis are still unknown. Furthermore, recent data with green cell cultures of *Lupinus polyphyllus* have documented that the synthesis of the quinolizidine alkaloids lupanine and sparteine are associated with the chloroplasts (Wink and Hartmann, 1980).

Perhaps photosynthetic/photoautotrophic plant cell culture used as an experimental tool will contribute to a better understanding of the regulatory parameters displayed in chloroplast-associated secondary metabolism.

V. CONCLUSIONS

The photoautotrophic growth of higher plant cells in culture has confirmed Haberlandt's assumption. Photoautotrophically cultured higher plant cells display physiological totipotency and provide a favorable tool for studying cytodifferentiation, i.e., the factors controlling chloroplast differentiation.

Photoautotrophic cell cultures offer the possibility to separate leaf structure from function. Therefore, they will aid in understanding whether C_4 photosynthesis and crassulacean acid metabolism (CAM) may function in isolated cells cultured under photoautotrophic conditions.

Photoautotrophic cell cultures will aid in a better understanding of the regulation of photosynthetic CO_2 assimilation, subsequent carbon metabolism, and interrelationships between the assimilative and dissimilative metabolic pathways in the cells at different growth stages.

Continuous photoautotrophic cell cultures are an outstanding tool for measuring photosynthetic yields and balanced growth depending on nutrient supply and physicochemical factors limiting growth and cell development. Possibly, the full expression of the biochemical properties of the chloroplasts in cells grown under controlled conditions will provide a better insight in those metabolic pathways that are related to these organelles.

Finally, photoautotrophic cell cultures may represent a favorable system for studying the biochemical and physiological actions of xenobiotics and phytotoxins on growth, differentiation, and metabolism of the "green" cell.

ACKNOWLEDGMENT

I wish to thank Dr. Kumar D. Mukherjee for helpful discussions and for critically reading the manuscript. The authors' studies on photoautotrophic cell cultures have been supported by the Deutsche Forschungsgemeinschaft.

REFERENCES

Azcon-Bieto, J. (1983). Inhibition of photosynthesis by carbohydrates in wheat leaves. *Plant Physiol.* **73**, 681–686.

Barz, W., and Hüsemann, W. (1982). Aspects of photoautotrophic cell suspension cultures. In "Plant Tissue Culture 1982" (A. Fujiwara, ed.), pp. 245–248, Maruzen, Tokyo.

Barz, W., Herzbeck, H., Hüsemann, W., Schneiders, G., and Mangold, H. (1980). Alkaloids and lipids of heterotrophic, photomixotrophic, and photoautotrophic cell suspension cultures of *Peganum harmala*. *Planta Med.* **40**, 127–136.

Beauchesne, G., and Poulain, M. C. (1966). Influence des éclairements approximativement monochromatique sur le dévelopment des tissus de moelle de tabac cultivés *in vitro* en presence d'auxine et de kinitine. *Photochem. Photobiol.* **5**, 157–167.

Bender, L., Kumar, A., and Neumann, K. H. (1981). Photoautotrophe pflanzliche Gewebekulturen in Laborfermentern. In "Fermentation" (R. M. Lafferty, ed.), pp. 193–203. Springer-Verlag, Berlin and New York.

Berger, C., and Bergmann, L. (1967). Farblicht und Plastidendifferenzierung im Speichergewebe von *Solanum tuberosum*. *Z. Pflanzenphysiol.* **56**, 439–445.

Bergmann, L. (1967). Wachstum grüner Suspensionskulturen von *Nicotiana tabacum* var. "Samsun" mit CO_2 als einziger Kohlenstoffquelle. *Planta* **74**, 243–249.

Bergmann, L., and Berger, C. (1966). Farblicht und Plastidendifferenzierung in Zellkulturen von *Nicotiana tabacum* var. "Samsun." *Planta* **69**, 58–69.

Berlyn, M. B., and Zelitch, I. (1975). Photoautotrophic growth and photosynthesis in tobacco callus cells. *Plant Physiol.* **56**, 752–756.

Berlyn, M. B., Zelitch, I., and Beaudette, P. D. (1978). Photosynthetic characteristics of photoautotropically grown tobacco callus cells. *Plant Physiol.* **61**, 606–610.

Böcher, M., and Kluge, M. (1977). Der C_4-Weg der C-Fixierung bei *Spinacea oleracea*. I. [^{14}C]Markierungsmuster suspendierter Blattstreifen unter dem Einfluss des Suspensionsmediums. *Z. Pflanzenphysiol.* **80**, 81–86.

Brangeon, J., and Nato, A. (1981). Heterotrophic tobacco cell cultures during greening. I. Chloroplast and cell development. *Physiol. Plant.* **53**, 327–334.

Büchner, K. H., Zimmermann, U., and Bentrup, F. W. (1981). Turgor pressure and water transport properties of suspension-cultured cells of *Chenopodium rubrum*. *Planta* **151**, 95–102.

Chandler, M. T., de Marsac, N. T., and de Kouchkovsky, Y. (1972). Photosynthetic growth of tobacco cells in liquid suspensions. *Can. J. Bot.* **50**, 2265–2270.

Corduan, G. (1970). Autotrophe Gewebekulturen von *Ruta graveolens* und deren [$^{14}CO_2$] Markierungsprodukte. *Planta* **91**, 291–301.

Dalton, C. C. (1980a). Photoautotrophy of spinach cells in continuous culture: Photosynthetic development and sustained photoautotrophic growth. *J. Exp. Bot.* **31**, 791–804.

Dalton, C. C. (1980b). The biotechnology of green-cell cultures. *Biochem. Soc. Trans.* **8**, 475–477.

Dalton, C. C. (1983). Photosynthetic development of *Ocimum basilicum* cells on transition from phosphate to fructose limitation. *Physiol. Plant.* **59**, 623–626.

Dalton, C. C. (1984). The effect of sugar supply rate on photosynthetic development of *Ocimum basilicum* (sweet basil) cells in continuous culture. *J. Exp. Bot.* **35**, 505–516.

Dalton, C. C., and Peel, E. (1983). Product formation and plant cell specialization: A case study of photosynthetic development in plant cell cultures. *Prog. Ind. Microbiol.* **17**, 109–166.

Dalton, C. C., and Street, H. E. (1976). The role of the gas phase in the greening and growth of illuminated cell suspension cultures of *Spinacia oleracea*. *In Vitro* **7**, 485–493.

Dalton, C. C., and Street, H. E. (1977). The influence of applied carbohydrates on the growth and greening of cultured spinach (*Spinacia oleracea*) cells. *Plant Sci. Lett.* **10**, 157–164.

Dalton, C. C., Iqbal, K., and Turner, D. A. (1983). Iron phosphate precipitation in Murashige and Skoog media. *Physiol. Plant.* **57**, 472–476.

Davey, M. R., Fowler, M. M., and Street, H. E. (1971). Cell clones contrasted in growth, morphology and pigmentation isolated from a callus culture of *Atropa belladonna* var. Lutea. *Phytochemistry* **10**, 2559–2575.

Edelman, J., and Hanson, A. D. (1971). Sucrose suppression of chlorophyll synthesis in carrot callus cultures. *Planta* **98**, 150–156.

Gautheret, R. J. (1959). "La Culture des Tissus Végétaux: Techniques et Réalisation," pp. 341–343. Masson, Paris.

Gogarten, J. P., and Bentrup, F.-W. (1984). Properties of a hexose carrier at the plasmalemma of green suspension cells from *Chenopodium rubrum* L. *In* "Membrane Transport in Plants" (W. J. Cram, K. Janacek, R. Rybova, and K. Sigler, eds.), pp. 183–189. Czechoslovak Academy of Sciences, Prague.

Gross, M., and Richter, G. (1982). Influence of sugars on blue light-induced synthesis of chlorophyll in cultured plant cells. *Plant Cell Rep.* **1**, 288–290.

Haberlandt, G. (1902). Culturversuche mit isolierten Pflanzenzellen. *Sitzungsber. Akad. Wiss. Wien, Math.-Naturwiss. Kl. Abt. 1*, **111**, 69–92.

Hagimori, M., Matsumoto, T., and Mikami, Y. (1984). Photoautotrophic culture of undifferentiated cells and shoot-forming cultures of *Digitalis purpurea* L. *Plant Cell Physiol.* **25**, 1099–1102.

Herold, A., Mc Gee, E. E. M., and Lewis, D. H. (1980). The effect of orthophosphate concentration and exogenously supplied sugars on the distribution of newly fixed carbon in sugar beet leaf discs. *New Phytol.* **85**, 1–13.

Herzbeck, H. (1979). Untersuchungen über Anlage, Wachstum und Inhaltsstoffe in Zellsuspensionskulturen von *Peganum harmala*. Thesis, Münster Univ., W. Germany.

Herzbeck, H., and Hüsemann, W. (1985). Photosynthetic carbon metabolism in photoautotrophic cell suspension cultures of *Chenopodium rubrum.In* "Primary and Secondary Metabolism of Plant Cell Cultures," 1984. Springer-Verlag, Berlin and New York.

Hildebrandt, A. C., Wilmar, J. C., Johns, H., and Riker, A. J. (1963). Growth of edible chlorophyllous plant tissue *in vitro*. *Am. J. Bot.* **50**, 248–254.

Horn, M. E., and Widholm, J. M. (1984). Aspects of photosynthetic plant tissue cultures. *In* "Application of Genetic Engineering to Crop Improvement" (G. B. Collins, and J. F. Petolino, eds.), pp. 113–161. M. Nijhoff/Dr. W. Junk Publ.

Horn, M. E., Sherrard, J. H., and Widholm, J. M. (1983). Photoautotrophic growth of soybean cells in suspension culture. *Plant Physiol.* **72**, 426–429.

Hundrieser, J., and Richter, G. (1982). Blue light-induced synthesis of ribulosebisphosphate carboxylase. *Plant Cell Rep.* **1**, 115–118.

Hüsemann, W. (1970). Der Einfluss verschiedener Lichtqualitäten auf Chlorophyllgehalt und Wachstum von Gewebekulturen aus *Crepis capillaris* (L.). *Plant Cell Physiol.* **11**, 315–322.

Hüsemann, W. (1981). Growth characteristics of hormone and vitamin independent photoautotrophic cell suspension cultures from *Chenopodium rubrum*. *Protoplasma* **109**, 415–431.

Hüsemann, W. (1982). Photoautotrophic growth of cell suspension cultures from *Chenopodium rubrum* in an airlift fermenter. *Protoplasma* **113**, 214–220.

Hüsemann, W. (1983). Continuous culture growth of photoautotrophic cell suspensions from *Chenopodium rubrum*. *Plant Cell Rep.* **2**, 59–62.

Hüsemann, W. (1984). Photoautotrophic cell cultures. *In* "Cell Culture and Somatic Cell Genetics of Plants" (I. K. Vasil, ed.), Vol. I, Laboratory Techniques, pp. 182–191. Academic Press, New York.

Hüsemann, W., and Barz, W. (1977). Photoautotrophic growth and photosynthesis in cell suspension cultures of *Chenopodium rubrum*. *Physiol. Plant.* **40**, 77–81.

Hüsemann, W., Plohr, A., and Barz, W. (1979). Photosynthetic characteristics of photomixotrophic and photoautotrophic cell suspension cultures of *Chenopodium rubrum*. *Protoplasma* **100**, 101–112.

Hüsemann, W., Radwan, S. S., Mangold, H. K., and Barz, W. (1980). The lipids in photoautotrophic and heterotrophic cell suspension cultures of *Chenopodium rubrum*. *Planta* **147**, 379–383.

Hüsemann, W., Herzbeck, H., and Robenek, H. (1984). Photosynthesis and carbon metabolism in photoautotrophic cell suspensions of *Chenopodium rubrum* at different phases of batch growth. *Physiol. Plant* **62**, 349–355.

Igbavboa, U., Sieweke, H.-J., Leistner, E., Röwer, I., Hüsemann, W., and Barz, W. (1985). Alternative formation of anthraquinones and lipoquinones in heterotrophic and photoautotrophic cell suspension cultures of *Morinda lucida* Benth. *Planta* (in press).

Kamiya, A., Ikegami, I., and Hase, E. (1983). Effects of light on chlorophyll formation in cultured tobacco cells. II. Blue light effect on 5-aminolevulinic acid formation. *Plant Cell Physiol.* **24**, 799–809.

Katoh, K. (1983). Kinetics of photoautotrophic growth of *Marchantia polymorpha* cells in suspension culture. *Physiol. Plant.* **59**, 242–248.

Katoh, K., Ohta, Y., Hirose, Y., and Iwamura, T. (1979). Photoautotrophic growth of *Marchantia polymorpha* L. cells in suspension culture. *Planta* **144**, 509–510.

Katoh, K., Ishikawa, M., Miyake, K., Ohta, Y., Hirose, Y., and Iwamura, T. (1980). Nutrient utilization and requirement under photoheterotrophic growth of *Marchantia polymorpha:* Improvement of the culture medium. *Physiol. Plant.* **49**, 241–247.

Kaul, K., and Sabharwal, P. S. (1971). Effects of sucrose and kinetin on growth and chlorophyll synthesis in tobacco tissue cultures. *Plant Physiol.* **47**, 691–695.

Laetsch, W. M. (1971). Plant tissue culture. *In* "Methods in Enzymology" (A. San Pietro, ed.), Vol. 23, pp. 96–109. Academic Press, New York.

Laetsch, W. M., and Stetler, D. A. (1965). Chloroplast structure and function in cultured tobacco tissue. *Am. J. Bot.* **52**, 798–804.

LaRosa, P. C., Hasegawa, P. M., and Bressan, R. A. (1984). Photoautotrophic potato cells: Transition from heterotrophic to autotrophic growth. *Physiol. Plant.* **61**, 279–286.

Lescure, A.M. (1978) Choloroplast differentiation in cultured tobacco cells: *In vitro* protein synthesis efficiency of plastids at various stages of their evolution. *Cell Differ.* **7**, 139-152.

McLaren, I., and Thomas, D. R. (1967). CO_2 fixation, organic acids and some enzymes in green and colorless tissue cultures of *Kalanchoe crenata. New Physiol.* **66**, 683–695.

Martin, S. M. (1980). Mass culture systems for plant cell suspensions. In "Plant Tissue Culture as a Source of Biochemicals" (E. J. Staba, ed.), pp. 149–166. CRC Press, Boca Raton, Flordia.

Martin, A. A., Horn, M. E., Widholm, J. M., and Rinne, R. W. (1984). Synthesis, composition and location of glycerolipids in photoautotrophic soybean cell cultures. *Biochim. Biophys. Acta* **796**, 146–154.

Miginiac-Maslow, M., Mathieu, Y., Nato, A., and Hoarau, A. (1981). Contribution of photosynthesis to the growth and differentiation of cultured tobacco cells. The regulatory role of inorganic phosphate. *In* "Phytosynthesis" (G. Akoyunoglou, ed.), V. Chloroplast Development, pp. 977–984. Balaban International Science Service, Philadelphia, Pennsylvania.

Murashige, T., and Skoog, F. (1962). A revised medium for rapid growth and bioassay with tobacco cultures. *Physiol. Plant.* **15**, 473–479.

Nato, A., and Mathieu, Y. (1978). Changes in phosphoenolpyruvate carboxylase and ribulosebisphosphate carboxylase activities during the photoheterotrophic growth of *Nicotiana tabacum* (cv. Xanthi) cell suspensions. *Plant Sci. Lett.* **13**, 49–56.

Nato, A., Bazetoux, S., and Mathieu, Y. (1977). Photosynthetic capacities and growth characteristics of *Nicotiana tabacum* cell suspension cultures. *Physiol. Plant.* **41**, 116–123.

Nato, A., Mathieu, Y., and Brangeon, J. (1981). Heterotrophic tobacco cell cultures during greening. II. Physiological and biochemical aspects. *Physiol. Plant.* **53**, 335–341.

Neumann, K. H., and Rafaat, A. (1973). Further studies on the photosynthesis of carrot tissue cultures. *Plant Physiol.* **51**, 685–690.

Nishida, K., Sato, F., and Yamada, Y. (1980). Photosynthetic carbon metabolism in photoautotrophically and photomixotrophically cultured tobacco cells. *Plant Cell Physiol.* **21**, 47–55.

Ohkawa, T., Köhler, K., and Bentrup, F. W. (1981). Electrical membrane potential and resistance in photoautotrophic suspension cells of *Chenopodium rubrum* L. *Planta* **151**, 88–94.

Pamplin, E. J., and Chapman, J. M. (1975). Sucrose suppression of chlorophyll synthesis in tissue culture: Changes in the activity of the enzymes of the chlorophyll biosynthetic pathway. *J. Exp. Bot.* **26**, 212–220.

Peel, E. (1982). Photoautotrophic growth of suspension cultures of *Asparagus officinalis* L. cells in turbidostats. *Plant Sci. Lett.* **24**, 147–155.

Radwan, S. S., Mangold, H. K., Hüsemann, W., and Barz, W. (1979a). Lipids in plant tissue cultures. VII. Heterotrophic and mixotrophic cell cultures of *Chenopodium rubrum. Chem. Phys. Lipids* **24**, 79–84.

Radwan, S. S., Mangold, H. K., Barz, W., and Hüsemann, W. (1979b). Lipids in plant tissue cultures. VIII. Reversible changes in the composition of lipids and their constituent fatty acids in response to alternative shifts in the mode of carbon supply. *Chem. Phys. Lipids* **25**, 101–109.

Richter, G., Reihle, W., Wietoska, B., and Beckmann, J. (1980). Blue light-induced development of thylakoid membranes in isolated roots and cultured plant cells. In "The Blue Light Syndrome" (H. Senger, ed.), pp. 465–472. Springer-Verlag, Berlin and New York.

Richter, G., Beckmann, J., Gross, M., Hundrieser, J., and Schneider, C. (1982). Blue-light induced synthesis of chloroplast proteins in cultured plant cells. In "Progress in Clinical and Biological Research, Vol. 102B: Cell Function and Differentiation." FEBS Vol. 65, pp. 267–276. Liss, New York.

Richter, G., Hundrieser, J., Gross, M., Schulz, S., Botländer, K., and Schneider, C. (1984). Blue light effects in cell cultures. In "Blue Light Effects in Biological System" (H. Senger, ed.), pp. 387–396. Springer-Verlag, Berlin and New York.

Sato, F., Asada, K., and Yamada, Y. (1979). Photoautotrophy and the photosynthetic potential of chlorophyllous cells in mixotrophic cultures. *Plant Cell Physiol.* **20**, 193–200.

Sato, F., Nishida, K., and Yamada, Y. (1980). Activities of carboxylation enzymes and products of $^{14}CO_2$ fixation in photoautotrophically cultured cells. *Plant Sci. Lett.* **20**, 91–97.

Sato, F., Nakagawa, N., Tanio, T., and Yamada, Y. (1981). An improved medium for the photoautotrophic culture of *Cytisus scoparius* Link cells. *Agric. Biol. Chem.* **45**, 2463–2467.

Seeni, S., and Gnanam, A. (1981). Relationship between chlorophyll concentration and photosynthetic potential in callus cells. *Plant Cell Physiol.* **22**, 1131–1135.

Seeni, S., and Gnanam, A. (1982). Carbon assimilation in photoheterotrophic cells of peanut (*Arachis hypogaea* L.) grown in still nutrient medium. *Physiol. Plant.* **70**, 823–826.

Seyer, P., Marty, D., Lescure, A. M., and Péaud-Lenoel, C. (1975). Effect of cytokinin on chloroplast cyclic differentiation in cultured tobacco cells. *Cell Differ.* **4**, 187–197.

Steinmüller, F., and Bentrup, F.-W. (1981). Amino acid transport in photoautotrophic suspension cells of *Chenopodium rubrum* L.: Stereospecificy and interaction with potassium ions. *Z. Pflanzenphysiol.* **102**, 353–361.

Stetler, D. A., and Laetsch, W. M. (1965). Kinetin-induced chloroplast maturation in cultures of tobacco tissue. *Science* **149**, 1387–1388.

Storck, M. (1979). Untersuchungen zur Photorespiration in photoautotrophen Zellsuspensionskulturen von Chenopodium rubrum. Thesis, Münster Univ., Münster.

Sunderland, N., and Wells, B. (1968). Plastid structure and development in green callus tissues of *Oxalis dispar*. *Ann. Bot.* **32**, 327–346.

Tamiya, H., Hase, E., Shibata, K., Miyata, A., Iwamura, T., Nihei, T., and Sasa, T. (1953). Kinetics of growth of *Chlorella* with special reference to its dependence on quality of available light and on temperature. In "Algal Culture from Laboratory to Pilot Plant" (J. S. Burley, ed.), pp. 204–234. Carnegie Institute Washington Publication 600, Carnegie Inst., Washington, D.C.

Tsuzuki, M., Miyachi, S., Sato, F., and Yamada, Y. (1981). Photosynthetic characteristics and carbonic anhydrase activity in cells cultured photoautotrophically and mixotrophically and cells isolated from leaves. *Plant Cell Physiol.* **22**, 51–57.

Vasil, I. K., and Hildebrandt, A. C. (1966). Growth and chlorophyll production in plant callus tissues grown *in vitro*. *Planta* **68**, 69–82.

Venketeswaran, S. (1965). Studies on the isolation of green pigmented callus tissue of tobacco and its continued maintenance in suspension cultures. *Physiol. Plant.* **18**, 776–789.

Wilson, G. (1980). Continuous culture of plant cells using the chemostat principle. In "Advances in Biochemical Engineering" (A. Fiechter, ed.), Vol. 16, Plant Cell Cultures I, pp. 1–25. Springer-Verlag, Berlin and New York.

Wink, M., and Hartmann, T. (1980). Production of quinolizidine alkaloids by photomixotrophic cell suspension cultures: Biochemical and biogenetic aspects. *Planta Med.* **40**, 149–155.

Yamada, Y., and Sato, F. (1978). The photoautotrophic culture of chlorophyllous cells. *Plant Cell Physiol.* **19,** 691–699.

Yamada, Y., Sato, F., and Hagimori, M. (1978). Photoautotrophism in green cultured cells. *In* "Frontiers of Plant Tissue Culture 1978" (T. Thorpe, ed.), pp. 453–462. University of Calgary Press, Calgary, Alberta, Canada.

Yamada, Y., Imaizumi, K., Sato, F., and Yasuda, T. (1981). Photoautotrophic and photomixotrophic culture of green tobacco cells in a jar-fermenter. *Plant Cell Physiol.* **22,** 917–922.

Yamada, Y., Sato, F., and Watanabe, K. (1982). Photosynthetic carbon metabolism in cultured photoautotrophic cells. *In* "Plant Tissue Culture 1982" (A. Fujiwara, ed.), pp. 249–250. Maruzen, Tokyo.

Yasuda, T., Hashimoto, T., Sato, F., and Yamada, Y. (1980). An efficient method of selecting photoautotrophic cells from cultured heterogenous cells. *Plant Cell Physiol.* **21,** 929–932.

Yeoman, M. M., and Aitchison, P. A. (1973). Growth patterns in tissue (callus) cultures. *In* "Plant Tissue and Cell Culture" (H. E. Street, ed.), Botanical Monographs Vol. 11, pp. 240–268. Blackwell, Oxford.

Zenk, M. H. (1978). The impact of plant cell culture on industry. *In* "Frontiers of Plant Tissue Culture 1978," (T. Thorpe, ed.), pp. 1–14. Univ. of Calgary Press, Calgary, Alberta, Canada.

CHAPTER 7

Cryopreservation of Cultured Cells and Meristems

Lyndsey A. Withers

Department of Agriculture and Horticulture
School of Agriculture
University of Nottingham
Loughborough, Leicestershire, England

I. INTRODUCTION

A. Germplasm in Agriculture and Biotechnology

The limits of improvement in agricultural production are defined by the genetic building blocks—genes—with which the breeder and the agronomist have to work. The demands made of these agricultural technologists are increasing with time, toward the achievement of greater yields, susceptibility to mechanization, and utilization of more marginal land. Such pressures are leading to a search for varieties containing superior genes and then a concentration upon their use to the potential exclusion of other genotypes.

Several historical examples can be drawn upon to demonstrate the dangers inherent in this trend toward the use of a small number of favored crop genotypes. The Irish potato famine of the 1840s was a result of vulnerability to pathogen attack in a narrow genetic base. More recently, wheat stem rust and southern corn leaf blight caused serious crop failures in the United States in 1954 and 1970, respectively. The lesson to be drawn from these is that genetic diversity provides an insurance against changes in extrinsic circumstances which reveal "genetic vulnerability" in crops (see Wilkes, 1983).

The conventional view on genetic diversity of crop plants is that we should seek it in the wild forms and relatives of the crops by exploration in their centers of diversity and sequester the material thus collected in genebanks. For many crops these genebanks take the form of seed stores wherein representative seed samples are maintained under controlled conditions of temperature and humidity. Extensive studies into loss of viability in storage, genetic instability, and physiological deterioration have helped to provide a good understanding of the optimal requirements for efficient storage (Cromarty et al., 1982). In recent years, attempts to improve stability in storage have led to the investigation of cryopreservation as an alternative conservation method (see Stanwood, 1985; Styles et al., 1982; Withers, 1985). It remains to be seen whether a shift toward ultralow temperature storage will occur in genebanks; one would speculate not.

Although "seed bank" and "genebank" are often used synonymously, it should be recognized that seed stores provide for only one form of ex situ genetic conservation. In addition there are in situ reserves, ex situ reserves—plantations—and ex situ stores of vegetative material such as tubers and stakes (Frankel and Soulé, 1981; Ingram and Williams, 1984; Martin, 1975). These latter two types of store are crucial elements of conservation policy for vegetatively propagated crops and those producing recalcitrant (short-lived) seeds. Acknowledgment of the special problems in storing genetic diversity for these crops has led to a concentration of interest on potential "technological fixes" which might resolve the difficulties in handling material which is neither dormant nor actively growing and therefore seriously prone to deterioration in storage.

A prime candidate for such a "fix" is seen in the storage of *in vitro* cultures. Initial views of its potential involved the storage of meristems (valued particularly as virus-free units of germplasm) in cultured isolation for periods of years (Henshaw, 1975; Morel, 1975). However, this approach is seen now more as a fallback position, with cryopreservation in liquid nitrogen as holding the greatest promise (IBPGR, 1983a). Opinions on the most suitable tissue for cryopreservation vary (see Withers, 1980a), but meristems/shoot tips are prominent in experimentation and planning, as will be seen in later sections (II,B, IV, V,B,C).

Culture storage is a subject which bridges the two very different fields of genetic conservation and biotechnology. It finds applications in both and, furthermore, *in vitro* storage is itself seen as a biotechnological discipline. Biotechnology in the broadest sense mirrors conventional agriculture by embracing *in vitro* propagation, genetic manipulation (i.e., *in vitro* breeding), and *in vitro* synthesis of plant products. In consequence, biotechnological agriculture must be seen to have its own genetic conservation needs to preserve genetic diversity. Whereas with the earlier situation we were dealing with seeds and some vegetative, somatic tissues, the biotechnological subjects requiring preservation will range from protoplasts, cells, and calluses to meristems, shoot tips, embryos, and plantlets. Both here and in the conventional context, there is some interest in preserving pollen.

Plant breeding utilizing preexisting germplasm encompasses a vast range of genotypes but does present a finite number of subjects for conservation. Furthermore, conservation policy may be defined to require that only limited, strategically selected genotypes, including some cultivars, be stored. In contrast, the activities of the scientist involved in plant genetic manipulation may yield a limitless number of diverse genotypes of potential value in the future, thereby requiring safe conservation. Until the mechanisms underlying somaclonal variation (Scowcroft *et al.*, 1984) are more fully understood, this process alone will be capable of presenting unmanageably large numbers of variants which, like the products of other procedures with a random element such as protoplast fusion and mutagenesis, will require storage until they can be adequately screened.

In examining in detail the methodology of cryopreservation of *in vitro* cultures, emphasis will be placed upon cell cultures and meristems/shoot tips, but the underlying principles, alternative storage methods, and storage procedures for a wider range of culture systems will be addressed in

Section II. The interested reader is directed to other reviews on *in vitro* conservation (Henshaw and O'Hara, 1983; Withers, 1982a; 1983a,b; 1984a; 1985; Withers and Williams, 1985).

B. The Cryobiology of *in Vitro* Plant Cultures

Cryopreservation is one aspect only of cryobiology and merits the attention it receives more by virtue of its potential applications than its biological significance. The process of ultralow temperature storage in liquid nitrogen at $-196°C$ (or at a similarly low temperature) is for the most part an artificial one involving artificially protective treatments. Other cryobiological studies involving *in vitro* cultures have a potential input at the level of understanding to the development of successful cryopreservation procedures but should not be confused with the latter. They are mentioned here to provide a biological perspective, to permit access to the wider literature on low-temperature phenomena, and for their relevance to improvement in cryopreservation procedures by cold hardening and facilitation of recovery from cryoinjury.

Freeze-drying has not been attempted with higher plant somatic tissues, although it has been used widely for the preservation of certain microbes (Ashwood-Smith, 1980). Questions relating to the potential risks of genetic instability have been raised in the latter context. Higher plant pollen of many species can be freeze-dried (see Akihama *et al.*, 1978, and references therein). Although fertility appears to be retained, genetic stability has not been examined adequately.

Recent studies involving the use of emulsions for the suspension of yeast and higher plant cells during cooling to intermediate temperatures (ca. $-30°C$ or lower) indicate that undercooling (supercooling) may provide an alternative means of suspending metabolism in organisms and their consequent preservation (Franks *et al.*, 1983). Little information is available as yet on the physical and genetic stability of such systems or the range of tissue types and sizes which may be subjected to undercooling in this way.

The cryomicroscope is a valuable tool in studies of the freezing process. Protoplasts and suspension cultured cells provide ideal subjects for examination in this instrument. Examination of these units after pretreatments such as cold hardening and in different freezing milieu will undoubtedly throw light on the process of freezing and thawing under putative conditions of cryopreservation (see, e.g., Steponkus *et al.*, 1983, and references therein). However, such studies should not be substituted for those involving aseptic material which can be returned to culture after freezing and tested for the only true criterion of success—recovery. The same potential value and criticisms could be attributed to physical studies of the freezing process as measured by differential thermal analysis, nuclear magnetic resonance, etc. They are only of true value when used alongside culture experiments (e.g., Chen *et al.*, 1984a).

Some attention has been given to the induction of cold hardening in cultures, particularly calluses. For example, Bannier and Steponkus (1972, 1976) have shown that callus of *Chrysanthemum morifolium* can be hardened and can also be exposed to supercooling, subzero temperatures for periods of weeks (see Section IIB,3). However, it is critical to note that such regimes will not provide for stable long-term storage as viability declines progressively with time. The risks of selection for more cold-tolerant variants from within a heterogeneous population far outweigh any perceived advantages of this approach. Further information on cold hardening and freezing stress can be found in Li and Sakai (1982) and references therein.

In the context of plant improvement, *in vitro* cultures are subjected to various stresses (cold, salt, herbicide, etc.) with the objective of selecting for tolerant (spontaneous or induced) mutants (Bright *et al.*, 1986). The success achieved in this, and the caution given above, would indicate the general need for awareness of potential risks of instability in any *in vitro* system, for continual monitoring of stability, and for avoidance of conditions (both environmental and morphological) known or thought to be conducive to the generation of instability (see Scowcroft *et al.*, 1984). The objective of all work associated with the generation of variation *in vitro* should be control of the phenomenon, rather than total elimination of variability. In this connection, cryopreservation has a valuable application as one of the few means whereby control may be exerted over time-related processes occurring *in vitro*.

C. A Decade of Progress in Cryopreservation

During the mid-1970s, cryopreservation as a means of genetic conservation for or by means of plant tissue cultures was little more than a fairly realistic expectation based upon microbiological and zoological precedents (Ashwood-Smith and Farrant, 1980) and a prospect of dramatic progress in all aspects of higher plant culture technology. Occasional early reports indicated the feasibility of cryopreservation *sensu stricto*. These include the partially successful freezing in liquid nitrogen of excised seedlings of *Pisum sativum* by Sun (1958), followed by two studies of the exposure of cell cultures to temperatures which, although not as low as that of liquid nitrogen, are likely to have involved the critical event of intracellular freezing. In 1968, Quatrano froze cells of *Linum usitatissimum* to −50°C, after which a

TTC viability test (see Section II,D) response of 14% was recorded. Three years later, Latta (1971) reported the regeneration of cells of *Daucus carota* and *Ipomoea* sp. after exposure to −40°C. While the temperature was higher than in the earlier sample, achievement of regeneration was a significant advance.

As yet, true cryopreservation of an *in vitro* culture in the sense of regeneration after exposure to liquid nitrogen had not been achieved. However, in 1973, two notable studies were reported in the literature. Sakai and Sugawara (1973), examining callus of *Populus euramericana*, and Nag and Street (1973), examining a cell suspension of *Daucus carota*, froze their experimental material to −196°C and achieved subsequent recovery growth. Nag and Street's findings were particularly significant in that the culture was embryogenic and demonstrated this totipotency both before and after preservation. It is arguable that the study of the cryopreservation of higher plant culture systems, as opposed to their cryodamage, dates from this point.

During the decade following Nag and Street's (1973) report, cryopreservation of cell suspensions has virtually become a matter of routine, whereas callus cultures have remained problematic (see Section II,B). Organized cultures—meristems, shoot tips, embryos, and plantlets—eluded successful preservation until 1976, when Seibert demonstrated callus and shoot regeneration at a moderate level (up to 33%) in shoot tips of *Dianthus caryophyllus*. Progress with this system led within one year to very successful (up to 100%) recovery and plant regeneration (Seibert and Wetherbee, 1977).

In 1978, Grout and Henshaw, examining shoot tips of *Solanum goniocalyx*, reported up to 60% survival and regeneration. In both this and the work with *Dianthus caryophyllus*, some impractical elements in the rapid freezing procedure and the fact that specimens dissected from independently growing *in vivo* plants were preserved and then recovered in culture led to the continued search for a slow freezing method which could be applied to material originating in culture.

A significant contribution was then made by Kartha and colleagues. Shoot tips dissected from germinated seeds of *Pisum sativum* were, in a report made in 1979 (Kartha *et al.*), cryopreserved by a slow freezing method. A similar method was then applied to *in vitro* propagated meristems of *Fragaria* × *ananassa* (Kartha *et al.*, 1980).

Over the last 4–5 years, reports have accumulated on the successful preservation of shoot tip material by either of the above rapid or slow freezing approaches. Some reports include small, but apparently critical, modifications to the handling of the material during freezing. However, the general level of success, reproducibility, and applicability of techniques have not always matched progress made in the preservation of cell cultures.

Other organized cultures have received disproportionately little attention, and, therefore, routine methods have not emerged for embryo (somatic and zygotic) cultures. This remark can also be made for callus and protoplast cultures, pollen plantlets, and cultured roots. Considerable research remains to be carried out before their preservation can be considered routine, as will be confirmed in Section II,B.

A very welcome trend in relatively recent studies has been an emphasis on the preservation of material with a genuine conservation need. Examples in the context of global germplasm conservation include examination of *Manihot esculenta* (cassava; Kartha *et al.*, 1982b), a number of temperate fruits (Katano *et al.*, 1983; Sakai and Nishiyama, 1978), and palms (Engelmann *et al.*, 1984; Grout, 1985; Grout *et al.*, 1983). In biotechnology, equivalent notable studies have involved the preservation of mutant cell lines (Hauptmann and Widholm, 1982) and secondary product synthesizing or biotransforming cultures (Chen *et al.*, 1984b; Diettrich *et al.*, 1982; Dougall and Whitten, 1980; Watanabe *et al.*, 1983; Seitz *et al.*, 1983). The latter studies have, of necessity, involved verification of stability in storage. It is hoped that awareness of this critical aspect of preservation will become more widespread.

Major contributions to progress in the development of methodology have been empirical studies. However, mention should be made of descriptive studies which have examined cellular responses to pregrowth conditions, freezing injury at the fine-structural level, and impairment of physiological competence (e.g., Cella *et al.*, 1982; Pritchard *et al.*, 1982; Withers, 1978a; Withers and Davey, 1978; Zavala and Finkle, 1983).

II. PRESERVATION PROCEDURES

A. Biological Requirements and Technical Options

In examining potential approaches to germplasm storage *in vitro*, it is necessary to define the demands which will be made of the storage system and their feasibility at a given stage in the development of methodology. We have already identified two major user groups: the genetic conservationist and the biotechnologist. In some respects, their requirements overlap in that both will expect maximal genetic stability in storage and reproducibility, a high level of viability, and minimal physiological impairment such that recovery of preserved material may proceed as rapidly as possible.

Differences in requirements may come in at a number of stages. The conservationist will, at present, tend to have a greater interest in the storage of organized cultures, these being the more appropriate for a rapid return to field conditions and utilization in breeding and, arguably, genetically more stable, whereas the biotechnologist will be interested in all types of culture, each for different reasons. In genetic manipulation programs, cells, protoplasts, and regenerated protoclones at the microcallus stage will require storage in order to avoid loss through genetic instability and to circumvent logistic problems in handling large quantities of potential variants. In situations where somaclonal variation is likely to be occurring, control material from stages prior to, during, and after variation-inducing stages will require storage.

Where plant regeneration is the end point of the project, clonal samples of organized or meristematic/embryogenic cultures should be stored for reference and for the production of further examples of the genotype. Where a biomass producing a secondary product synthesizing culture is the end point of the project, the cultures themselves will need to be stored as inocula for production runs. Thus it is possible to justify investment in development of storage protocols for all types of *in vitro* plant culture.

A primary aim in all aspects of *in vitro* storage of germplasm is freeing the scientist from time-related changes in his experimental material. It is important to assess the time scales involved in different applications. If we are to look upon stored cultures as an *in vitro* parallel to seeds or vegetative propagules, then there will be a year-by-year requirement for material of a high standard in all respects at "feed-stock" for laboratory experimental procedures, breeding programs, and crop or metabolite production. The requirement to ensure availability of identical genotypes for the long-term future may stretch into decades and centuries in all contexts, but particularly so in the case of irreplaceable germplasm rescued from threatened and diminishing natural environments.

A number of practical requirements can be attached to the biological prospectus. The preserved material should demand minimal attention during storage. The storage facility should be compact, inexpensive, and as maintenance free as possible and should have a high degree of buffering against environmental instability. Finally, material in storage should be susceptible to transportation without provision of elaborate supporting systems.

Having identified user needs, what can be offered to meet them? Cryopreservation is, for obvious reasons, a prime candidate for the favored approach to storage, since it offers a suspension of metabolism and, therefore, effectively, time. Furthermore, cryopreservation appears to be applicable to all types of culture (albeit with varying degrees of success, as will be described in Section II,B). Liquid nitrogen refrigerators are relatively unobtrusive, are mechanically unsophisticated, and provide a very stable storage environment. Aspects of freezing apparatus will be considered further in Section II,C.

With all of the above advantages, it might seem unnecessary to seek alternative approaches to storage. However, in addition to incomplete development of methodology for certain types of specimen, cryopreservation has other drawbacks if seen as the *sole* focus of storage policy. Storage in a small enclosure (ampoule, for example) in liquid nitrogen is not compatible with direct transfer to growth under normal conditions. Also, frozen material is not easily transported without provision of a special low-temperature-maintaining container. Critically, temporary warming up even to temperatures well below 0°C can be completely lethal.

On the scale of most *in vitro* culture operations, cryopreservation is not among the more costly. However, in circumstances in which other factors compete, the cost may be prohibitive, necessitating the adoption of a storage method which uses existing facilities. We are thus required to consider a second technical option in culture storage based upon slow growth through its limitation rather than its total suspension at ultralow temperatures. Growth may be limited by one of several means: a reduction in growth temperature, a reduction in available oxygen, application of osmotic inhibitors, and application of hormonal retardants. Desiccation is another, little explored, way of approaching storage at normal growth temperatures. In the following section each of these possibilities will be examined in the context of the various culture systems requiring storage. A list of species which have been subjected to cryopreservation is given in Table I; see also Withers, 1983b, 1984a; Withers and Williams, 1985).

B. A Survey of the Current Status of Preservation Procedures for a Range of Culture Systems

1. Protoplast Cultures

The ephemeral nature of protoplasts renders them inappropriate for storage by any means other than cryopreservation. Setting aside a number of studies wherein protoplasts have been involved in investigations of cryoinjury to membranes and other temperature-related cellular phenomena, the first success in the cryopreservation of protoplasts was reported by Mazur and Hartmann (1979). They were able to record a viability level of up to 68% in protoplasts of *Bromus inermis* and *Daucus carota* but,

TABLE I

Summary of Higher Plant Species and Tissue Culture Systems Subjected to Cryopreservation in Liquid Nitrogen

Species	Culture system	References
Acer pseudoplantanus	Cell suspension	Nag and Street (1975a,b); Sugawara and Sakai (1974); Withers (1983a, 1984a)[a]; Withers and King (1980); Withers and Street (1977b)
Arachis hypogaea	Shoot tip	Bajaj (1979)
Arachis hypogaea	Pollen embryo	Bajaj (1983b)
Arachis villosa	Pollen embryo	Bajaj (1983b)
Asparagus officinalis	Shoot tip	Seibert (1977)
Atropa belladonna	Cell suspension	Nag and Street (1975a,b)
Atropa belladonna	Anther	Bajaj (1978a)
Atropa belladonna	Pollen embryo	Bajaj (1977a, 1978a,b)
Berberis dictyophylla	Cell suspension	Withers (1983a, 1984a)[a]
Brassica campestris	Pollen embryo	Bajaj (1983b)
Brassica napus	Cell suspension	Weber *et al.* (1983)
Brassica napus	Shoot tip	Withers (1982b)
Brassica napus	Pollen embryo	Bajaj (1983b)
Catharanthus roseus	Cell suspension	Chen *et al.* (1984a,b); Kartha *et al.* (1982a); Withers (1983a, 1984a)[a]
Cicer arietinum	Shoot tip	Bajaj (1979)
Corydalis sempervirens	Cell suspension	Withers (1983a, 1984a)[a]
Datura innoxia	Cell suspension	Hauptmann and Widholm (1982); Weber *et al.* (1983)
Datura innoxia	Protoplast	Hauptmann and Widholm (1982)
Datura stramonium	Cell suspension	Bajaj (1976)
Daucus carota	Cell suspension	Dougall and Wetherell (1974); Dougall and Whitten (1980); Hauptmann and Widholm (1982); Nag and Street (1973, 1975a,b); Popov *et al.* (1978); Weber *et al.* (1983); Withers (1983a, 1984a)[a]; Withers and Street (1977b)
Daucus carota	Somatic embryo/clonal plantlets	Withers (1979)
Daucus carota	Protoplast	Hauptmann and Widholm (1982); Mazur and Hartmann (1979); Takeuchi *et al.* (1980, 1982); Withers (1980c)
Dianthus caryophyllus	Shoot tip	Anderson (1979); Seibert (1976, 1977); Seibert and

TABLE I (*Continued*)

Species	Culture system	References
		Wetherbee (1977); Uemura and Sakai (1980)
Digitalis lanata	Cell suspension	Diettrich *et al.* (1982); Seitz *et al.* (1983)
Elaeis guineensis	Embryogenic callus	Engelmann *et al.* (1984)
Elaeis guineensis	Zygotic embryo	Grout (1986); Grout *et al.* (1983)
Fragaria × ananassa	Shoot tip	Kartha *et al.* (1980); Sakai *et al.* (1978)
Glaucium flavum	Cell suspension	Withers (1983a, 1984a)[a]
Glycine max	Cell suspension	Bajaj (1976); Withers (1983a, 1984a)[a]
Glycine max	Protoplast	Weber *et al.* (1983); Takeuchi *et al.* (1982)
Gossypium arboreum	Callus	Bajaj (1982)
Gossypium hirsutum	Callus	Bajaj (1982)
Hordeum vulgare	Zygotic embryo	Withers (1982b)
Hordeum vulgare	Protoplast	Takeuchi *et al.* (1982)
Hyoscyamus muticus	Cell suspension	Withers (1983a, 1984a)[a]; Withers and King (1980)
Lactuca sativa	Shoot tip	Seibert (1977)
Lavandula vera	Callus	Watanabe *et al.* (1983)
Lycopersicon esculentum	Seedling	Grout *et al.* (1978)
Lycopersicon esculentum	Zygotic embryo	Grout (1979)
Malus domestica	Shoot tip	Katano *et al.* (1983); Sakai and Nishiyama (1978)
Manihot esculenta	Shoot tip	Henshaw *et al.* (1980b); Kartha *et al.* (1982b); Stamp (1978)
Manihot utilissima	Shoot tip	Bajaj (1977b)
Medicago sativa	Callus	Finkle *et al.* (1979, 1983)
Nicotiana plumbaginifolia	Cell suspension	Maddox *et al.* (1982)
Nicotiana sylvestris	Cell suspension	Maddox *et al.* (1982); Shillito (1978)
Nicotiana tabacum	Cell suspension	Bajaj (1976); Hauptmann and Widholm (1982); Maddox *et al.* (1982); Withers (1983a, 1984a)[a]
Nicotiana tabacum	Anther	Bajaj (1978a,b)
Nicotiana tabacum	Pollen embryo	Bajaj (1977a, 1978a,b)
Onobrychis viciifolia	Cell suspension	Withers (1983a, 1984a)[a]
Oryza sativa	Cell suspension	Cella *et al.* (1982); Sala *et al.* (1979); Withers (1983a, 1984a)[a]
Oryza sativa	Protoplast	Bajaj (1983a)[b]
Oryza sativa	Callus	Finkle *et al.* (1979, 1983)
Oryza sativa	Anther	Bajaj (1980)
Oryza sativa	Zygotic embryo	Bajaj (1981a)
Pennisetum americanum	Cell suspension	Withers (1983a, 1984a)[a]

(*continued*)

TABLE I (*Continued*)

Species	Culture system	References
Petunia hybrida	Anther	Bajaj (1978a)
Phoenix dactylifera	Embryogenic callus	Finkle *et al.* (1979, 1983); Tisserat *et al.* (1981)
Pisum sativum	Protoplast	Bajaj (1983a,)[b]
Pisum sativum	Shoot tip	Kartha *et al.* (1979)
Populus euramericana	Callus	Sakai and Sugawara (1973)
Populus sp.	Cell suspension	Binder and Zaerr (1980a)
Primula obconica	Anther	Bajaj (1981b)
Pseudotsuga menziesii	Cell suspension	Binder and Zaerr (1980b)
Rhazya orientalis	Cell suspension	Withers (1983a, 1984a)[a]
Rhazya stricta	Cell suspension	Withers (1983a, 1984a)[a]
Ribes sp.	Shoot tip	Sakai and Nishiyama (1978)
Rosa 'Paul's Scarlet'	Cell suspension	Withers and King (1980); Withers (1983a, 1984a)[a]
Rubus sp.	Shoot tip	Sakai and Nishiyama (1978)
Saccharum spp.	Cell suspension	Chen *et al.* (1979); Finkle and Ulrich (1979); Finkle *et al.* (1983); Ulrich *et al.* (1979)
Saccharum spp.	Callus	Ulrich *et al.* (1979)
Sambucus racemosa	Shoot tip	Sakai and Nishiyama (1978)
Solanum etuberosum	Shoot tip	Towill (1981a)
Solanum goniocalyx	Shoot tip	Grout and Henshaw (1978); Henshaw *et al.* (1980a,b)
Solanum melongena	Cell suspension	Withers (1983a, 1984a)[a]
Solanum tuberosum	Shoot tip	Bajaj (1978c, 1981c); Benson *et al.* (1984); Henshaw *et al.* (1980a,b, 1985); Jarvis (1983); O'Hara *et al.* (1984); Towill (1981b, 1983); Withers *et al.* (1985)
Sorghum bicolor	Cell suspension	Withers (1983a, 1984a)[a]; Withers and King (1980)
Triticum aestivum	Protoplast	Bajaj (1983a)[b]; Takeuchi *et al.* (1982)
Triticum aestivum	Zygotic embryo	Withers (1982b)
Triticum aestivum	Pollen embryo	Bajaj (1983b)
Triticum monococcum	Cell suspension	Withers (1983a, 1984a)[a]
Zea mays	Cell suspension	Withers (1983a, 1984a)[a]; Withers and King (1979, 1980)
Zea mays	Protoplast	Withers (1980b)
Zea mays	Callus	Withers (1980b)
Zea mays	Zygotic embryo	Withers (1980c)

[a] Cell suspensions in storage at the Friedrich Miescher Institute, Basel; line designations and special features are listed in Withers (1983a, 1984a).

[b] Species pairs frozen as fused protoplasts.

critically, did not demonstrate recovery growth. This was first described by Takeuchi *et al.* (1980) in a study of the preservation of protoplasts of *Daucus carota* and the liverwort *Marchantia polymorpha*. This work has been consolidated by successful preservation of protoplasts of several other crop species (Takeuchi *et al.*, 1982).

More recently, Bajaj (1983a) has achieved a low level of recovery in freshly isolated protoplasts of *Pisum sativum* fused with either *Triticum aestivum* or *Oryza sativa*, indicating that it may be possible to introduce the technique into somatic hybridization experimentation. In this and foregoing examples, the protoplasts were frozen at a rate which would bring about some cellular dehydration. In all but one reported case (Withers, 1980c), the protoplasts required cryoprotection. Rapid thawing and washing usually precede recovery on semisolid medium.

2. Cell Suspension Cultures

Like protoplasts, cultured cells are not easily stored other than by cryopreservation. Some anecdotal reports (e.g., Potrykus, 1978) suggest that storage of suspended cells in a laboratory refrigerator (ca. +4°C) may be feasible for relatively brief periods of time. Clearly, however, this would not be suitable for long-term storage.

Cryopreservation has, in contrast, developed to a high degree of refinement. It is now possible to preserve a wide range of species with excellent results. The protocols available are detailed and compared in Section III, but some common features are mentioned here: pregrowth of cultures in a medium supplemented with an osmoticum such as mannitol; cryoprotection with DMSO plus one or two other compounds; slow or stepwise freezing, rapid thawing; recovery on semisolid medium, usually without postthaw washing.

3. Callus Cultures

The greater bulk and different mode of culture of callus in comparison with cell suspensions are reflected in the available storage methods. Some success in growth limitation by oxygen limitation or a reduction in the culture temperature indicates the feasibility of developing routine short to medium-term storage methods.

In 1959 Caplin reported the successful maintenance of callus of *Daucus carota* under mineral oil. Growth was limited to one-quarter of controls by the reduced but adequate supply of oxygen regulated by the oil overlay. More recently, Bridgen and Staby (1981) have used oxygen limitation by reduced partial pressure of oxygen or reduced overall atmospheric pressure to control the growth of callus of *Nicotiana tabacum* among other culture systems and species.

As mentioned earlier, growth at reduced temperatures has been explored by Bannier and Steponkus (1972), who achieved survival (as estimated by the TTC test) in callus of *Chrysanthemum morifolium* exposed to a temperature of $-3.5°C$ for 28 days. Long-term studies to consolidate such work have yet to be carried out.

Callus cultures were among the first to be cryopreserved successfully (Sakai and Sugawara, 1973) but have received proportionately little attention in subsequent studies. The majority of such studies have been carried out by Finkle and colleagues (Finkle and Ulrich, 1979; Finkle *et al.*, 1979, 1983; Tisserat *et al.*, 1981; Ulrich *et al.*, 1979) examining a range primarily of tropical species. Generally, slow freezing after application of a cryoprotectant mixture consiting of DMSO, glucose, and polyethylene glycol, rapid thawing, washing, and return to culture on semisolid medium give success.

A number of features of this work are noteworthy. An effect of cryoprotectant temperature (at the time of application) and postthaw washing temperature has implications for other culture systems (for further consideration, see Section II,B). Plant regeneration has generally been reported in cryopreserved callus but sometimes with impairment in this capacity (e.g., Ulrich *et al.*, 1979), suggesting a structural element to freezing injury rather than an effect of freezing on totipotency per se.

4. Meristems, Shoot Tips, and Shoot Cultures

The three types of culture described here are related in that shoot tips can be dissected from shoot cultures prior to preservation and that specimens are often referred to as meristems when in fact they are more accurately described as shoot tips. The difference is in part a matter of size.

It is in this group of cultures that we find the most fully developed range of short- and long-term storage methods. Shoot cultures of a wide range of species can be maintained in slow growth for periods of years, with subculturing being carried out at approximately annual intervals. Reports fall into two categories: growth at a reduced temperature and growth in the presence of inhibitors. Examples are listed in Table II, from which storage procedures for other species may be derived (see also CIAT, 1979).

Few serious problems are encountered in this approach to storage, other than those familiar in any culture maintenance program, i.e., renewal of culture medium when necessary, monitoring for deterioration in health of the culture or appearance of microbial contamination, etc. Two studies involving shoot cultures of *Vitis* spp. (Barlass and Skene, 1983) and *Solanum* spp. (Henshaw *et al.*, 1980a,b; Westcott, 1981a,b) indicate that intergenotypic variations in response to growth-limiting conditions may require that careful exploration of exact requirements be carried out for each genotype under examination.

TABLE II

Examples of Slow Growth Procedures for Shoot Cultures

Species	Normal culture conditions	Storage conditions	Storage period	Reference
Dioscorea alata *Dioscorea rotundata* *Ipomoea batatas*	26°C; Standard medium	16–20°C; 3% mannitol	8–24 months	IITA (1981); Ng and Hahn (1985)
Solanum stenotomum *Solanum goniocalyx* *Solanum chaucha* *Solanum juzepcjukii* *Solanum tuberosum* *Solanum curtilobum*	22°C 16-hr photoperiod; 3% sucrose	12°C/16-hr day 6°C/8-hr night; 8% sucrose	≥12 months[a]	Westcott (1981a)
Solanum spp. (as above)	22°C; Standard medium; 20 ml medium	22°C; 20 mg/liter ABA; 60 ml medium	12 months[a]	Westcott (1981b)
Solanum spp. (as above)	22°C; Standard medium	22°C; 6% mannitol	12 months[a]	Westcott (1981b)
Solanum tuberosum	20–22°C; 16-hr photoperiod 4000 lx; Standard medium	10°C; 16-hr photoperiod 2000 lx; 50 mg/liter B-nine[c]	24 months[b]	Mix (1982, 1984)
Trifolium repens	25°C; 16-hr photoperiod	5°C; dark	10 months	Bhojwani (1981)
Vitis amurensis *Vitis berlandieri* *Vitis caribea* *Vitis champini* *Vitis labrusca* *Vitis rupestris* *Vitis vinifera* *Vitis* hybrids	26°C/15-hr day 20°C/9-hr night	9.5°C; 15-hr photoperiod	>6 months[d]	Barlass and Skene (1983)

[a] Interspecific differences in performance were noted.

[b] Cultures transferred to fresh medium and placed for 3–4 weeks at 20–22°C could be returned to the slow growth conditions for a further 2 years.

[c] N-dimethylsuccinamic acid.

[d] Differences were noted between species and types of culture (proliferating shoots or single-rooted shoots).

At the current state of development of techniques and experimental investment, it would seem that growth limitation/slow growth should permit the maintenance of shoot cultures for, say, a decade given adequate periodic attention and replacement of cultures lost for various reasons (deterioration, distribution, etc.). However, care should be taken to avoid selection for vigor over extended periods of time. This tendency should be resisted by either systematic or random assignment of cultures for transfer at each subculturing interval.

Despite the relatively optimistic picture described above, there is still a clear need for truly long-term storage under the minimal-input conditions offered by cryopreservation. Studies to date would suggest that a fairly narrow range of routine methodologies should soon be developed for shoot tips. Common features are likely to include a pregrowth period intervening between dissection and freezing, cryoprotection with DMSO (usually prepared in culture medium with or without other cryoprotectant additives), and recovery on semisolid medium. Major divergencies will lie in the freezing rate, between rapid and slow or stepwise, and exact method of handling at the freezing stage.

In contrast to the situation with cell suspension cultures, survival rates are erratic within and highly divergent between different studies, suggesting that much remains to be done in the development of preservation procedures. An additional dimension to the preservation of organized cultures such as shoot tips is the possibility of influencing the level and pattern of recovery by modification of the postthaw hormonal environment. These aspects are enlarged upon in Section IV.

5. Embryos

There is interest in preserving both zygotic and somatic embryos. Since embryos have a natural storage phase of dormancy in the seed, it might be expected that such specimens would be especially amenable to storage, particularly by growth limitation. This approach is, however, reflected in only one study of storage, involving the desiccation of somatic embryos of *Daucus carota* (Jones, 1974). The embryos were stored on semisolid medium for 1 year, during which time they progressively desiccated. Rehydration by provision of fresh sucrose solution brought about "germination" of the embryos. Parallel cultures maintained in growth at an uninhibited rate lost embryogenic potential during the storage period, thus illustrating the stabilizing potential of culture preservation.

Future studies of embryo storage might usefully draw on investigations of germination inhibition in recalcitrant seeds, involving osmotic inhibitors such as polyethylene glycol and growth substances (see King and Roberts, 1979; Roberts *et al.*, 1984).

Zygotic embryos in imbibed orthodox seeds such as those of *Lycopersicon esculentum* have been used as a model system to aid understanding of the freezing process in recalcitrant seeds. Using such an approach, Grout (1979) has been able to cryopreserve (by rapid freezing) seeds rehydrated to 23% water content without adding a cryoprotectant, and to 42% when treated with 15% DMSO. Rehydration to 72% destroyed survival potential in the entire seed, but, significantly, about 20% of shoot meristems dissected from the frozen and thawed seeds cryoprotected with 15% DMSO could be recovered in culture (for further information on the cryopreservation of seeds, see references cited in Withers, 1985).

A critical study of freezing of zygotic embryos of *Elaeis guineensis* (Grout, 1986; Grout *et al.*, 1983) indicates that such apparently recalcitrant seeds may, in fact, be orthodox at the embryo level. Thus, excised embryos can be subjected to desiccation and then frozen by rapid immersion in liquid nitrogen.

Immature zygotic embryos provide important inoculum material for genetic manipulation and other studies involving cereal *in vitro* cultures. In view of the expense of providing year-round growth facilities for seed-head production, there is an obvious application for cryopreservation of embryos. Accordingly, Withers (1982b) has examined the freezing requirements of immature seed heads and embryos of *Hordeum vulgare* and *Triticum aestivum*. Seed heads, individual seeds, and dehusked seeds respond to slow freezing after cryoprotection with DMSO, whereas embryos dissected from seeds respond to both rapid and slow freezing. Rapid freezing tends to favor recovery by germination of the embryonic axis, whereas slow freezing favors formation of callus, probably of scutellar origin. In a similar study on mature seeds of *Oryza sativa*, Bajaj (1981a) demonstrated survival of seeds and portions of the seed (dehusked seed, excised embryos, and endosperm) when cultured after thawing.

Somatic embryos have received relatively little attention in cryopreservation studies. Embryogenic callus derived from excised immature cereal embryos as described earlier and somatic embryos and embryogenic callus of *Theobroma cacao* are currently under investigation in the author's laboratory. These will help to determine whether survival potential is more influenced by genotype or culture system. Other studies of the preservation of "embryogenic callus" have probably really involved proembryonic structures (e.g., *Phoenix dactylifera*, Tisserat *et al.*, 1981; *Elaeis guineensis*, Engelmann *et al.*, 1984). They may throw some light upon the requirements of more advanced embryos.

However, only one study involving *Daucus carota* has so far indicated the feasibility of cryopreserving individual somatic embryos (Withers, 1979). Embryos ranging from globular stage to torpedo stage and plantelet failed to respond to a freezing method suitable for the cell suspensions giving rise to the embryos, although groups of cells within individual embryos might recover by production of secondary embryos. A very high percentage recovery could be achieved, however, by slowly freezing the DMSO cryoprotected embryos after draining free of the cryoprotectant solution and enclosing in an alumininum foil envelope. Either rapid or slow thawing was successful, the latter having advantages with large and brittle organized tissues. Recovery on semisolid medium proceeded by either callusing or meristematic development according to the composition of the culture medium.

6. Pollen, Pollen Embryos, and Anthers

Pollen has been the subject of preservation studies for some years in view of the potential convenience of being able to carry out fertilization between geographically and temporally separated parent plants. Both freeze-drying and cryopreservation have proved useful in this respect and the interested reader is referred to other publications for further information (Akihama *et al.*, 1978; Withers, 1985).

Pollen embryos and anthers have not been preserved successfully by growth limitation, but a number of studies by Bajaj of cryopreservation indicate that, while problematic, storage should eventually be possible (1977a, 1978a,b, 1980, 1981b, 1982). There is no clear advantage in either rapid or slow freezing, and cryoprotectant treatments are not optimized. Bisection of the pollen sacs may be advantageous (Bajaj, 1978a), but this would prevent the preservation of freshly harvested anthers of most species, since the pollen would not have embarked upon androgenesis and probably would not do so once the anther had been damaged.

In view of the difficulties experienced with handling anthers and the lack of an absolute necessity to preserve the intact anther, it may be more fruitful to pursue preservation of pollen-derived embryos isolated from anthers in which androgenesis has been induced. This precise approach has not been followed as yet, but in a similar procedure in which embryos were released from anthers after freezing and then assayed for viability Bajaj (1977a) demonstrated preferential survival of early stage (globular and less well heart-shaped) embryos of *Atropa belladonna* and *Nicotiana tabacum*.

C. Apparatus

The apparatus requirements for culture preservation by cryopreservation fall into those associated with (1) culture maintenance, (2) freezing, and (3) storage. The requirements for culture maintenance in no way differ from those normally found in the tissue culture laboratory. It is likely, however, that a slightly greater emphasis might be placed on static culture and culture on semisolid medium.

While it is undeniably preferable and convenient to use a purpose-built unit for freezing such as marketed by a number of suppliers as detailed in various research papers, unavailability of such facilities should not be taken as an absolute impediment to carrying out cryopreservation. Once the cooling requirements of the culture in question are determined (best by using a unit as just described), all that is required is a means of applying either a reproducible rate of cooling or a constant subzero temperature. The former may be achieved by suspending specimens in liquid nitrogen vapor, in an insulated container within a refrigerator, or in a low-melting-point liquid bath cooled by a cooling coil (Withers, 1984b; Withers and King, 1980).

Additionally, a suitable cooling environment may be improvised within a low-temperature, electrically cooled refirgerator by supplying heat at a controlled and decreasing rate to a specimen holder. Constant subzero conditions can be established in an alcohol–dry ice bath, in a low-melting-point liquid held at the required temperature by a thermostated cooling source, or in such a bath cooled by a cooling mixture. Formulas for preparing cooling mixtures which are very thermally stable and nonhazardous to users have been published (Kaufmann and Berthier, 1983). The above cooling facilities should have the capacity to cool to and hold a temperature of −40°C or lower.

A simple vacuum flask holding liquid nitrogen is all that is required to carry out rapid freezing as used for the preservation of some organized cultures (see Section IV,A) and for the quenching of slow- or stepwise-cooled specimens once they have reached the appropriate subzero temperature.

For convenience, a range of Dewar vessels is useful for collection, transportation, and dispensing of liquid nitrogen. A large vessel with inner containers designed to hold specimen ampoules will be required for storage. Various models are commercially available, which usually balance ease of access against liquid nitrogen consumption. A reliable supply of liquid nitrogen, if not available from liquified gas producers, may be ensured by installation of an air compressor. While this might be prohibitively expensive for a small-scale storage facility, once plant culture gene-banking becomes a routine, necessary safeguard for valuable genetypes, the cost might not be so daunting.

Certain consumable items and chemicals differ from those used in conventional tissue culture. Specimen containers for cooling and storing should be chosen with care. Glass ampoules (screw-top or flame-sealed) have been used in some studies but, despite their excellent thermal conductivity, are not generally recommended due to the risk of cross-contamination or even injury to the operator resulting from explosion caused by liquid nitrogen penetration. A safe alternative can be found in polypropylene screw-top or flame-sealed ampoules of various capacities. These are supplied presterilized.

In some preservation procedures, other containers are used, such as heat-sealed, plastic-lined aluminium envelopes (Takeuchi *et al.*, 1980), foil envelopes (Withers, 1979), a foil sheet within a petri dish (Kartha *et al.*, 1982b), and, most fragile, a piece of lens tissue (J. F. O'Hara, personal communication). However, use of the latter two in particular and the freezing of naked specimens (Grout and Henshaw, 1978; Sakai *et al.*, 1978; Withers, 1982b; Uemura and Sakai, 1980) introduce risks of contamination and present difficulties in handling. Accordingly, alternative plastic materials such as employed in the freezing of blood are under investigation by the author (L. A. Withers, unpublished observations) to combine the advantages of rapid heat transfer with secure enclosure (see Section IV,C and Fig. 11).

D. Viability Assays

The only truly satisfactory indicator of survival after cryopreservation is recovery growth, and a procedure should not be considered satisfactory until rapid recovery can be demonstrated. However, in the development of methodology, it can be invaluable to have recourse to a simple, objective assay of viability. A number of these can be found in the literature, but only two are recommended here, even then not without reservation.

1. Fluorescein Diacetate Staining

This test (Widholm, 1972) is particularly suitable for cell suspension cultures, protoplasts, and other small specimens (small callus pieces or embryos) which can be observed in a compound microscope. It is unsuitable for large, opaque specimens. A solution of fluorescein diacetate (2% dilution in distilled water) is prepared fresh from a 1% w/v stock solution of solid in acetone. One drop of the reagent is added to one drop of cell suspension, for example, on a microscope slide. The specimen is then viewed under the microscope with both bright field and ultraviolet (UV) illumination. "Living" cells fluoresce brightly in UV light. These differences are especially clear if a number of hours (preferably overnight) elapse before examination.

Several staining techniques are complementary to fluorescein diacetate staining in that they preferentially stain lethally damaged cells. Evans' blue (Gaff and O'Kong'o-Ogola, 1971; Withers, 1978b) and phenosaffranine (Widholm, 1972) have both been used in cryopreservation studies. In the absence of UV microscope facilities, they would suffice. Microscopic observation alone is taxing and unreliable as a sole means of estimating viability levels. Cyclosis can be difficult to detect, although the cytoplasmic appearance of surviving cells is, to the experienced eye, usually quite different from plasmolyzed or shrunken, lethally damaged cells.

2. TTC Test

Reduction of 2,3,5-triphenyl tetrazolium chloride has been used for some years to test viability in seed samples. Adaptation of the test for application to plant tissue cultures (Towill and Mazur, 1974) has provided an invaluable means of assaying freezing injury in both cell suspensions and larger unit specimens. (There are no reports of its application to protoplasts.) The test indicates the level of respiration in the specimen. Since freshly thawed material may give an unrepresentatively low reaction whereas seriously damaged tissues can give a very strong, transient reaction, the test should be applied after a suitable interval, again preferably overnight.

The reagent solution which can be prepared in advance and stored in the dark at $+4°C$ for several weeks consists of 1% w/v tetrazolium salt in $0.05 M$ phosphate buffer (pH 7.5). For the assay of cell cultures, the reagent is added to an equal volume of cells suspended in culture medium. Bulk specimens are immersed in the reagent, diluted to 50% strength. In both cases, narrow-necked, covered vessels are most suitable for incubation, which is carried out for about 20 hr at 20°C in the dark.

After incubation, the specimens should be washed three times in distilled water and then evaluated either by observation (microscopic or with the naked eye) or by extraction of the red coloration (formazan) formed in respiring cells with 95% ethanol and spectrophotometric measurements (485 nm). In the latter case, the mass of specimens and the volume of alcohol should be standardized, and comparison made with a calibration curve constructed using mixtures of live and dead cells. Large numbers of samples can be processed rapidly by the TTC test, and it is generally highly reliable. The only difficulties encountered by the author have been in the case of assaying very small volumes of cell suspension and green, photosynthesizing tissues, neither of which gave consistent positive reactions.

Both of the aforementioned viability assays can be used to determine initial postthaw viability levels and to monitor recovery growth manifested as proportionate increases in viable cell counts. Corroboration by measurement of parameters of growth is recommended wherever possible. Microscopic studies of thick or thin sections to reveal tissue organization and cellular ultrastructure are useful in determining qualitative aspects of recovery (see later sections; III,D, IV,D) but are not appropriate for the evaluation of the many specimens generated in the development of cryopreservation protocols.

III. CRYOPRESERVATION OF CULTURED CELLS

A. Developing a Routine Protocol

Over the 10 years during which the cryopreservation of cultured cells has been explored, studies at first followed a very similar logic in which cryoprotection, freezing, and thawing conditions were optimized, but then attention turned to the pregrowth stage preceding cryoprotection and the recovery stage following thawing. It now appears that there is little latitude in the treatments applied at any of these stages if preservation is to result in a high-percentage survival. However, fortunately we now know a great deal about likely requirements at each stage and find that interspecific variations in requirements are small.

In one research program (Withers and King, 1979, 1980; Withers, 1980b) attention was given to the development of a simple routine protocol that could be applied to a wide range of species. The stages in this protocol are as follows: pregrow cells for 3–4 days in standard medium or medium supplemented with 10% proline or 6% mannitol; cryoprotect with 0.5 M DMSO, 0.5 M glycerol, 1 M sucrose or proline (final concentrations) for ca. 1 hr at 4°C; dispense in 1-ml aliquots into 2-ml ampoules; freeze at 1°C/mm to −35°C; hold at −35°C for 40 min; plunge into liquid nitrogen; store in liquid nitrogen (−196°C) or liquid nitrogen vapor (ca. −150°C); thaw in warm water (+40°C); pour onto semisolid medium of standard composition; return cells to liquid medium as soon as healthy growth is reestablished (7–14 days). Over 30 species have responded favorably to one or more of the permutations in the procedure given above (Withers, 1983a, 1984a; Withers and Williams, 1985).

A number of other recent publications have reported similar approaches to cryopreservation and may offer potential improvements to make the procedure even more widely applicable. These aspects will be considered in the following sections under appropriate headings.

B. Pregrowth and Cryoprotection

Pregrowth and cryoprotection have been combined here and in Section IV,B since they are becoming increasingly interrelated due to the development in some studies of pregrowth treatments involving compounds with cryoprotectant properties. Various pregrowth procedures and cryoprotectant formulations are listed in Table III.

It has been appreciated for some years that the recent history of a cell

TABLE III

Examples of Pregrowth and Cryoprotection Treatments Applied to Cell Suspension Cultures[a]

Species	Pregrowth	Cryoprotection	Reference
Daucus carota	Cells harvested 2 weeks after transfer to maintenance medium containing 10 mM myoinositol and 0.1% casein hydrolysate	10% DMSO applied at room temperature; no period of equilibration	Dougall and Whitten (1980)
Daucus carota *Nicotiana tabacum*	Cells harvested after 4 days growth in standard medium and in exponential growth	10% DMSO plus 10% glycerol applied at 0°C; no period of equilibration	Hauptmann and Widholm (1982)
Nicotiana tabacum	Cells harvested after 3 days growth in the presence of 6% mannitol and in exponential growth	0.5 M DMSO plus 0.5 M glycerol plus 1 M sucrose applied at ice temperature; 1 hr equilibration	Maddox *et al.* (1983)
Nicotiana sylvestris	Cells harvested after 3–5 days growth in the presence of 6% sorbitol	0.5 M DMSO plus 0.5 M glycerol applied at ice temperature; 1 hr equilibration	Maddox *et al.* (1983)
Zea mays and many other species	Cells harvested after 4–7 days growth in standard medium or in the presence 6% mannitol or 10% proline	0.5 M DMSO plus 0.5 M glycerol plus 1 M sucrose or proline applied at room temperature or on ice; 1 hr equilibration	Withers and King (1980); Withers (1983a, 1984a,b)
Catharanthus roseus	Cells harvested after 4 days growth in standard medium and cultured for 20 hr in the presence of 1 M sorbitol	1 M sorbitol plus 0.5 M DMSO applied at ice temperature; 1 hr equilibration	Chen *et al.* (1984b)

[a] The treatments are listed in order of increasing modification of pregrowth conditions and integration of pregrowth and cryoprotection.

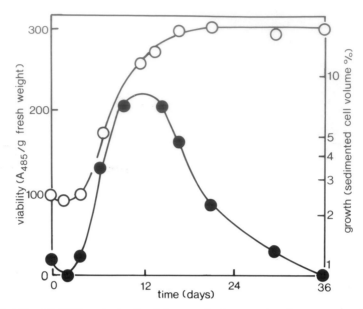

Fig. 1. Viability (●; TTC test) of cells of *Oryza sativa* harvested and cryopreserved during a growth cycle (○). Cells in rapid growth have the highest freeze tolerance. (Redrawn from Sala *et al.*, 1979.)

culture could have a profound effect upon its freeze tolerance. Cells in early lag phase and stationary phase are particularly sensitive to freezing injury (Fig. 1). This is a relfection upon their size and degree of vacuolation. Metabolic factors may also be involved. Accordingly, cells should be harvested for cryopreservation in late lag or exponential phase. This contrasts, for example, with findings for algal cells which become more freeze tolerant as they age and accumulate storage products (Morris, 1980).

Superimposed upon the cell age effect may be one of cell cycle stage, but this is probably usually masked by asynchrony in most cultures. The first suggestion of an involvement of the cell cycle appeared in a study of cells of *Acer pseudoplatanus* by Withers and Street (1977b), who found that frequent subculturing led to an oscillation in the percentage survival during the first few days of successive passages. Examination of a culture synchronized intentionally by nitrate starvation revealed clear changes in freeze tolerance with passage through each cell cycle (Withers, 1978b). By combining Feulgen staining with prior vital staining with Evans' blue, the individual surviving cells could be identified as predominantly in early G_1 or G_0 phase. Later studies have achieved much higher overall percentage survivals than in the synchronized culture, thus relegating this observation to one of academic interest rather than a pointer to a general means of improving success in cryopreservation.

Supplementation of the culture medium in the days preceding cryo-preservation with various compounds can have a dramatic effect upon survival potential. These compounds include mannitol, sorbitol, proline, and DMSO. The action of the first two of these compounds is most readily appreciated in that they bring about substantial reduction in mean cell size (e.g., 50% volume reduction by application of 3% mannitol; Withers and Street, 1977b), thus reducing the likely requirement for protective dehydra-tion in the early stages of freezing. Pritchard and colleagues (1982) have investigated the cellular phenomena accompanying cell size reduction. A redistribution of the reduced vacuolar volume into a number of small vesi-cles was observed (see also Withers, 1978a). Levels of soluble proteins and lipids increased and respiration decreased.

Mannitol and sorbitol at levels of about 0.075–0.3 M achieve benefits with no apparent detrimental effects when applied over days or weeks. More recently, however, sorbitol has been applied at a much higher level of 1 M for 6–20 hr before cryoprotection and freezing (Chen et al., 1984b). It may operate by a combined dehydrative and physiological effect (e.g., resembling cold hardening) upon individual cells rather than by providing an environment within which cells grow and reproduce to a higher degree of freeze-tolerance. During the first 10 hr of pregrowth, survival drops by ca. 20% and continues to decline slowly thereafter (Fig. 2). An exposure

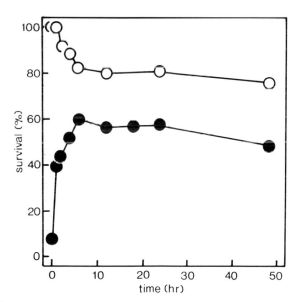

Fig. 2. Effect of the duration of pregrowth in the presence of 1 M sorbitol on the viability (TTC test) of cells of *Catharanthus roseus* before (○) and after (●) cryopreservation. (Redrawn from Chen et al., 1984b.)

period of 20 hr was chosen for practical convenience; it is unlikely that a concentration of 1 M could be tolerated for extended periods.

The two remaining compounds listed earlier as effective pregrowth additives, DMSO and proline, also have cryoprotective properties and therefore may be exerting a dual effect during the period of their relatively prolonged application in the examples studied—3–4 days. DMSO has been applied effectively at a level of 5% (ca. 0.54 M) to cells of *Catharanthus roseus* (Kartha *et al.*, 1982a). Changes in cell size and vacuolar distribution similar to those described for mannitol-treated cells (above) were observed. Other than a tendency to darkening in color suggestive of an incipient detrimental effect (L. A. Withers, unpublished observation), no detailed observations have been made of proline pregrowth treatment.

Although, as pointed out earlier, pregrowth treatments can involve potential cryoprotectant compounds, most successful cryopreservation protocols almost invariably involve a distinct cryoprotection phase of some hours' duration immediately prior to freezing (one possible exception being the study of Weber *et al.*, 1983, in which 1 M sorbitol was applied for 16 hr prior to freezing without further cryoprotection, thus combining pregrowth and cryoprotection). A further generalization that can usefully be made is that mixtures of cryoprotectants are usually more effective than single compounds for cell cultures, although, of course, the culture medium in which cells are suspended and in which cryoprotectants are prepared will contain sucrose and salts which are likely to contribute to the overall protective effect.

Earliest studies with cell cultures concentrated upon the exploration of DMSO and glycerol, alone or in combination. The former is effective alone for some species, including *Daucus carota* (Nag and Street, 1973) and *Catharanthus roseus* (Kartha *et al.*, 1982a). However, Glycerol is of little value alone as a cryoprotectant and may even be detrimental. There is some evidence from other biological systems that glycerol can have an effect of predisposing cells to deplasmolysis injury (Jenkins and Weed, 1977; Withers, 1980c). Chen and colleagues (1984b) have found that a mixture of 5% DMSO plus 1 M sorbitol (i.e., avoiding the application of gycerol) is effective for the cryoprotection of cells of *Catharanthus roseus* pregrown in 1 M sorbitol. Higher levels of DMSO proved toxic.

Finkle and colleagues have carried out extensive investigations into the benefit of combining cryoprotectants for application to callus cultures (Finkle and Ulrich, 1979; Finkle *et al.*, 1983). These revealed the efficacy of a mixture of polyethylene glycol, DMSO, and glucose totaling a 1.9 M solution overall. A range of similarly highly concentrated mixtures was explored by Withers and King (1980) and Withers (1980b). While both of these studies involved Gramineous species (*Saccharum* sp. and *Zea mays*), it is encouraging that a much wider efficacy has been demonstrated making possible the development of the routing protocol mentioned at the beginning of Section III.

Cryoprotectant mixtures have been sterilized by various procedures, but it is strongly recommended that filter sterilization be used to avoid caramelization. By convention, cryoprotectants have been prepared at double strength in culture medium, chilled on ice, and then added slowly to an equal volume of chilled cell suspension, with a period of equilibration (e.g., 1 hr) preceding further cooling. Although this procedure is used in current studies (e.g., Chen et al., 1984b), there is some evidence to suggest that this may not be an optimal procedure.

P. J. King (personal communication) now applies DMSO, glycerol, and sucrose or DMSO, glycerol, and proline mixtures after chilling but in one operation, since slow uptake is assured by the reduced temperature of incubation. Hauptmann and Widholm (1982) observed on interaction between duration of exposure to cryoprotectants above and below freezing temperatures; i.e., a short exposure to cryoprotectants before commencing cooling could be compensated for by a longer period of time at an intermediate temperature before freezing in liquid nitrogen. Most significant of all is the observation by Finkle and Ulrich (1982) that there are interactions between the temperature of applying and removing cryoprotectants. The optimal combination involves cold (ca. 0°C) application and warm (ca. +22°C) removal, although warm application and warm removal are significantly less damaging than cold application and cold removal (although see comments on cryoprotectant removal and washing in Section III,D).

C. Freezing and Thawing

Cryoprotected cells, dispensed into containers at a suitable cell density (e.g., 20% by volume; Kartha et al., 1982a), should first be exposed to conditions conducive to extracellular freezing before quenching in liquid nitrogen and transfer to the storage temperature. Either slow or stepwise freezing is appropriate, or, as in fact occurs especially in a number of improvised freezing units, a combination of slow freezing to an intermediate temperature and holding that for a period of time. In addition to the cooling program recommended in Section III,A, cooling at 0.5–1°C/min to a temperature between −40 and −100°C are offered as alternatives deriving from successful reports in the literature (e.g., Chen et al., 1984b; Dietrich et al., 1982; Dougall and Whitten, 1980; Kartha et al., 1982a; Watanabe et al., 1983). Figure 3 shows the differing responses of two culture systems to freezing rate. The differences are likely to be attributable to both the species and their pregrowth and cryoprotectant treatments.

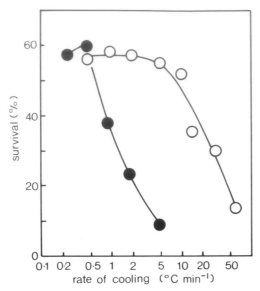

Fig. 3. Effect of cooling rate on the survival of cells of *Catharanthus roseus* (●; TTC test) cryoprotected with 1 *M* sorbitol plus 5% DMSO and *Digitalis lanata* (○; phenosaffranine staining) cryoprotected with 20% sucrose plus 20% glycerol. (Redrawn from Chen *et al.*, 1984b, and Diettrich *et al.*, 1982, respectively.)

While it is clearly necessary to continue slow cooling to an adequately low temperature, cooling to too low a temperature may be deleterious in some cases. Figure 4 demonstrates one case in which cooling could be continued beyond the threshold of an optimum response and one in which this was deleterious. Prolonged exposure to suboptimal temperatures may be injurious by causing either excessive cellular dehydration or damaging ice crystallization once intracellular freezing ensues. The observations of Chen *et al.* (1984a) that the use of a mixture of sorbitol and DMSO as cryoprotectants after pregrowth in 1 *M* sorbitol may help survival by preventing removal of an excessive amount of intracellular water yet permitting sufficiently slow cooling with its advantages of more gentle dehydration are relevant here.

Clearly, no one cooling program can be recommended to the exclusion of all others due to the inevitable interactions between stages in the procedure. Observation of such interactions can reveal the potential for improving survival by, for example, application of thawing conditions which counteract potentially damaging treatments at the freezing stage. Conventionally, and with a firm experimental backing, thawing is carried out rapidly by agitating the frozen specimen in warm water at about +40°C. This avoids the risk of ice recrystallization damage when freezing has terminated at too high a temperature (Fig. 5; Withers, 1980b). A reversal of this effect is seen at much lower temperatures (e.g., −70°C; Fig. 5), where-

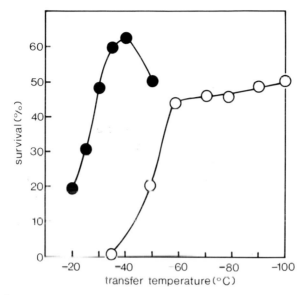

Fig. 4. Effect of temperature of transfer to liquid nitrogen on survival of cells of *Catharanthus roseus* (●) cooled at 0.5°C/min and *Digitalis lanata* (○) cooled at 1°C/min. Viability tests and cryoprotectants as detailed in legend to Fig. 3. (Redrawn from Chen *et al.*, 1984b, and Diettrich *et al.*, 1982, respectively.)

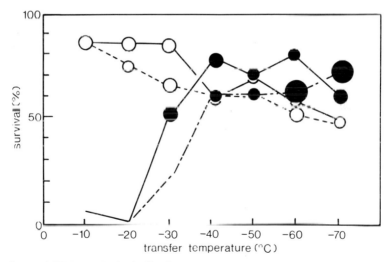

Fig. 5. Survival (FDA staining) of cells of *Zea mays* cryoprotected with 0.5 *M* DMSO plus 0.5 *M* glycerol plus 1 *M* sucrose, cooled at 1°C/min to various subzero temperatures before thawing rapidly (solid lines) or slowly (broken lines) with (●) or without (○) first quenching in liquid nitrogen. Presence of circle indicates recovery of cell colonies on plates, and the size of the circles indicates relative efficiency of recovery in terms of number of colonies formed and their size. Cells thawed directly from transfer temperatures of −30°C and above escape ice damage. (Redrawn from Withers, 1980b.)

upon cells survive as well, if not better, when thawed slowly in air at room temperature.

D. Postthaw Treatments and Recovery

Early examinations of cell cultures involved a postthaw washing stage as a matter of routine to avoid deleterious effects of cryoprotectants. However, a number of subsequent studies have shown that washing is itself often far more damaging than prolonged exposure to cryoprotectants. Three options are available in the immediate postthaw handling of cryopreserved cells: (1) washing before returning to culture, (2) dispensing into liquid medium without washing but with inevitable dilution, or (3) inoculating onto a prepared plate of semisolid medium. All of these have been attempted in various studies with some success, but few critical, comparative tests have been carried out.

Washing has been carried out using standard medium or medium supplemented with an osmoticum, sorbitol. There is clear evidence for the former's being damaging (Withers and King, 1979), probably resulting from loss of membrane during freeze-dehydration predisposing the cell to deplasmolysis injury. Washing with medium containing sorbitol is claimed to reduce such injury while conferring beneficial effects of removal of cryoprotectants (Maddox *et al.*, 1983). No doubt various other measures could be taken to similar effect (see also following). Examples can be found of cells suffering a prolonged lag period before resuming growth after thawing (e.g., Bajaj, 1976; Withers and Street, 1977b). This may have been caused, or at least exacerbated, by deplasmolysis injury.

In an attempt to study relative damage under different postthaw treatments, Withers and King (1979) returned thawed cells of *Zea mays* cryoprotected with various preparations to culture on semisolid medium after washing with distilled water, culture medium, culture medium supplemented with proline (one of the cryoprotectants under examination), and without washing. In all cases, unwashed cells were able to resume growth (as estimated by increase in percentage viability) more rapidly than washed cells. This difference was particularly marked in cells cryoprotected with DMSO and glycerol and least in cells cryoprotected with proline. Furthermore, it was noted that cells benefited by remaining in contact with the solution in which they were frozen rather than being drained free of it. It would appear that the medium is taken up by the cells as growth resumes.

The simple postthaw treatment of spreading over semisolid medium without washing or removal of the suspending liquid was adopted as part of the routine procedure of Withers and King (1980). Two refinements of detail have been added by P. J. King (personal communication) involving (1) pushing cells together into a mound for the first few days of recovery and (2) cutting a depression into the semisolid medium to drain away remaining unincorporated medium after the initial recovery period.

An alternative means of achieving rapid postthaw recovery while avoiding deplasmolysis injury and cryoprotectant toxicity has been explored by Chen et al. (1984b). They layered thawed cells of various Catharanthus roseus genotypes onto a filter paper disk overlying semisolid medium. After 4–5 hr, the filter paper bearing the cells was transferred to a plate of fresh medium, at the same time giving the opprotunity for weighing the cells to determine the rate of recovery. Despite the several reports of successful recovery of cells after washing or dilution, none have clearly been demonstrated to recover more rapidly than unwashed cells. The converse has, however, been demonstrated. Therefore, a strong recommendation is made that washing should be avoided unless absolutely essential because of sensitivity to washing, in which case the use of a warm wash (Finkle and Ulrich, 1982) or one supplemented with an osmoticum (Maddox et al., 1982) might be tried.

Maddox et al. (1983) adopted a procedure involving recovery growth in liquid medium as they found that this avoided callusing on semisolid medium. However, in the experience of the author and as evidenced by successful reports elsewhere, cells can readily form a suspension again when transferred from plates to liquid medium after 7 days (Diettrich et al., 1982) or unspecified periods (Chen et al., 1984b; Seitz et al., 1983) Little attention has been given to the effect of medium composition upon recovery, although feeder plates have been used successfully by Hauptmann and Widholm (1982) and a liquid medium designed to permit recovery from very low cell densities has also proved successful (Maddox et al., 1983).

E. Structural and Genetic Stability

Cells at various stages in the cryopreservation procedure have been examined ultrastructurally. These studies have been valuable in revealing the nature of cellular injury, particularly under suboptimal cryopreservation conditions. For the reason that observations of pathological conditions, can involve problems of poor fixation and susceptibility to deterioration during the very early stages of fixation, the caution is given that these observations should not be overinterpreted.

A useful comparison can, however, be made between cells (1) frozen by a procedure which yields a fairly high percentage of living cells capable of

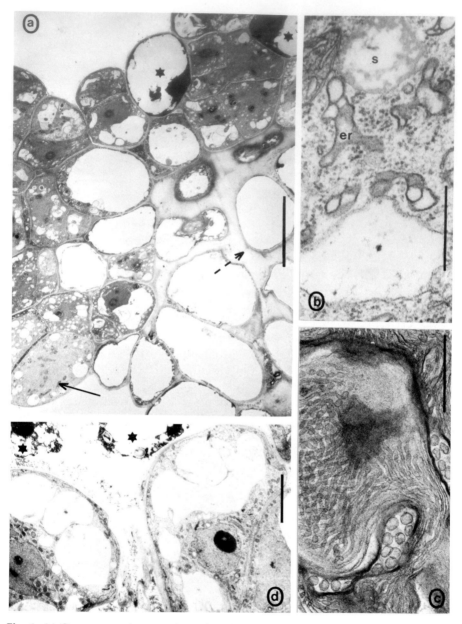

Fig. 6. (a) Cryopreserved suspension cultured cells of *Daucus carota* observed 10 days after returning to culture. Some cells have highly electron-dense contents (∗). Others have produced large amounts of wall material (--→). A minority are undergoing recovery and cell division (→). (Bar = 5 μm.) (b) Detail from a preparation as in (a) showing a spherosome (s; probably formed from degraded membrane material) undergoing erosion at the same time as new endoplasmic reticulum (er) membrane forms. (Bar = 500 nm.) (c) Detail from a prepara-

regrowing after a long period, probably of weeks (e.g., Withers, 1980d; Withers and Street, 1977a,b), and (2) frozen by an improved procedure of cooling, cryoprotection, and postthaw treatment which permits rapid recovery within days (Withers, 1980d; Withers and King, 1980). In the former, cells examined shortly after thawing, breakage damage is observed to the plasmalemma, plasmodesmata, and tonoplast. Cellular organelles—mitochondria, plastids, Golgi body/dictyosome vesicles, and endoplasmic reticulum—all display swelling suggestive of a disturbance in control of water movement in and out of membrane-bound structures. This is, of course, consistent with leakiness of freeze-injured cells and tissues and the observed influence of the postthaw osmotic environment upon survival and recovery.

Viability test data suggest that viability loss can continue for days after thawing, and this is borne out by ultrastructural changes over the same period (Figs. 6, a–c). Recovery would appear to be occurring in only a proportion of the initial surviving cells. Others remain capable of metabolism directed toward various degenerative processes but incapable of cell division and contribution to the regenerating population. Some appear to produce considerable quantities of cell wall material, thus isolating themselves from neighboring cells. Others accumulate large quantities of cytoplasmic and vacuolar osmiophilic material—probably phenolic in nature. Browning in the appearance of fresh, unfixed material bears this out.

Within the recovering cells, a number of changes take place which involve redistribution of membrane material from damaged membranes, via accumulation in spherosomes/lipid droplets, to regenerated membrane masses and then functional organelle membranes. These recovering cells appear to be occupied in a process of recycling damaged cellular building blocks before attention can be given to cell division and recovery of the culture as a whole. While no studies of the potentially toxic effects of dead, dying through injury, and senescing cells have been carried out, it is highly likely that the cells capable of recovery are, at best, retarded by the presence of more seriously injured cells. (Note that Diettrich et al., 1982, have shown that in artificial mixtures of living and dead cells, 30% viability is necessary to permit recovery.)

The resumption of a relatively normal cellular appearance takes about 2 weeks in the circumstances described above, whereas, in the case of cells preserved by the recommended routine procedure described earlier, struc-

---

tion as in (a) showing a mass of membrane material forming in a recovering cell. (Bar = 500 nm.) (d) Suspension cultured cells of Zea mays cryopreserved by an improved method (cf. Figs. a–c) observed 2 days after returning to culture. Recovering cells show little sign of freezing injury and are clearly distinguishable from dead cells (*). (Bar = 5 μm.) (From Withers, 1980c,d.)

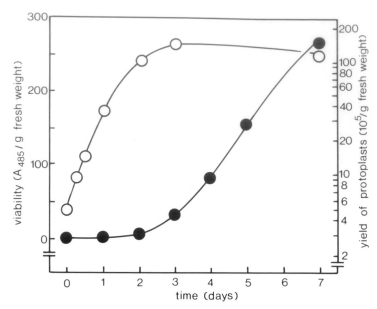

Fig. 7. Cell growth (●) and yield of protoplasts (○) during the first 7 days of recovery growth of a cryopreserved cell suspension culture of *Oryza sativa*. (Redrawn from Cella *et al.*, 1982.)

tural symptoms of injury are greatly reduced and within about 2 days cell divisions are noted. Further, the gradation of relatively healthy through increasingly injured to dead cells is not observed. Within 2 days, cells are clearly alive and recovering or dead and breaking down (Fig. 6d).

Freezing-induced weakness of the plasmalemma manifested as susceptibility to deplasmolysis injury is further evidenced by observations of Cella *et al.* (1982) involving cells of *Oryza sativa* subjected to enzyme digestion for protoplast isolation. From freshly thawed cells, very few stable protoplasts could be isolated, whereas within 2 days the cells were capable of yielding stable protoplasts virtually at the control rate (Fig. 7).

Rates of recovery from cryopreservation vary from study to study, and as yet too few comprehensive investigations have been carried out to deduce controlling factors. However, drawing from available evidence, slow recovery would appear to be due in part to effectively reduced inoculum densities and in part to physiological impairment. A specific recovery period in physiological competence, as earlier suggested by ultrastructural observations, susceptibility to deplasmolysis injury, and sensitivity to protoplast isolation, is borne out by measurement of a range of metabolic parameters. In a wide-ranging study into such aspects of recovery in cells of *Oryza sativa*, Cella *et al.*, (1982) revealed that glucose uptake, respiration, functional integrity of the plasmalemma, and transport across membranes

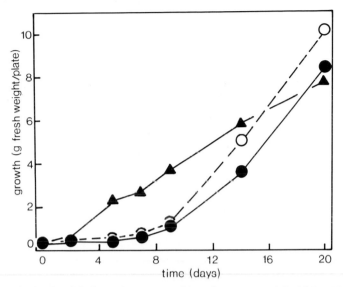

Fig. 8. Growth of cells of *Catharanthus roseus* subjected to pregrowth in 1 *M* sorbitol, cryoprotection in 1 *M* sorbitol plus 5% DMSO, and then inoculation onto filter paper over plates of recovery medium before (○) and after (●) cryopreservation. Control cells (▲) were plated without any treatment. (Redrawn from Chen *et al.*, 1984b.)

are impaired. As a result, intracellular pools of K^+, ATP, G-6-P, and pyruvate were depleted.

Within 2–7 days of thawing, recovery is normally under way in cells preserved by a satisfactory procedure. Increase in cell number can proceed rapidly and close to the control rate. However, a slight increase in growth rate in cells of *Catharanthus roseus* after cryopreservation (Fig. 8) has been attributed to elimination of more fragile fractions from the population of cells (Chen *et al.*, 1984b). It is important that this point be pursued in order to demonstrate satisfactorily that the apparent selection be phenotypic only.

Most investigations to date have explored technical aspects of cryopreservation with emphasis on achievement of a high-percentage survival and, where relevant, demonstration of totipotency. In the latter category one can cite the study of Nag and Street (1973), for example, wherein cells of *Daucus carota* were capable of regenerating somatic embryos after cryopreservation and storage in liquid nitrogen. However, in one of the most important potential applications of cryopreservation—the conservation of cell lines with certain, valuable biochemical or synthetic properties—totipotency in the sense of plant regeneration is not a character of great relevance. Yet stability of genotype is of the utmost importance. Cryopreservation presents two major opportunities for the destabilization of genotypes: (1) by selection for freeze tolerant variants, and (2) by direct damage to the genome as a result of the manipulations involved in the cryopreservation procedure. Fortunately, no serious evidence has been generated to date to suggest that either of these forces is operating.

Six particular studies have addressed the question of biochemical stability in preservation, and their findings are very encouraging. Hauptmann and Widholm (1982) have investigated the preservation of wild-type and variant cells of *Nicotiana tabacum*. They were able to demonstrate that cryopreservation does not impair the auxotrophic response of a 5-methyltryptophan–resistant line which retains its dose–response and growth characteristics after freezing and thawing (Fig. 9). A second aspect of the demonstration of stability through preservation involves cell cultures capable of synthesizing valuable secondary products or carrying out biotransformations.

Dougall and Whitten demonstrated in 1980 that 25 different lines of wild carrot (*Daucus carota*) produce approximately the same amount of anthocyanins as unfrozen controls. Similarly, Seitz *et al.* (1983) have demonstrated that in cells of *Digitalis lanata* the ability to transform β-methyldigitoxin to β-methyldigoxin is unchanged after cryopreservation. Diettrich *et al.* (1982) have investigated the preservation of cells of *Digitalis*

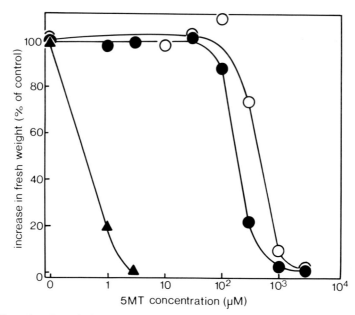

Fig. 9. Effect of DL-5 methyltryptophan (5MT) concentration on the growth of wild type (▲) and 5MT resistant cells of *Nicotiana tabacum*. The latter were tested before (○) and after (●) cryopreservation. (Redrawn from Hauptmann and Widholm, 1982.)

lanata, which are capable of transforming cardenolides such as digitoxin. In addition to retaining this capacity after cryopreservation, the cells were demonstrated to have an unchanged frequency distribution of nuclear DNA contents when compared with unfrozen controls. Watanabe and colleagues (1983) have shown that frozen cells of *Lavandula vera* retain the capacity to produce biotin as well as the ability to differentiate chloroplasts and regenerate plantlets.

In a study by Chen *et al.* (1984b) which demonstrated the retention of the capacity to synthesize a range of alkaloids by cells of *Catharanthus roseus,* some suggestion is made of a slight qualitative shift in synthesis which cannot be satisfactorily explained but may suggest a phenomenon of selection within the population of cells (see also earlier). This selection might be on a morphological or genotypic basis. Either would give cause for concern if tending to depress synthesis and clearly warrants further investigation. It is interesting that the alkaloid-producing cell lines required a refined cryopreservation procedure, responding badly to one which could preserve nonproducing cells (Kartha *et al.,* 1982a). Differences in morphology and cell size between cell lines of this species (see Chen *et al.,* 1984b; Kartha *et al.,* 1982a) may account for the variation in requirements.

As yet, little attention has been given to other markers of genetic stability in cell suspension cultures. However, stability in isozyme patterns has been demonstrated in a cryopreserved culture of *Phoenix dactylifera* (Finkle *et al.,* 1983).

IV. CRYOPRESERVATION OF MERISTEMS

A. A Divergence in Procedures

Despite differences in technical detail in the various reports of successful cryopreservation of cell suspension cultures, a common methodological theme can be recognized. However, in the case of meristems and shoot tips this is not the case; two significantly different approaches are taken. One of these is similar to that for cell cultures in that slow or stepwise freezing is involved, whereas the other involves rapid freezing. Success in the latter depends upon a mechanism whereby cellular water freezes very rapidly into microscopically small crystals which remain innocuous provided that thawing is carried out sufficiently rapidly to avoid recrystallization.

The first reports of success in the cryopreservation of meristems involved rapid freezing, and this approach was prominent in a number of subsequent publications. However, with few exceptions (e.g., *Brassica napus;* Withers, 1982b), the successes in rapid freezing have been matched by ones involving slow or stepwise freezing: *Solanum tuberosum* (Grout and Henshaw, 1978; Towill, 1983), *Dianthus caryophyllus* (Seibert and Wetherbee, 1977; Anderson, 1979), and *Fragaria* sp. (Sakai *et al.,* 1978; Kartha *et al.,* 1980). Overall, successes with methods which rely upon extracellular freezing and protective dehydration to minimize freezing injury outnumber others. In the future, it is likely that some general methods will emerge on this basis, but ultrarapid freezing should not be ruled out, especially for certain recalcitrant specimens. An interesting recent example of successful rapid freezing can be found in the study by Engelmann *et al.* (1984), who found that somatic embryos of *Elaeis guineensis* could survive rapid freezing after cryoprotection with 0.75 M sucrose.

B. Pregrowth and Cryoprotection

Specimens within this category can have one of several histories. They may be isolated from seedlings/plants (e.g., *Lycopersicon esculentum;* Grout *et al.,* 1978; *Malus domestica;* Katano *et al.,* 1983). Unless already under aseptic conditions, they must be sterilized to remove surface contaminants. Second, specimens may derive from shoot cultures maintained *in vitro* (e.g., *Brassica napus;* Withers, 1982b; *Solanum tuberosum;* Towill, 1983). These are structurally very similar to those just mentioned but have the advantage of being more numerous, thus providing clonal replicates and permitting more satisfactory experimentation and duplication of samples in storage.

The third type of specimen, again, is a truly *in vitro* system but one which involves rapid proliferation by multimeristems or adventitious means (e.g., *Fragaria* × *ananassa;* Kartha *et al.,* 1980). These share the advantages of the other two types but have the potential disadvantage of a greater susceptibility to somaclonal variation (Scowcroft *et al.,* 1984), possibly exacerbated by their capacity to regenerate by adventitious means. This point brings a timely reminder that in considering the cryopreservation and recovery of organized structures such as meristems, there is a very important qualitative aspect to recovery as well as the bare statistic of percentage plant regeneration. This is discussed in more detail in Section IV,D.

In contrast to the cell culture situation, meristems and shoot tips must be dissected from the plant or culture before freezing (the whole specimen would be much too large to cryopreserve intact). Here the third category of specimen may be attractive in that multiple samples for freezing may be produced far more quickly and easily. An appropriate unit specimen for freezing consists of the apical dome plus two to four leaf primordia and measures 0.5–1 mm in length depending upon the species. After dissection the specimen may be processed directly for freezing or inoculated into a pregrowth phase.

Pregrowth in shoot tips is usually of 1–2 days' duration. Henshaw et al. (1980b) have investigated the progressive increase and subsequent decrease in specimen of Solanum tuberosum over a period of 72 hr. Survival potential is maximized at about 48 hr and may be related to healing of the dissection wound and/or achievement of an optimum unit size. In support of the former suggestion, Withers (1982b) has found in the case of Brassica napus that supplementation of the first of 2 days' pregrowth medium with hormones conducive to the induction of callusing can increase percentage recovery of plants.

Requirement for a pregrowth period ranges from virtually essential in some reports (e.g., Henshaw et al., 1980b; Withers, 1982b; Benson et al., 1984) through advantageous (e.g., Kartha et al., 1979) to inconsequential or of unexplored value (Kartha et al., 1982b; Katano et al., 1983; Seibert and Wetherbee, 1977). In developing a protocol for an untried species, this would be a priority area for examination of alternate treatments. O'Hara and colleagues (O'Hara et al., 1984; J. F. O'Hara, personal communication) have combined a 24-hr pregrowth period under normal culture conditions with a subsequent overnight period at a reduced temperature (ca. 4°C) for shoot tips of Solanum tuberosum. L. A. Withers (unpublished observations) has found that pregrowth at 4°C can partially substitute for the effects of pregrowth in the presence of DMSO in the case of shoot tips of Brassica napus (see later).

Some reports describe supplementation of the pregrowth medium with DMSO at 5%. DMSO was first used as a pregrowth additive by Nag and Street (1975a) for suspension cultured cells of Atropa belladonna. In the case of Pisum sativum and Fragaria × ananassa, pregrowth in the presence of this compound increases percentage recovery significantly from 28 to 73%, and from 5 to 95% respectively, when compared with the performance of freshly dissected specimens (Kartha et al., 1979, 1980). (No data are available on specimens pregrown on unsupplemented medium.) In the case of shoot tips of Brassica napus, DMSO affects both the frequency and the nature of recovery growth, presumably by an effect upon survival of critical regions of the shoot tip (Benson et al., 1984; Withers, 1982b). Two days of pregrowth on a filter paper soaked in standard culture medium leads to regeneration in up to 40% of specimens, usually involving leaf primordial tissues only, whereas supplementation of the medium with DMSO at 5% increases overall recovery (up to 90%) and promotes survival and outgrowth of the shoot apical meristem.

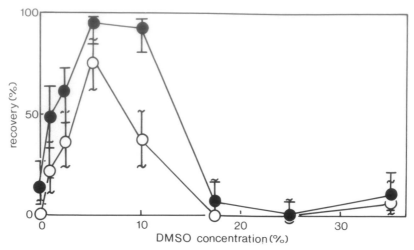

Fig. 10. Effect of DMSO concentration upon the level and nature of recovery in shoot tips of *Dianthus caryophyllus.* Survival as indicated by greening, callusing, and shoot regeneration is indicated by ●; shoot regeneration alone is indicated by ○. (Redrawn from Seibert and Wetherbee, 1977.)

Cryoprotection conditions are remarkably uniform in most reports, involving DMSO at levels of 5–15%. Glycerol has been found to be far less effective (Kartha *et al.,* 1979, 1980). Exceptionally, some specimens, particularly dormant buds dissected from fruit trees, may survive freezing without any cryoprotection (Katano *et al.,* 1983). Cryoprotectants have, conventionally, been applied at a reduced temperature (ca. +4°C), but the author has found no serious deleterious effects by application at the normal culture temperature (ca. +25°C; L. A. Withers, unpublished observations). An exposure period of approximately 1 hr prior to freezing appears to be generally adopted, although Towill (1983) notes that exposure as brief as 10 sec at room temperature can confer protection. There is some evidence that DMSO may have a beneficial effect on total survival levels while causing excessive callusing (Seibert and Wetherbee, 1978; Fig. 10; L. A. Withers and E. E. Benson, unpublished observations), indicating that prolonged exposure and/or the use of other than minimal effective concentrations should be avoided.

C. Freezing and Thawing

As mentioned at the beginning of this section, both rapid and slow freezing have proved successful in certain applications. Each will be examined further here. For rapid freezing, the specimen may be exposed directly to the coolant without enclosure in a vial (Fig. 11a). Grout and Henshaw (1978) first described this approach for the preservation of shoot tip of *Solanum goniocalyx*. The specimens were collected from the cryoprotectant solution on the tip of a hypodermic needle and then plunged into liquid nitrogen. More recent reports of this method have standardized the cooling treatment by using a depth of 6 cm for the immersion (Henshaw *et al.*, 1980a). Withers and colleagues have used this method for *S. tuberosum* and *Brassica napus* (Benson *et al.*, 1984; Withers, 1982b; Withers *et al.*, 1985).

An alternative means of direct exposure to the coolant is described by Sakai and colleagues (Sakai *et al.*, 1978; Uemura and Sakai, 1980), who collected meristems of *Fragaria* × *unanassa* or *Dianthus caryophyllus* on a cover glass held in forceps and plunged this into liquid nitrogen (Fig. 11b). Some survival was achieved thus but was improved upon by prefreezing (see later).

It is difficult to record accurately the rates of cooling involved in operations such as described above. Further, accurate reproduction of freezing conditions can be problematic. However, it is likely that the specimen may be cooling at a rate of up to $10^5°C/min$. The necessity to maintain such a rapid rate of heat withdrawal may thwart attempts to overcome some of the problems in rapid freezing. Naked specimens present difficulties in handling, especially during freezing, when they may pick up microbial contaminants, and during storage, when they may become detached from the supporting needle or cover glass, lost within the refrigerator, or cross-contaminated.

In some exploratory experiments, the author (L. A. Withers, unpublished observation) has been attempting to combine maintenance of adequately rapid cooling with enclosure of the specimen (Fig. 11c). Shoot tips of *Brassica napus* and *Solanum tuberosum* have been placed within small envelopes of aluminium foil and various plastic materials. Most of the suspending liquid was removed, and the envelopes were sealed (by heat in the case of plastic). Survival rates after immersion in liquid nitrogen and rapid thawing were lower than in the case of naked specimens, but by cooling the envelopes to -10 or $-15°C$ *without* ice nucleation in the remaining cryoprotectant solution (and therefore no opportunity for protective dehydration) some improvement was noted. In all cases, results were superior to those achieved by enclosing the specimens in polypropylene ampoules. This investigation is continuing.

Seibert and Wetherbee (1977) reported a very high rate of success (approaching 100% callusing or plant regeneration) in the cryopreservation of meristems dissected from cold-hardened plants of *Dianthus caryophyllus* and frozen in open ampoules (Fig. 11d). While this is a very good result, it has not been used successfully for other species and would appear to

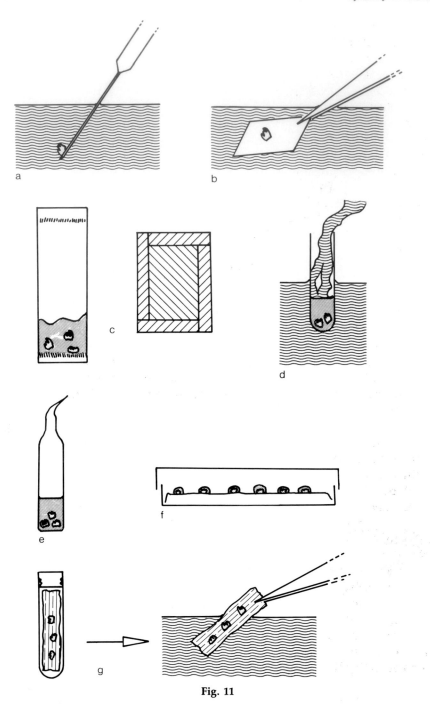

Fig. 11

combine the worst practical aspects of freezing within and without an enclosing ampoule, since the nitrogen coolant was poured into the open top of the ampoule at the same time as it was plunged into liquid nitrogen.

Before addressing the use of slow freezing, one other report should be noted. Grout and colleagues (1978) successfully employed an intermediate rate of freezing (20–50°C/min) for the preservation of shoot tips dissected from seedlings of *Lycopersicon esculentum.* The specimens were suspended in the vapor above liquid nitrogen and then plunged into the liquid nitrogen to complete cooling. A recovery rate of up to 45% was recorded, but this again is an isolated example and cannot set any precedents for wider application.

Slow and stepwise freezing are attractive in that a high degree of re producibility is possible and considerable control can be exerted over the freezing conditions. Two approaches differing in practicalities but both relying upon protective dehydration during the early stages of cooling can be described: (1) continuous slow cooling and (2) prefreezing. Among the most impressive examples of continuous slow cooling are the reports of Kartha and colleagues (1979, 1980, 1982b) covering species as diverse as *Pisum sativum, Fragaria* × *ananassa,* and *Manihot esculenta.* Freezing rates of 0.6, 0.85, and 0.5°C per minute, respectively, proved optimal, suggesting a considerable sensitivity to under- or overdehydration in these specimens. Survival rates range widely, with the highest being recorded for *Fragaria* × *ananassa* (95%) and the lowest for *Manihot esculenta,* in which considerable variation was found in terms of absolute number of specimens recovering (16–80%) and plant regeneration (0–42%). The latter species has proved particularly recalcitrant to efforts to preserve it satisfactorily (cf. Bajaj, 1977b; Henshaw et al., 1980b), and this result should be taken as encouraging.

A noteworthy feature of the work with *Manihot esculenta* is the handling of the meristems during freezing. Whereas those of the other two species mentioned above were suspended in about 1.2 ml of cryoprotectant solution in a flame-sealed glass vial (Fig. 11e), in this case they were each dispensed into a 2- to 3-µl droplet of cryoprotectant solution on an alumi-

Fig. 11. Various methods for handling shoot tips during cryopreservation. (a) Supported on a hypodermic needle (Grout and Henshaw, 1978; Withers, 1982b), (b) Supported on a cover glass (Sakai et al., 1978; Uemura and Sakai, 1980), (c) In plastic or foil bags/envelopes (Withers, 1979; unpublished observations), (d) In an ampoule into which liquid nitrogen is poured as the ampoule is dropped into the liquid nitrogen. Such plastic ampoules can also be used closed. (Seibert and Wetherbee, 1977), (e) In a flame-sealed glass ampoule (e.g., Chen et al., 1982b; Kartha et al., 1982a), (f) In droplets on a foil sheet inside a Petri dish (Kartha et al., 1982b), (g) On a lens tissue carrier inside an ampoule. After slow cooling the carrier is dropped into liquid nitrogen. Storage is in an ampoule with perforated lid to permit liquid nitrogen entry and pressure equalization (J. F. O'Hara, personal communication; Henshaw et al., 1985).

num foil sheet in a sterile plastic petri dish (Fig. 11f). The entire dish was then placed within the cooling chamber of a controlled freezer and subsequently transferred to liquid nitrogen. In another study, O'Hara and colleagues (J. F. O'Hara, personal communication; Henshaw *et al.*, 1985) modified the handling of shoot tips of *Solanum tuberosum* at the cryoprotection, freezing, and thawing stages. The shoot tips were supported on a lens tissue carrier which capillated cryoprotectant solution toward the specimens. Freezing was carried out by placing the carrier inside an ampoule, which was then cooled slowly at 0.3°C/min to −30°C. To achieve the highest rate of survival, the carrier was then removed from the ampoule and dropped into liquid nitrogen (Fig. 11g). In both the study of Kartha and colleagues (1982b) and the latter example, there is a risk of microbial contamination via the liquid nitrogen. The aforementioned method of Withers might overcome this.

An alternative approach to methodological development with the aim of optimizing protective dehydration involves the use of a consistent freezing rate but varying the temperature of transfer to liquid nitrogen. Two such studies using this approach for the cryopreservation of specimens as diverse as potato and apple illustrate the dramatic effect that transfer temperature can make to survival potential (see also Withers, 1979). (Note that the term "transfer temperature" is used here but in some reports "prefreezing temperature" may be found.)

Towill (1983) has investigated the cryopreservation of shoot tips of *Solanum tuberosum* dissected from shoot cultures. After cryoprotection with DMSO these were cooled to −5°C, and after 10 min ice was seeded with cold forceps. After a further 10 min the specimens were cooled at 0.2–0.3°C/min to either −30, −35, −40, or −45°C. At the transfer temperatures duplicate specimen tubes were either plunged into liquid nitrogen or thawed directly in warm water at +37°C. In the two varieties examined, transfer at −35°C gave the highest rate of recovery, with total survival percentages being 75 and 92% and, within these totals, regeneration of shoots about 33 and 20%, respectively (Fig. 12). The data illustrate considerable further scope for improvement, especially in shoot regeneration (see Section IV,D). This material shows a clear sensitivity to overdehydration and/or the risk of intracellular ice damage by slow crystal growth in the period between the optimum and actual transfer temperatures (cf. Fig. 13).

A rather different situation is illustrated in the second example chosen here, preservation of dormant shoot tips of *Malus domestica* (Katano *et al.*, 1983). The specimens were surface sterilized after dissection from the parent plant and then frozen in specimen tubes with a suspending milieu of distilled water. Initial cooling was to −2.5°C at which temperature ice was seeded, followed by continued cooling at −2.5°C/5 min (0.5°C/min) by the addition of ice to the cooling ethanol bath. Transfer to liquid nitrogen

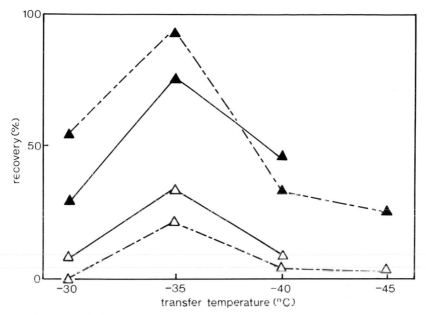

Fig. 12. Recovery of shoot tips of *Solanum tuberosum* cv. Norlond (broken lines) and cv. Pontiac (solid lines) cryoprotected with 10% DMSO and frozen at 0.2–0.3°C/min to various subzero temperatures before plunging into liquid nitrogen and thawing rapidly. Regrowth including callusing is indicated by ▲; formation of new shoots is indicated by △. (Plotted from data in Towill, 1983.)

occurred from 0 (unseeded) to −70°C. Specimens were then thawed either rapidly as above or slowly in air at 0°C. A dramatic increase in survival occurred at −5°C for the rapidly thawed and −10°C for the slowly thawed shoot tips, thus indicating achievement of a critical, minimal degree of protective dehydration between −5 and −10°C (Fig. 13). It is noteworthy that overdehydration/intracellular ice damage does not appear to be occurring. All samples which were capable of undergoing recovery in subsequent regenerated shoots at no less than about 70%. [Interactions between freezing rate and thawing rate have also been noted by Uemura and Sakai (1980) working with shoot tips of *Dianthus caryophyllus*.]

A final point of interest in comparison of the two examples above is the striking difference between the critical transfer temperatures for survival: −35°C *optimum* in the case of *Solanum tuberosum* and −5°C *threshold* in the case of *Malus domestica*. No doubt there is the likelihood of genotypic and physiological effects (*in vitro* versus *in vivo*, actively growing versus dormant) here, but it is equally likely that the freezing-point differences between the suspending milieu (plus and minus cryoprotectant, respectively) will have influenced the onset of protective dehydration.

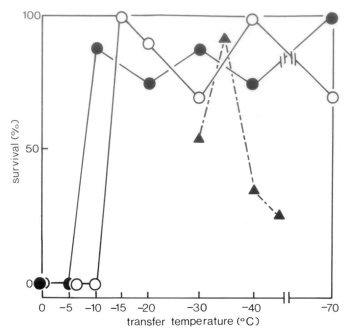

Fig. 13. Survival of shoot tips of *Malus domestica* suspended in distilled water and frozen to various temperatures before transfer to liquid nitrogen and subsequent rapid (●) thawing in warm water or slow (○) thawing in air. Thawed shoots grew and multiplied *in vitro*. Data relating to the survival of shoot tips of *Solanum tuberosum* cv. Norlond are plotted for comparison (▲; see Fig. 12 and Towill, 1983). (Redrawn from Katano *et al.*, 1983.)

D. Postthaw Treatments and Recovery

Washing in the early postthaw period is commonly carried out (and, indeed, is inevitable in the case of rapidly frozen, naked specimens) with no apparent deleterious effects. However, a comparative study would be useful to confirm the necessity for washing and also to test the possibility of a washing temperature effect (see Finkle and Ulrich, 1982). Kartha (1984) advised slow dilution of cryoprotectants to avoid deplasmolysis, which would obviously be most critical in slowly frozen material. Little is known of the effects of other features of the physical environment during the postthaw period, although the use of a low level of lighting has been recommended (e.g., O'Hara *et al.*, 1984).

Probably the most important aspect of the handling of shoot tips and meristems at the postthaw stage is the composition of the medium (invariably semisold) onto which they are inoculated. The requirements of this medium are (1) that it should sustain recovery in a high percentage of individual specimens, (2) that within each individual specimen a sufficiently high level of cell survival should be maintained, and (3) that recovery should proceed in an organized manner minimizing adventitious activity. Unfortunately, the first two of these may conflict with the third.

Several studies on solanaceous species have revealed the potential for a high degree of manipulation at the recovery stage. In the case of shoot tips dissected from seedlings of *Lycopericon esculentum* (Grout *et al.*, 1978), it was shown that, whereas cryoprotected but unfrozen control specimens callused in culture but were able to proceed to continued organized growth, frozen and thawed specimens required GA_3 in the recovery medium to minimize callusing and adventitious bud formation permitting outgrowth of the apical meristem.

A subsequent study involving several *Solanum* species revealed a strong genotypic effect necessitating modification of the postthaw medium for a high net rate of recovery (Henshaw *et al.*, 1980a). Shoot tips of *Solanum goniocalyx* could recover at a rate of 54% on standard medium containing BAP at 1 mg/liter. However, *Solanum tuberosum* subspecies *tuberosum* recovered at a level of 2% under the same conditions. Supplementation of the recovery medium with NAA at 0.5 mg/liter and GA_3 at 0.2 mg/liter instead of BAP raised the level of recovery dramatically to 32%.

This phenomenon has been investigated further (Benson *et al.*, 1984; Jarvis, 1983; Withers *et al.*, 1985). Recovery on the standard medium, while occurring at a low level, follows normal nonadventitious development, whereas the modified medium promotes callusing and variable shoot recovery. Light and electron microscopic investigations have revealed that the surviving shoot apex may be overgrown by callus and lose in the competition for vigorous growth. Furthermore, meristematic development may involve more than one primary center of growth. In order to control the pattern of recovery, the application of the supplemented recovery medium has been limited to the first 4 or 7 days postthawing, after which the shoot tips were transferred to plain medium. This procedure appears to control callusing while taking advantage of the boost to recovery afforded by the hormonal supplements. (Also note that DMSO may influence the level of callusing; see Section IV,B).

Towill (1983) has investigated recovery medium requirements in slowly frozen shoot tips of *Solanum tuberosum* (cf. the above mentioned rapidly frozen specimens). Early experiments (Towill, 1981) involved a medium containing 0.5 mg/liter IAA, 1 mg/liter GA_3, and 0.04 mg/liter kinetin, which promoted single shoot growth in unfrozen controls. Replacement of kinetin by zeatin at 0.5 mg/liter (Towill, 1983) led to formation of a multiple shoot mass in these controls. The modified medium improved rates of recovery in frozen specimens, the first signs of recovery being development of nodular callus from which shoots progressively emerged after about 1 month in culture. Not all calluses regenerated shoots, but some of the unresponsive ones could be induced to develop by transfer to fresh medium. Towill considers that zeatin is necessary for recovery by promoting callus formation from a few remaining meristematic cells.

O'Hara et al. (1984), working with the same species, find that hormones are essential to recovery but that the auxin and cytokinin involved should be at a very low level (GA$_3$ at 5 mg/liter, NAA at 1 μg/liter, BAP at 10 μg/liter). Recovery is variable in terms of overall levels and morphology, but direct, apparently nonadventitious recovery is observed in some cases.

These observations of a necessity for carefully optimized recovery media indicate that the determination between degeneration and recovery may be very finely balanced, leading to problems in reproducibility of techniques and genotypic stability. One means of overcoming this impediment to wider use of cryopreservation for shoot tips would lie in increasing the percentage cell survival within individual specimens. In this connection, it is regrettable that so few studies, outlined below, have been carried out into fine-structural aspects of shoot freezing, thawing, and recovery.

Grout and Henshaw (1980) have used scanning and transmission electron microscopy to observe recovery in shoot tips of Solanum goniocalyx. No serious abnormalities in regeneration were observed except for a tendency to form gross, malformed leaves at the expense of the shoot apex in a proportion of the specimens. [This observation is mirrored in the performance of shoot tips of Brassica napus (Benson et al., 1984) and Fragaria × ananassa (Haskins and Kartha, 1980).] Within the shoot apical region, localized cellular damage was observed, indicating that meristematic development must, at best, be interrupted for a period of time.

Benson et al. (1984) have investigated the development of shoot tips of Brassica napus rapidly frozen, as in the above example. Within 3 days of thawing, cells in the meristematic regions of the shoot tip are clearly zoned (Fig. 14). Many leaflet and leaf primordial cells and cells in the region of the developing vascular tissues can be seen to be recovering with few signs of cellular injury other than development of large lipid droplets and occlusion of some vacuoles with electron-dense granules. The outer two or three cell layers of the apical dome are generally similar in appearance except for localized lesions involving a few dead cells. Beneath the superficial cell layers lies another clear 6- to 8-cell-deep layer of damaged but living cells which are lighter in electron density and optical staining properties. These cells exhibit considerable membrane damage, including loss of vacuolar and cytoplasmic compartmentalization, and show no signs of the operation of recovery mechanisms. The remaining areas of tissue, remnants of the short stem of the shoot tip and regions of the leaflets, consist of lethally damaged cells.

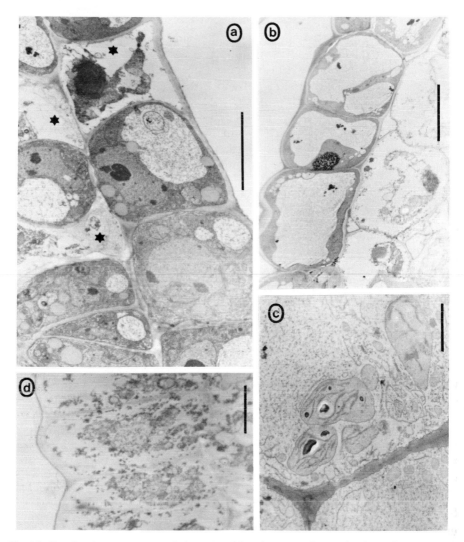

Fig. 14. Details of a cryopreserved shoot tip of *Brassica napus* observed 3 days after returning to culture. (a) Apical dome region: Groups of surviving cells are separated by lethally damaged cells (*). (Bar = 5 μm.) (b) Region of a leaf primorium showing surviving epidermal cells and lethally damaged subepidermal cells. (Bar = 5 μm.) (c) Detail of a dying cell in the middle region of the shoot tip showing deterioration of organelles. (Bar = 2 μm.) (d) Dead cells in the basal region of the shoot tip. (Bar = 2 μm.) L. A. Withers and E. E. Benson, unpublished observations.)

Over the subsequent 3–4 days, degeneration continues in the intermediate zone of damaged cells. Often, large fleshy leaves developed from original surviving primordia. The well-defined apical dome of the unfrozen specimen appears, however, to be replaced by a larger meristematic "plate" with localized regions of more intense activity. This work is continuing.

Haskins and Kartha (1980) have studied the origin of recovery in shoot tips of *Pisum sativum* cryopreserved by a constrasting (to the above) slow freezing regime. Perhaps surprisingly, recovery often proceeds by adventitious development from leaf primordial cells and cells in regions to the periphery of the apical dome. The dome cells themselves are lethally damaged in many cases.

Drawing together the above observations, some general points can be made.

1. Postthaw recovery in shoot tips often requires modification of the culture medium.

2. The shoot apical dome region is a primary target for damage and may be entirely destroyed.

3. Much of the supporting stem tissue will be lethally damaged, whereas leaf tissues may be relatively resistant to damage.

4. Callusing of potentially recovering tissues may be a prerequisite for the subsequent development of organized structures, possibly leading to problems of genetic stability.

5. The two serious problems of callusing and apical dome damage, both leading to the risk of adventitious development, may be most serious in shoot tips preserved by a slow or stepwise freezing method.

Clearly this is rich territory for further useful investigation. Another area which demands attention is the implication of tissue damage to the integrity of chimeral structures. This would be relevant to some *Solanum* species, ornamentals, and other crops.

The point concerning relative damage under different freezing conditions relates to observations made of cryopreservation of immature zygotic embryos of cereals. Withers (1982b) found that rapid freezing tended to favor organized development (germination) of the embryonic axis, whereas slow freezing favored callusing from the axial tissues and the scutellum. Thus, useful information relevant to the cryopreservation of shoot tips may be derived from studies of other organized tissues such as somatic and zygotic embryos, seedlings, and other bulky tissues such as callus (see Table I). Studies of structural survival should address individual cell survival, tissue survival, intercellular connections essential to maintenance of the symplasm, and influence of damaged cells upon neighboring survivors.

An important, alternative route to recovery of an independently growing plant from the frozen shoot tip or meristem involves grafting. The feasibility of this has been demonstrated by Sakai and Nishiyama (1978) with cryopreserved vegetative buds of *Malus domestica* grafted onto 2-year-old seedlings. This indicates a considerable degree of structural integrity in the bud region.

V. FROM THE PRESENT TO THE FUTURE

A. A Summary of Achievements

On any scale of evaluation, progress over recent years in the development of cryopreservation techniques has been considerable, especially when reviewed against the level of activity and small number of individual workers involved. In the context of cell cultures, a very satisfactory level of technical refinement has been achieved such that few, if any, species are likely to be recalcitrant to known procedures. For any such problem species, we have a wide knowledge of the technical modifications at the pregrowth, preservation, and recovery stages to permit systematic exploration of suitable alternative cryopreservation procedures.

Central interests in studies of the cryopreservation of cell cultures have moved from cryoprotection, freezing and thawing, through physiological aspects of pregrowth, injury, and recovery, to evaluation of stability in storage and application of techniques to species of biotechnological interest rather than to model systems. This is a very healthy progression which has not really been matched in the case of meristems/shoot tips.

Despite the latter remark, the past 10 years have seen progress with the cryopreservation of shoot tips, with interest being concentrated increasingly on good, clonally propagated experimental systems, on the nature of recovery growth, and on promotion of organized recovery. Increasingly, studies have tended to concentrate on species of interest in the context of germplasm storage (e.g., *Solanum* spp., *Manihot esculenta*, and *Malus domestica*).

Overall some 60 species have succumbed to cryopreservation (see Table I; others are recorded in the IBPGR *In Vitro* Database—see Section D). Both cultures and meristems/shoot tips are well represented, as are both monocotyledonous and dicotyledonous families. Whereas temperate species feature most prominently, tropical crops are represented (e.g., *Elaeis guineensis*, *Manihot esculenta*, *Phoenix dactylifera*, and *Saccharum* spp.). In the general development of *in vitro* culture as a whole, cereals and woody crops are probably considered among the more problematic groups. These are fairly well represented in cryopreservation studies.

B. Technical Shortfalls

The fact that technical development for organized cultures lag behind that for unorganized cell cultures is really a reflection of the nature of the culture systems and their amenability to routine handling rather than a criticism of research workers in relevant fields. Furthermore, there is underinvestment in appropriate research. Nonetheless, whereas cryopreservation could now be recommended for the conservation of genotypes in the form of cell cultures, it could not for shoot tips. This is particularly consequential for the conservation of naturally occurring genotypes collected by germplasm exploration projects, and therefore global genetic conservation efforts.

The following pinpoint specific problems or neglected ones as in the cryopreservation of shoot tips.

1. Pregrowth and cryoprotection conditions are underexamined.

2. Little attention has been given to the structural nature of the individual specimen: overall size, number of lead primorida, and length of stem attachment.

3. Similarly, little attention has been given to environmental factors at all cultural stages: lighting, temperature, use of liquid versus semisolid medium, and use of osmotically active supplements during the pregrowth and recovery periods.

4. The conflict, if there be one, between the two approaches of rapid and slow freezing (do they differentially favor certain regions of the specimen?) should be resolved.

5. The efficient handling of large numbers of specimens requires attention.

6. The consequences in terms of genetic stability resulting from disruptions in tissue organization are unknown.

C. Tackling Problems with Meristematic Tissues

All practical endeavors must have as their target improvements in the maintenance of structural integrity in the preserved shoot tip or meristem. Looking ahead to the possible situation wherein no more than a small proportion of the meristematic cells survives, we may have to readdress the question of the suitability of the apical meristem as a specimen for cryo-preservation. Proponents of this culture system for faithful genetic conservation (e.g., Henshaw, 1975; Henshaw and O'Hara, 1983) have presented very cogent supporting arguments, but these may have to be seen against the background of persistent failure to preserve the intact apical meristem. This leads to the question of whether we are encumbering the cryopreservation technique with an inappropriate subject.

Accordingly, the following scheme is proposed as an alternative scenario for the development of conservation techniques based upon cryopreservation of organized specimens. Shoot apical tissue has been found to be very suitable for clonal propagation, especially through multimeristem development. Such activity may not necessarily lead to unacceptable levels of instability through somaclonal variation (Scowcroft et al., 1984). In any case, we shall probably have to accept a certain amount of disruption in the freezing process; better that it be at a stage at which the operator has more control. If, during the pregrowth stage, development in the shoot apex could be diverted toward the establishment of many, localized meristematic regions—rather than in the form of a truncated stem in which apical dominance has been lost—or, perhaps, a multiple floral-type meristem, these individual meristematic regions might more closely approximate to the critical size of the surviving unit within meristems observed to date.

Taking the argument one stage further, if somatic embryogenesis could be induced before freezing, the individual embryo or proembryo structures might be even more resistant to freezing damage as a consequence of their greater structural autonomy. Development of cryopreservation procedures for such specimens might derive a considerable number of technical leads from studies with somatic embryo and cell cultures.

To complete this piece of speculation, the author suggests that in the future two approaches are likely to be taken. The first is the scheme proposed above, which should be particularly appropriate for germplasm involved in clonal propagation for crop production wherein adventitious development may already be an acceptable feature and wherein problems of genetic or phenotypic stability may be traded off against practical convenience in propagation. Indeed, in these cases, techniques may eventually tend toward those used for cells and involve rapidly proliferating liquid suspension cultures.

A second approach may be taken in which the preservation of an exact genotype is absolutely critical and in which individual specimens may be precious, having been collected one by one in the field. Here, the approaches adopted in current work may be more appropriate, with emphasis being placed on the integrity of the apical tissue. Recovery by nonadventitious means of one plant from one preserved specimen and recovery by postthaw grafting may be more prominent here also. In both of these approaches, considerable attention will have to be given to procedural refinement if success is to be achieved. The magnitude and importance of this task should not be underestimated.

D. Evaluation and Documentation of *in Vitro* Germplasm

Studies carried out, particularly over recent years, into the characterization of cryopreserved cell cultures are testimony to the increasing recognition of the importance of stability through storage (e.g., Chen *et al.*, 1984b; *Diettrich et al.*, 1982; Dougall and Whitten, 1980; Hauptmann and Widholm, 1982; Nag and Street, 1973; Watanabe *et al.*, 1983). When cryopreservation becomes a routine procedure in the cell culture laboratory (and this day cannot be far away), the availability of critical data upon the stored cultures will be essential. These should include exact taxonomic classification, culture history, morphogenic potential, synthetic or biotransformation capabilities, genetic manipulations, somaclonal variations, growth kinetics, culture medium and environmental requirements, phytopathological status, and any other important features such as isozyme data, autotrophy, karyotype, presence of chimeral structures (for organized specimens), or population composition and distribution (for cell cultures). Unlike in the case of growing stock cultures, some of these features will not be readily apparent or susceptible to testing for stored material.

Within the individual laboratory, much of this information will be well known if not documented, but in the future the exchange of cultures and their sale as feed-stocks for propagation and biosynthesis programs will become increasingly common. At such a time exact, tangible documentation will be essential both for descriptive purposes and patent certification. It is therefore recommended here that scientists should become more aware of the documentable features of their cultures and the necessity for recording of such data, preferably in computerized form. A second area of documentation relates to technical aspects of *in vitro* propagation and storage of plant species, particularly crops of interest from the point of view of global plant genetic conservation.

In view of the critical importance of promoting contacts between scientists, making available valuable but unpublished information, and collating disparate pieces of information which could lead to significant technical progress, the International Board for Plant Genetic Resources is sponsoring an information project on *in vitro* conservation (IBPGR, 1983b; Wheelans and Withers, 1984a,b). To date, this project has led to the compilation of two editions of a report on appropriate work deriving from the literature and questionnaires completed by scientists active in appropriate areas of research (Withers, 1981, 1982c). Now, as a continuation of the project, original and newly submitted information is being computerized to provide an on-request data base searching service.

E. Transportation of *in Vitro* Germplasm

In the past, considerations of transportation have involved the exchange of cultures, usually shoots growing under normal conditions or under growth limitation. Examples of very successful transportation can be cited (e.g., Roca *et al.*, 1979, 1984), and there is no reason why tubes of *in vitro* plantlets should not be exchanged by post, air freight, or carrier on a routine basis in the future. In such operations, care should be taken regarding adequately safe packaging, instructions to the recipient, and phytosanitary requirements.

Unfortunately, however, the exchange of frozen specimens has not been explored. Although semen of farm animals is transported nationally and internationally, no such procedures exist for plant material. Perhaps with the development of other aspects of *in vitro* technology in animal husbandry such as embryo cloning, artificial implantation postfertilization, and surrogate pregnancies, more thought will be given to general aspects of the carriage of cryogenic samples.

During the last 1 or 2 years, consideration of transporting plant germ plasm *in vitro* has extended back to the original collecting stage in which germplasm is acquired in the natural environment. Several crucial world crops present collecting problems in that their germplasm deteriorates or even dies before it can be established in genebanks. *In vitro* culture has been proposed as a possible solution to these difficulties (IBPGR, 1984; Withers and Williams, 1985). A pilot project has commenced in the author's laboratory to investigate the practical aspects of collecting one such problem crop, *Theobroma cacao*. Aspects under investigation include the nature of the unit tissue collected, surface sterilization, simple inoculation procedures compatible with the forest/field environment, culture media, and subsequent processing of collected samples (Yidana and Withers, 1984; Yidana *et al.*, 1985). As yet, cryopreservation has not been introduced into the experimentation but must be considered as an obvious option.

F. Education and Implementation

Takeup of germplasm storage techniques, other than the simpler approaches of slow growth/growth limitation, has been regrettably slow. However, it is hoped that, as the central importance of unique germplasm to all aspects of modern *in vitro* culture becomes recognized, this situation will change. A crucial factor in the successful implementation of cryopreservation will be the availability of accurate, clear information on procedures and the provision of appropriate training.

To date, one specialist training course has been given on cryopreservation techniques (P. J. King and L. A. Withers, Friedrich Miescher Institute, Basel, 1982, under the auspices of the European Molecular Biology Organization), and it is hoped that similar courses will be run in the future. With the standardization of cryopreservation procedures and the use of relatively inexpensive, improvised equipment, it is suggested that tuition in cryopreservation techniques could, more often, form part of international training courses.

As in so many aspects of *in vitro* culture, the methodology is relatively unsophisticated and involves no technical "wizardry." However, an understanding of the often complex physiological basis for the behavior observed in cryopreserved material is invaluable, and nothing can substitute for learning basic techniques and appreciating the importance of apparent idiosyncracies of cryopreservation procedures in a direct teaching situation. This must offer the best way of ensuring efficient, widespread use of *in vitro* germplasm storage and realization of the great benefits it can offer to plant genetic conservation and biotechnology alike.

ACKNOWLEDGMENT

The author holds an Advanced Fellowship of the U.K. Science and Engineering Research Council. The support of the S.E.R.C. and of the International Board for Plant Genetic Resources (IBPGR) is gratefully acknowledged.

REFERENCES

Akihama, T., Omura, M., and Kozaki, I. (1978). Further investigation of freeze-drying for deciduous tree pollen. *In* "Long Term Preservation of Favourable Germplasm" (T. Akihama and K. Nakajima, eds.), pp. 1–7. Fruit Tree Research Station, MAF, Japan.

Anderson, J. O. (1979). Cryopreservation of apical meristems and cells of carnation (*Dianthus caryophyllus*). *Cryobiology* **16**, 583.

Ashwood-Smith, M. J. (1980). Preservation of microorganisms by freezing, freeze-drying and desiccation. *In* "Low Temperature Preservation in Medicine and Biology" (M. J. Ashwood-Smith and J. Farrant, eds.), pp. 219–252. Pitman Medical, Tunbridge Wells, England.

Ashwood-Smith, M. J., and Farrant, J. (1980). "Low Temperature Preservation in Medicine and Biology." Pitman Medical, Tunbridge Wells, England.

Bajaj, Y. P. S. (1976). Regeneration of plants from cell suspensions frozen at −20°, −70° and −196°C. Physiol. Plant. 37, 263–268.

Bajaj, Y. P. S. (1977a). Survival of Nicotiana and Atropa pollen embryos frozen at −196°C. Curr. Sci. 46, 305–307.

Bajaj, Y. P. S. (1977b). Clonal multiplication and cryopreservation of cassava through tissue culture. Crop Improv. 4, 198–204.

Bajaj, Y. P. S. (1978a). Effect of super-low temperature on excised anthers and pollen-embryos of Atropa, Nicotiana and Petunia. Phytomorphology 28, 171–176.

Bajaj, Y. P. S. (1978b). Regeneration of plants from pollen-embryos frozen at ultra-low temperatures: A method for preservation of haploids. IV. Int. Palynol. Conf., 4th Lucknow (1976–1977) 1, 343–346.

Bajaj, Y. P. S. (1978c). Tuberization in potato plants regenerated from freeze-preserved meristems. Crop Improv. 5, 137–141.

Bajaj, Y. P. S. (1979) Freeze-preservation of meristems of Arachis hypogaea and Cicer arietinum. Indian J. Exp. Biol. 17, 1405–1407.

Bajaj, Y. P. S. (1980). Induction of androgenesis in rice anthers frozen at −196°C. Cereal Res. Commun. 8, 365–369.

Bajaj, Y. P. S. (1981a). Growth and morphogenesis in frozen (−196°C) endosperm and embryos of rice. Curr. Sci. 50, 947–948.

Bajaj, Y. P. S. (1981b). Regeneration of plants from ultra-low frozen anthers of Primula obconica. Sci. Hortic. (Amsterdam) 14, 93–95.

Bajaj, Y. P. S. (1981c). Regeneration of plants from potato meristems freeze-preserved for 24 months. Euphytica 30, 141–145.

Bajaj, Y. P. S. (1982). Survival of anther and ovule derived cotton callus frozen in liquid nitrogen. Curr. Sci. 51, 139–140.

Bajaj, Y. P. S. (1983a). Survival of somatic hybrid protoplasts of wheat × pea and rice pea subjected to −196°C. Indian J. Exp. Biol. 21, 120–122.

Bajaj, Y. P. S. (1983b). Regeneration of plants from pollen-embryos of Arachis, Brassica and Triticum spp. cryopreserved for one year. Curr. Sci. 52, 484–486.

Bannier, L. J., and Steponkus, P. L. (1972). Freeze-preservation of callus cultures of Chrysanthemum morifolium Ramat. HortScience 7, 194.

Bannier, L. J., and Steponkus, P. L. (1976). Cold acclimation of Chrysanthemum callus cultures. J. Am. Soc. Hortic. Sci. 101, 409–412.

Barlass, M., and Skene, K. G. M. (1983). Long-term storage of grape in vitro. FAO/IBPGR Plant Genet. Resour. Newsl. 53, 19–21.

Benson, E. E., Marshall, H., and Withers, L. A. (1984). A light and electron microscopical study of the cryopreservation of cultured shoot-tips of Brassica napus and Solanum tuberosum. In Abstracts: "Plant Tissue Culture and Its Agricultural Applications," p. 95. Univ. of Nottingham School of Agriculture, Nottingham, England.

Bhojwani, S. S. (1981). A tissue culture method for propagation and low temperature storage of Trifolium repens genotypes. Physiol. Plant. 52, 187–190.

Binder, W., and Zaerr, J. B. (1980a). Freeze-preservation of suspension cultured cells of a hardwood poplar. Cryobiology 17, 625.

Binder, W., and Zaerr, J. B. (1980b). Freeze-preservation of suspension cultured cells of a gymnosperm, Douglas-fir. Cryobiology 17, 624.

Bridgen, M. P. and Staby, G. L. (1981). Low pressure storage and low oxygen storage of plant tissue cultures. Plant Sci. Lett. 22, 177–186.

Bright, S. W. J., Ooms, G., Foulger, D., Evans, N., and Karp, A. (1986). Isolation and characterization of mutants from tissue culture. In "Plant Tissue Culture and Its Agricultural Application" (L. A. Withers and P. G. Alderson, eds.). Butterworth, London. (In press.)

Caplin, S. M. (1959). Mineral-oil overlay for conservation of plant tissue cultures. *Am. J. Bot.* **46**, 324–329.

Cella, R., Colombo, R., Galli, M. G., Nielsen, E., Rollo, F., and Sala, F. (1982). Freeze-preservation of *Oryza sativa* L. cells: A physiological study of freeze-thawed cells. *Physiol.Plant.* **55**, 279–284.

Centro Internacional de Agricultura Tropical (CIAT) (1979). "Annual Report: Cassava Programme." CIAT, Cali, Colombia.

Chen, T. H. H., Kartha, K. K., Constabel, F., and Gusta, L. V. (1984a). Freezing characteristics of cultured *Catharanthus roseus* (L.) G. Don cells treated with dimethylsulphoxide and sorbitol in relation to cryopreservation. *Plant Physiol.* **75**, 720–725.

Chen, T. H. H., Kartha, K. K., Leung, N. L., Kurz, W. G. W., Chatson, K. B., and Constabel, F. (1984b). Cryopreservation of alkaloid producing cell cultures of periwinkle (*Catharanthus roseus*). *Plant Physiol.* **75**, 726–731.

Chen, W. H., Cockburn, W., and Street, H. E. (1979). Preliminary experiments on freeze-preservation of sugarcane cells. *Taiwania* **24**, 70–74.

Cromarty, A. S., Ellis, R. H., and Roberts, E. H. (1982). "The Design of Seed Storage Facilities for Genetic Conservation," IBPGR Publication AGP IBPGR/82/23. International Board for Plant Genetic Resources, Rome.

Diettrich, B., Popov, A. S., Pfeiffer, B., Neumann, D., Butenko, R., and Luckner, M. (1982). Cryopreservation of *Digitalis lanata* cell cultures. *Planta Med.* **46**, 82–87.

Dougall, D. K., and Wetherall, D. F. (1974). Storage of wild carrot cultures in the frozen state. *Cryobiology* **11**, 410–415.

Dougall, D. K., and Whitten, G. H. (1980). The ability of wild carrot cell cultures to retain their capacity for anthocyanin synthesis after storage at −140°C. *Planta Med.* (1980 Suppl.), pp. 129–135.

Engelmann, F., Duval, Y., and Dereuddre, J. (1984). First successful cryopreservation of oilpalm somatic embryos. *In* Abstracts: "Plant Tissue Culture and Its Agricultural Applications," p. 96. Univ. of Nottingham School of Agriculture, Nottingham, England.

Finkle, B. J., and Ulrich, J. M. (1979). Effect of cryoprotectants in combination on the survival of frozen sugar cane cells. *Plant Physiol.* **63**, 589–604.

Finkle, B. J., and Ulrich, J. M. (1982). Cryoprotectant removal temperature as a factor in the survival of frozen rice and sugarcane cells. *Cryobiology* **19**, 329–335.

Finkle, B. J., Ulrich, J. M., Rains, D. W., Tisserat, B. B., and Schaeffer, G. W. (1979). Survival of alfalfa, rice and date palm callus after liquid nitrogen freezing. *Cryobiology* **16**, 583.

Finkle, B. J., Ulrich, J. M., Schaeffer, G. W.,and Sharpe, F. (1983). Cryopreservation of rice cells. *In* "Cell and Tissue Culture Techniques for Cereal Crop Improvement" (Academia Sinica/International Rice Research Institute), pp. 343–369. Science Press, Beijing.

Frankel, O. H., and Soulé, M. E. (1981). "Conservation and Evolution." Cambridge Univ. Press, London and New York.

Franks, F., Mathias, S. F., Galfre, P., Webster, S. D., and Brown, D. (1983). Ice nucleation and freezing in undercooled cells. *Cryobiology* **20**, 298–309.

Gaff, D. F., and Okong'o-Ogola, O. (1971). The use of non-penetrating pigments for testing survival of cells. *J. Exp. Bot.* **22**, 756–758.

Grout, B. W. W. (1979). Low temperature storage of imbibed tomato seeds: a model for recalcitrant seed storage. *Cryo-Lett.* **1**, 71–76.

Grout, B. W. W. (1986). Embryo culture and conservation of genetic resources of a species with recalcitrant seeds. *In* "Plant Tissue Culture and Its Agricultural Applications" (L. A. Withers and P. G. Alderson, eds.). Butterworth, London. (In press.)

Grout, B. W. W., and Henshaw, G. G. (1978). Freeze-preservation of potato shoot-tip cultures. *Ann. Bot. (London)* **42**, 1227–1229.

Grout, B. W. W., and Henshaw, G. G. (1980). Structural observations on the growth of potato shoot-tip cultures after thawing from liquid nitrogen. *Ann. Bot. (London)* **46**, 243–248.

Grout, B. W. W., Westcott, R. J. and Henshaw, G. G. (1978). Survival of shoot meristems of tomato seedlings frozen in liquid nitrogen. *Cryobioloby* **15**, 478–483.

Grout, B. W. W., Shelton, K., and Pritchard, H. W. (1983). Orthodox behaviour of oilpalm seed and cryopreservation of the excised embryo for genetic conservation. *Ann. Bot. (London)* **52**, 381–384.

Haskins, R. H., and Kartha, K. K. (1980). Freeze-preservation of pea meristems: Cell survival. *Can. J. Bot.* **58**, 833–840.

Hauptmann, R. M., and Widholm, J. M. (1982). Cryostorage of cloned amino acid analog-resistant carrot and tobacco suspension cultures. *Plant Physiol.* **70**, 30–34.

Henshaw, G. G. (1975). Technical aspects of tissue culture storage for genetic conservation. *In* "Crop Genetic Resources for Today and Tomorrow" (O. H. Frankel and J. G. Hawkes, eds.), pp. 349–358. Cambridge Univ. Press, London and New York.

Henshaw, G. G., and O'Hara, J. F. (1983). In vitro approaches to the conservation and utilization of global plant genetic resources. *In* "Plant Biotechnology" (S. H. Mantell and H. Smith, eds.), SEB Seminar Series 18, pp. 219–238. Cambridge Univ. Press, London and New York.

Henshaw, G. G., O'Hara, J. F., and Westcott, R. J. (1980a). Tissue culture methods for the storage and utilization of potato germplasm. *In* "Tissue Culture Methods for Plant Pathologists" (D. S. Ingram and J. P. Helgeson, eds.), pp. 71–76. Blackwell, Oxford.

Henshaw, G. G., Stamp, J. A., and Westcott, R. J. (1980b). Tissue culture and germplasm storage. *In* "Developments in Plant Biology," (F. Sala, B. Parisi, R. Cella, and O. Cifferi, eds.), Vol. 5: Plant Cell Cultures: Results and Perspectives, pp. 277–282, Elsevier, Amsterdam.

Henshaw, G. G., Keefe, P. D. and O'Hara, J. F. (1985). Cryopreservation of potato meristems. *In* "Proceedings EEC Symposium on In Vitro Technique—Propagation and Long-Term Storage," Nijhof, Amsterdam. (In press.).

Ingram, C. C. B., and Williams, J. T. (1984). In situ conservation of wild relatives of crops. *In* "Crop Genetic Resources: Conservation and Evaluation" (J. H. W. Holden and J. T. Williams, eds.), pp. 163–179. Allen and Unwin, London.

International Board for Plant Genetic Resources (IBPGR) (1983a). "IBPGR Advisory Committee on in Vitro Storage: Report of the First Meeting," IBPGR Publication AGP:IBPGR/82/84. IBPGR, Rome.

International Board for Plant Genetic Resources (IBPGR). (1983b). Information on in vitro culture. *FAO/IBPGR Plant Genet. Resour. Newsl.* **55**, 16.

International Board for Plant Genetic Resources (IBPGR). (1984). "The Potential for Using in Vitro Techniques for Germplasm Collection," IBPGR Publication AGP:IBPGR/83/108. IBPGR, Rome.

International Institute for Tropical Agriculture (IITA) (1981). "Annual Report." IITA, Ibadan, Nigeria.

Jarvis, R. V. (1983). The cryopreservation of *Solanum tuberosum* shoot-tip cultures. B.Sc. Dissertation, Univ. of Nottingham School of Agriculture, Nottingham, England.

Jenkins, E. C., and Weed, R. G. (1977). Slow and fast post-thaw dilutions in the cryoprotection of lymphocytes in whole blood. *Cryobiology* **14**, 696–697.

Jones, L. H. (1974). Long term survival of embryoids of carrot (*Daucus carota* L.). *Plant Sci. Lett.* **2**, 221–224.

Kartha, K. K. (1984). Freeze preservation of meristems. *In* "Cell Culture and Somatic Cell

Genetics of Plants'' (I. K. Vasil, ed.), Vol. 1: Laboratory Procedures and Their Applications, pp. 621–628. Academic Press, Orlando, Florida.

Kartha, K. K., Leung, N. L., and Gamborg, O. L. (1979). Freeze-preservation of pea meristems in liquid nitrogen and subsequent plant regeneration. *Plant Sci. Lett.* **15**, 7–15.

Kartha, K. K., Leung, N. L., and Pahl, K. (1980). Cryopreservation of strawberry meristems and mass propagation of plantlets. *J. Am. Soc. Hortic. Sci.* **105**, 481–484.

Kartha, K. K., Leung, N. L., Guadet LaPrairie, P. and Constabel, F. (1982a). Cryopreservation of periwinkle, *Catharanthus roseus* cells cultured *in vitro*. *Plant Cell Rep.* **1**, 135–138.

Kartha, K. K., Leung, N. L., and Mroginski, L. A. (1982b). *In vitro* growth responses and plant regeneration from cryopreserved meristems of cassava (*Manihot esculenta* Crantz). *Z. Pflanzenphysiol.* **107**, 133–140.

Katano, M., Ishihara, A., and Sakai, A. (1983). Survival of dormant apple shoot-tips after immersion in liquid nitrogen. *HortScience* **18**, 707–708.

Kaufmann, A., and Berthier, R. (1983). Simple thermostable baths for the two-step freezing technique and their use for cooling human bone marrow in tubes or bags. *Cryo-Lett.* **4**, 241–250.

King, M. W., and Roberts, E. H. (1979). "The Storage of Recalcitrant Seeds: Achievements and Possible Approaches," IBPGR Publication AGP:IBPGR/79/44. International Board for Plant Genetic Resources, Rome.

Latta, R. (1971). Preservation of suspension cultures of plant cells by freezing. *Can. J. Bot.* **49**, 1253–1254.

Li, P. H., and Sakai, A. (1982). "Plant Cold Hardiness and Freezing Stress." Academic Press, New York.

Maddox, A., Gonsalves, F., and Shields, R. (1983). Successful preservation of plant cell cultures at liquid nitrogen temperatures. *Plant Sci. Lett.* **28**, 157–162.

Martin, F. W. (1975). The storage of germplasm of tropical roots and tubers in the vegetative form. *In* "Crop Genetic Resources for Today and Tomorrow" (O. H. Frankel and J. G. Hawkes, eds.), pp. 369–378, Cambridge Univ. Press, London and New York.

Mazur, R. A. and Hartmann, J. X. (1979). Freezing of plant protoplasts. *In* "Plant Cell and Tissue Culture: Principles and Applications: (W. R. Sharp, P. O. Larsen, E. F. Paddock and V. Raghavan, eds.), p. 876. Ohio State Univ. Press, Columbus, Ohio.

Mix, G. (1982). *In vitro* preservation of potato material. *FAO/IBPGR Plant Genet. Resour. Newsl.* **51**, 6–8.

Mix, G. (1984). *In vitro* preservation of potato germplasm. *In* "Efficiency in Plant Breeding" (W. Lange, A. C. Zeven, and N. G. Hogenboom, eds.), pp. 194–195. Pudoc, Wageningen, The Netherlands.

Morel, G. (1975). Meristem culture techniques for the long-term storage of cultivated plants. *In* "Crop Genetic Resources for Today and Tomorrow" (O. H. Frankel and J. G. Hawkes, eds.), pp. 327–332. Cambridge Univ. Press, London and New York.

Morris, G. J. (1980). Cryopreservation of plant tissues. *In* "Low Temperature Preservation in Medicine and Biology". (M. J. Ashwood-Smith and J. Farrant, eds.), pp. 253–284. Pitman Medical, Tunbridge Wells, England.

Nag, K. K., and Street, H. E. (1973). Carrot embryogenesis from frozen cultured cells. *Nature (London)* **245**, 270–272.

Nag, K. K., and Street, H. E. (1975a). Freeze-preservation of cultured plant cells: I The pretreatment phase. *Physiol. Plant.* **34**, 254–260.

Nag, K. K., and Street, H. E. (1975b). Freeze-preservation of cultured plant cells: II The freezing and thawing phases. *Physiol. Plant.* **34**, 261–265.

Ng, S. Y., and Hahn, S. K. (1985). Application of tissue culture to tuber crops at IITA. *In* "Biotechnology in International Agricultural Research." Proceedings of the Inter-Center Seminar on International Agricultural Research Centers (IARCs) and Biotechnology,

IRRI, Los Banos, Philippines, April 1984, pp. 29–40. International Rice Research Institute, Manila.

O'Hara, J. F., Keefe, P. D., and Henshaw, G. G. (1984). An improved method for the cryopreservation of potato germplasm. In Abstracts: "Plant Tissue Culture and its Agricultural Applications," p. 95. Univ. of Nottingham School of Agriculture, Nottingham, England.

Popov, A. S., Butenko, R. G., and Glukhova, J. N. (1978). Effect of pretreatment and deep freezing conditions on renewal of a suspension culture of Daucus carota cells. Fiziol. Rast. (Moscow) 25, 1227.

Potrykus, I. (1978). Discussion comment. In "Production of Natural Compounds by Cell Culture Methods" (A. W. Alferman and E. Reinhard, eds.), p. 192. Gesellschaft für Strahlen und Umweltforschung mbH, Munich.

Pritchard, H. W., Grout, B. W. W., Reid, D. S., and Short, K. C. (1982). The effects of growth under water stress on the structure, metabolism and cryopreservation of cultured sycamore cells. In "The Biophysics of Water" (F. Franks and S. Mathias, eds.), pp. 315–318. Wiley, New York.

Quatrano, R. S. (1968). Freeze-preservation of cultured flax cells utilizing DMSO. Plant Physiol. 43, 2057–2061.

Roberts, E. H., King, M. W., and Ellis, R. H. (1984). Recalcitrant seeds: Their recognition and storage. In "Crop Genetic Resources: Conservation and Evaluation" (J. H. W. Holden and J. T. Williams, eds.), pp. 38–52. Allen and Unwin, London.

Roca, W. M., Bryan, J. E., and Roca, M. R. (1979). Tissue culture for the international transfer of potato genetic resources. Am. Potato J. 56, 1–10.

Roca, W. M., Rodriguez, J. A., Mafla, G., and Roa, J. (1984). "Procedures for Recovering Cassava Clones Distributed in Vitro." CIAT, Cali, Colombia.

Sakai, A., and Nishiyama, Y. (1978). Cryopreservation of winter vegetative buds of hardy fruit trees in liquid nitrogen. HortScience 13, 225–227.

Sakai, A., and Sugawara, Y. (1973). Survival of poplar callus at superlow temperatures after cold acclimation. Plant Cell Physiol. 14, 1201–1204.

Sakai, A., Yamakawa, M., Sakato, D., Harada, T., and Yakuwa, T. (1978). Development of a whole plant from an excised strawberry runner apex frozen to −196°C. Low Temp. Sci. Ser. B 36, 31–38.

Sala, F., Cella, R., and Rollo, F. (1979). Freeze-preservation of rice cells. Physiol. Plant. 45, 170–176.

Scowcroft, W. R., Ryan, S. A., Brettel, R. I. S., and Larkin, P. J. (1984). Somaclonal variation: A new genetic resource. In "Crop Genetic Resources: Conservation and Evaluation." (J. H. W. Holden and J. T. Williams, eds.), pp. 258–267. Allen and Unwin, London.

Seibert, M. (1976). Shoot initiation from carnation shoot apices frozen to −196°C. Science 191, 1178–1179.

Seibert, M. (1977). Process for storing and recovering plant tissue. U.S. Patent No. 4,052,817.

Seibert, M., and Wetherbee, P. M. (1977). Increased survival and differentiation of frozen herbaceous plant organ cultures through cold treatment. Plant Physiol. 59, 1043–1046.

Seitz, U., Alfermann, A. W., and Reinhard, E. (1983). Stability of biotransformation capacity in Digitalis lanata cell cultures after cryogenic storage. Plant Cell Rep. 2, 273–276.

Shillito, R. D. (1978). Problems associated with the regulation of auxotrophic mutants from plant cell tissue cultures. Ph.D. Thesis, Univ. of Leicester, Leicester, England.

Stamp, J. A. (1978). Freeze-preservation of shoot-tips of potato varieties. MSc. Thesis, Univ. of Birmingham, Birmingham, England.

Stanwood, P. C. (1985). Cryopreservation of seed germplasm for genetic conservation. In "Cryopreservation of Plant Cells and Organs," (K. K. Kartha, ed.), pp. 199–226. CRC Press, Boca Raton, Florida.

Steponkus, P. L., Dowgert, M. F., and Gordon-Kamm, W. J. (1983). Destabilization of the plasma membrane of isolated plant protoplasts during a freeze-thaw cycle: The influence of cold acclimation. *Cryobiology* **20**, 448–465.

Styles, E. D., Burgess, J. M., Mason, C., and Huber, B. M. (1982). Storage of seed in liquid nitrogen. *Cryobiology* **19**, 195–199.

Sugawara, Y., and Sakai, A. (1974). Survival of suspension cultured sycamore cells cooled to the temperature of liquid nitrogen. *Plant Physiol.* **54**, 772–774.

Sun, C. N. (1958). The survival of excised pea seedlings after drying and freezing in liquid nitrogen. *Bot. Gaz. (Chicago)* **19**, 234–236.

Takeuchi, M., Matsushima, H., and Sugawara, Y. (1980). Long-term freeze-preservation of protoplasts of carrot and *Marchantia*. *Cryo-Lett.* **1**, 519–524.

Takeuchi, M., Matsushima, H., and Sugawara, Y. (1982). Totipotency and viability of protoplasts after long-term freeze-preservation. *In* "Plant Tissue Culture, 1982" (A. Fujiwara, ed.), pp. 797–798. Maruzen, Tokyo.

Tisserat, B., Ulrich, J. M., and Finkle, B. J. (1981). Cryogenic preservation and regeneration of date palm tissue. *HortScience* **16**, 47–48.

Towill, L. E. (1981a). *Solanum etuberosum*: A model for studying the cryobiology of shoot-tips in the tuber bearing *Solanum* species. *Plant Sci. Lett.* **20**, 315–324.

Towill, L. E. (1981b). Survival at low temperatures of shoot-tips from cultivars of *Solanum tuberosum* group Tuberosum. *Cryo-Lett.* **2**, 373–382.

Towill, L. E. (1983). Improved survival after cryogenic exposure of shoot-tips derived from *in vitro* plantlet cultures of potato. *Cryobiology* **20**, 567–573.

Towill, L. E., and Mazur, P. (1974). Studies on the reduction of 2,3,5-triphenyltetrazolium chloride as a viability assay for plant tissue cultures. *Can. J. Bot.* **53**, 1097–1102.

Uemura, M., and Sakai, A. (1980). Survival of carnation (*Dianthus caryophyllus* L.) shoot apices frozen to the temperature of liquid nitrogen. *Plant Cell Physiol.* **21**, 85–94.

Ulrich, J. M., Finkle, B. J., Moore, P. H., and Ginoza, H. (1979). Effect of a mixture of cryoprotectants in attaining liquid nitrogen survival of callus cultures of a tropical plant. *Cryobiology* **16**, 550–556.

Watanabe, K., Mitsuda, H., and Yamada, Y. (1983). Retention of metabolic and differentiation potentials of green *Lavandula vera* callus after freeze preservation. *Plant Cell Physiol.* **24**, 119–122.

Weber, G., Roth, E. J., and Schweiger, H.-G. (1983). Storage of cell suspensions and protoplasts of *Glycine max* (L.) Merr., *Brassica napus* (L.), *Datura innoxia* (Mill.) and *Daucus carota* (L.) by freezing. *Z. Pflanzenphysiol.* **10**, 23–29.

Westcott, R. J. (1981a). Tissue culture storage of potato germplasm. 1. Minimal growth storage. *Potato Res.* **24**, 331–342.

Westcott, R. J. (1981b). Tissue culture storage of potato germplasm. 2. Use of growth retardants. *Potato Res.* **24**, 343–352.

Wheelans, S. K., and Withers, L. A. (1984a). The IBPGR/Nottingham University information project on *in vitro* conservation. *In* Abstracts: "Plant Tissue Culture and Its Agricultural Applications," p. 96. Univ. of Nottingham School of Agriculture, Nottingham, England.

Wheelans, S. K., and Withers, L. A. (1984b). Information on *in vitro* conservation. *FAO/IBPGR Plant Genet. Resour. Newsl.* **60**, 33–38.

Widholm, J. M. (1972). The use of fluorescein diacetate and phenosafranine for determining viability of cultured plant cells. *Stain Technol.* **47**, 189–194.

Wilkes, G. (1983). Current status of crop plant germplasm. *CRC Crit. Rev. Plant Sci.* **1** (2), 131–181.

Withers, L. A. (1978a). A fine-structural study of the freeze-preservation of plant tissue cultures: II The thawed state. *Protoplasma* **94**, 235–247.

Withers, L. A. (1978b). Freeze-preservation of synchronously dividing cells of *Acer pseudoplatanus*. *Cryobiology* **15**, 87–92.

Withers, L. A. (1979). Freeze-preservation of somatic embryos and clonal plantlets of carrot (*Daucus carota* L.). *Plant Physiol.* **63**, 460–467.

Withers, L. A. (1980a). "Tissue Culture Storage for Genetic Conservation," IBPGR Publication AGP:IBPGR/80/8. International Board for Plant Genetic Resources, Rome.

Withers, L. A. (1980b). The cryopreservation of higher plant tissue and cell cultures—an overview with some current observations and future thoughts. *Cryo-Lett.* **1**, 239–250.

Withers, L. A. (1980c). Low temperature storage of plant tissue cultures. *In* "Plant Cell Cultures II" (A. Fiechter, ed.), Advances in Biochemical Engineering, 18, pp. 102–150. Springer-Verlag, Berlin and New York.

Withers, L. A. (1980d). Preservation of germplasm. *Int. Rev. Cytol. Suppl.* **11B**, 101–136.

Withers, L. A. (1981). Institutes Working on Tissue Culture for Genetic Conservation, IBPGR Publication AGP.IBPGR/81/30. International Board for Plant Genetic Resources, Rome.

Withers, L. A. (1982a). Storage of plant tissue cultures. *In* "Crop Genetic Resources: The Conservation of Difficult Material" (L. A. Withers and J. T. Williams, eds.), IUBS Series B42, pp. 49–82. IUBS/IBPGR, Paris.

Withers, L. A. (1982b). The develoment of cryopreservation techniques for plant cell, tissue and organ cultures. *In* "Plant Tissue Culture 1982" (A. Fujiwara, ed.), pp. 793–794. Maruzen, Tokyo.

Withers, L. A. (1982c). "Institutes Working on Tissue Culture for Genetic Conservation," IBPGR Publication AGP:IBPGR/82/30. International Board for Plant Genetic Resources, Rome.

Withers, L. A. (1983a). Germplasm storage in plant biotechnology. *In* "Plant Biotechnology" (S. H. Mantell and H. Smith, eds.), SEB Seminar Series, 18, pp. 187–218. Cambridge Univ. Press, London and New York.

Withers, L. A. (1983b). Germplasm preservation through tissue culture: An overview. *In* "Cell and Tissue Culture Techniques for Cereal Crop Improvement" (Academia Sinica/International Rice Research Institute), pp. 315–341. Science Press, Beijing.

Withers, L. A. (1984a). *In vitro* techniques for germplasm storage. *In* "Efficiency in Plant Breeding" (W. Lange, A. C. Zeven, and N. G. Hogenboom, eds.), pp. 182–193. Pudoc, Wageningen, The Netherlands.

Withers, L. A. (1984b). Freeze-preservation of cells. *In* "Cell Culture and Somatic Cell Genetics of Plants" (I. K. Vasil, ed.), Vol. 1: Laboratory Procedures and Their Applications, pp. 608–620. Academic Press, Orlando, Florida.

Withers, L. A. (1984c). Germplasm conservation *in vitro:* Present state of research and its application. *In* "Crop Genetic Resources: Conservation and Evaluation" (J. H. W. Holden and J. T. Williams, eds.), pp. 138–157. Allen and Unwin, London.

Withers, L. A. (1985). The preservation of plant cell, tissue and organ cultures, and seed for genetic conservation and improved agricultural practice. *In* "The Effects of Low Temperature on Biological Systems" (B. W. W. Grout and J. G. Morris, eds.). Arnold, London. (In press.)

Withers, L. A., and Davey, M. R. (1978). A fine-structural study of the freeze-preservation of plant tissue cultures. I The frozen state. *Protoplasma* **94**, 207–219.

Withers, L. A., and King, P. J. (1979). Proline: A novel cryoprotectant for the freeze-preservation of cultured cells of *Zea mays* L. *Plant Physiol.* **64**, 675–678.

Withers, L. A., and King, P. J. (1980). A simple freezing unit and cryopreservation method for plant cell suspensions. *Cryo-Lett.* **1**, 213–220.

Withers, L. A., and Street, H. E. (1977a). Freeze-preservation of plant cell cultures. *In* "Plant Tissue Culture and Its Bio-technological Application" (W. Barz, E. Reinhard, and M.-H. Zenk, eds.), pp. 226–244. Springer-Verlag, Berlin and New York.

Withers, L. A., and Street, H. E. (1977b). Freeze-preservation of cultured plant cells: III The pregrowth phase. *Physiol. Plant.* **39**, 171–178.

Withers, L. A., and Williams, J. T. (1985). Research on long-term storage and exchange of *in vitro* plant germplasm. *In* "Biotechnology in International Agricultural Research." Proceedings of the Inter-Center Seminar on International Agricultural Research Centers (IARCs) and Biotechnology. IRRI, Los Banos, Philippines, April 1984, pp. 11–24. International Rice Research Institute, Manila.

Withers, L. A., Jarvis, R. V., and Marshall, H. (1985). Cryopreservation of potato shoot-tips: Effect of phasing hormonal treatments during recovery. (Submitted for publication.)

Yidana, J. A., and Withers, L. A. (1984). The development of a tissue culture based collecting method for germplasm of cocoa (*Theobroma cacao* L.). *In* Abstracts: "Plant Tissue Culture and its Agricultural Applications" p. 97. Univ. of Nottingham School of Agriculture, Nottingham, England.

Yidana, J. A., Withers, L. A. and Ivins, J. D. (1985). Development of a simple method for collecting and conserving cocoa germplasm *in vitro*. *Acta Hort.* (In press.).

Zavala, M. E., and Finkle, B. J. (1983). Ultrastructural changes in cultured sugarcane cells following the addition of cryoprotectants at low temperatures. *Cryo-Lett.* **4,** 371–380.

Index